Information-Processing Machines

Information-Processing Machines

Panos A. Ligomenides
University of California, Los Angeles

Holt, Rinehart and Winston
New York Chicago San Francisco Atlanta
Dallas Montreal Toronto London Sydney

Copyright © 1969 by Holt, Rinehart and Winston, Inc.
All rights reserved

Library of Congress Catalog Card Number: 78–77812
SBN: 03-075585-9

Printed in the United States of America

9 8 7 6 5 4 3 2 1

To my father, whose strong attachment to human
values taught me the peace-giving truth and
left me the inheritance of a holy life.
To my mother, who showed me how to make dreams.

Preface

Information processing is the conversion (transformation) of information in a prescribed way. When this is done by machines, information is presented in the form of binary-coded symbols. The use of only two symbols results in long and repetitious codes, and in reduced coding efficiency. However, this type of coding has an advantage in that the machine is required to recognize, produce, and measure only two types of signals or states of circuits and devices.

Manipulation of binary-coded combinations of symbols is done according to established rules and procedures of a special type of algebra—Boolean algebra. Information is processed when the coded binary combinations are transformed in a meaningful way. The commonly used symbols are 0 and 1. It should be pointed out that the symbols are not the information themselves, but they only represent the information. The same information may be represented in other coding systems by different symbols.

An information-handling device, which receives a certain combination of binary-coded symbols and delivers the same combination, although possibly by physically different means, is not performing data-processing as such, because no conversion of information takes place. The device performs a "memory" function, whereby it delivers on demand the same information received previously.

This is an introductory level treatment of the principles of operation, organization, and design of information-processing machines. The text evolved during four years of lecturing at San Jose State College and Santa Clara University. The single most important feature of this book is the unification and correlation of the most essential topics in the operation, the organization and the design of information-processing machines, in a simple, direct, and meaningful manner. Topics included are machine organization, arithmetic, logic and sequential circuit design, coding, and programming. The text is aimed to provide a valuable point of departure for engineering students, technical trainees, and scientifically-minded high school graduates. It should stimulate and facilitate a better understanding and further study in the field of information-processing machines. The scope of the book is such that it is also useful to engineering executives who need to have a well-founded understanding of computers, computer organization, and principles of operation and design, so that they may keep abreast of current trends and developments in technologies and in their profession.

This book is written explicitly for use as a textbook, and may be used at almost any college or technical school level consistently with local curricula. It may be used by engineering students, by computer technical trainees, or by scientifically-minded high school graduates. This is primarily so because the character of the subject matter is such that the principal academic background required is one which has largely been gained by the time the student leaves high school. Chapter 9 may be an exception, as the student is required to have an elementary understanding of electronic circuits and devices. However, the rest of the text does not depend on Chapter 9, which may be omitted.

In the presentation of the material, new concepts are first introduced at an appropriate point with a brief mention, while more complete discussions follow later when the reader is more adequately prepared. This inductive process of presentation has been found to be very effective in classroom teaching. In addition, it offers the instructor the opportunity to elaborate at different levels of sophistication.

The text progresses in a natural manner beginning with a general discussion of organization, operation, and programming in Chapters 1 and 2. Questions of coding and reliability are discussed next in Chapters 3 and 4. Logic design and techniques of minimization and logical circuit implementation are presented in Chapters 5 through 8. It was felt that the student should be familiarized with the elementary switching devices and circuits which make up the arithmetic and memory portions of an information processing machine. However, any attempt to discuss hardware in some detail would be severely curtailed if we are to maintain any measure of balanced text proportions. Thus, the most essential cases of the very important topics of computer hardware are presented in Chapter 9. Questions on the design, organization, and operation of the arithmetic portion of a machine are discussed in Chapter 10. Sequential and finite-state machines are defined and illustrated with examples in Chapter 11. The design of a small computer, in Chapter 12, brings together many of the principles and techniques discussed in previous chapters. While the major part of this book is accepted doctrine, the problem of explaining the concepts involved has been a challenging experience for the author. Historical presentation of the material may often become the enemy of logic and ludicity; consequently, the chronological evolution of theories is not followed, unless needed for clarity.

ACKNOWLEDGMENTS

I would like to acknowledge the encouragement and help extended to me by numerous friends and colleagues. In particular, I wish to thank Professor M. L. Dertouzos of M.I.T. for his help and encouragement; also, Professor E. J. McCluskey of Stanford University for his helpful comments on my manuscript. The encouragement of Dr. J. H. Smith of IBM and of Professor M. M. McWhorter of Stanford University is also gratefully acknowledged. I wish also to thank the IBM Corporation for allowing me time off to work on this textbook.

To my students at San Jose State College and Santa Clara University goes my deepest appreciation for their critical review of my class notes. My many thanks to Mr. John Minnick of IBM for editorial help, to Messrs. J. Scherer and V. Garbushian for help with programming and drawing, and to Miss P. Schotthoefer for expert typing and for her patience during the long hours of corrections and revisions. The assistance of Holt, Rinehart and Winston is gratefully acknowledged.

The text is my own responsibility. This is not to ignore the influence of many discussions with my colleagues. If they, or anyone else, feel that certain errors are too glaring, I hope that they will write and tell me about them.

Stanford University   P.A.L.
July 1969

Contents

1
Reflections on Information-Processing Machines 1

1.1 Machines 1
1.2 A Classification of Machines 2
1.3 Current Uses of Information-Processing Machines 4
1.4 EDP and Scientific Computers 5
1.5 Cybernetic and Heuristic Machines 6
1.6 Special-Purpose and General-Purpose Machines 7
1.7 Machine Logic, Programming and Timing 8
1.8 Information Structure 10
1.9 General Machine Organization 14
1.10 Peripheral Hardware 18
1.11 History and Classification of Digital Computers 24

2
Programming the Machine 28

2.1 Information Conveyance and Communication 29
2.2 Plugboard and External Programming 29
2.3 Preparation of an Algorithm 34
2.4 Preparation of a Program 35
2.5 Assembly Languages—Mnemonics 40
2.6 Temperature Conversion Example 44
2.7 Income Tax Example 45
2.8 Procedure-Oriented Languages 45
2.9 Compilers and Assemblers 48

- 2.10 Example of ALGOL Programming 49
- 2.11 Comments on ALGOL Programming 52
- 2.12 The DO-Statement 53
- 2.13 Error Detection by the Compiler 54
- 2.14 A Second Illustrative Example 55
- 2.15 FORTRAN Programming 56
- 2.16 Examples of FORTRAN Programming 59
- 2.17 Additional Comments on Programming 63

3
Number Systems and Coding 67

- 3.1 Sets 67
- 3.2 Number 68
- 3.3 Numerical Symbols 69
- 3.4 Equal Numbers 70
- 3.5 Counting 70
- 3.6 Systems of Numeration 70
- 3.7 Counting with Natural Numbers 72
- 3.8 Radix Conversion 72
- 3.9 Appropriate Numerical System for Mechanizing Arithmetic 74
- 3.10 The Pure Binary System 76
- 3.11 Binary Coded Numerical Systems—Machine Languages 77
- 3.12 Fixed- and Floating-Point Number Representation 82

4
Reliability 85

- 4.1 Reliability Demands 85
- 4.2 System Reliability 86
- 4.3 Reliability in the Nervous System 87
- 4.4 Redundancy and Error Detection in Codes 88
- 4.5 Error-Detecting Codes 90
- 4.6 Single-Error Detection 92
- 4.7 Single-Error Correction 95
- 4.8 The Hamming Code 96
- 4.9 Rounding-Off and Truncation 99
- 4.10 Comments 100

5
Logic Design 103

5.1 The Design Procedure 104
5.2 Calculus of Propositions and Networks of Switches 105
5.3 Gates 107
5.4 Logical Equations From Circuit Diagrams 109
5.5 Circuit Diagrams from Logical Equations 110
5.6 Formation of Logical Equations 112
5.7 Logical Diagrams for Resolution of Alternatives 116
5.8 Combinational and Sequential Logic 120

6
Symbolic Logic 125

6.1 The Use of Symbols 126
6.2 The Calculus of Classes—Definitions 127
6.3 Operations 128
6.4 Equations 130
6.5 Fundamental Laws 130
6.6 Calculus of Inference 131
6.7 Examples of Inference 132
6.8 Algebra of Propositions 135

7
Switching Algebra 140

7.1 Definitions 141
7.2 Axiomatic Definition of a Boolean Algebra 142
7.3 Theorems and Operation Rules 143
7.4 Combinations and Functions of Binary Variables 145
7.5 Equality of Boolean Functions—Boolean Equations 149
7.6 The Operation of Complementation 150
7.7 Canonical Forms 152
7.8 Simplification of Switching Circuits 156
7.9 The Karnaugh Maps 161
7.10 The Mapping Method 164

- 7.11 Adjacent Cells 167
- 7.12 Demapping 170
- 7.13 Examples of Karnaugh Map Simplifications 172
- 7.14 Boolean Versus Ordinary Algebra 176
- 7.15 Time Considerations—Boolean Calculus 178

8
Combinational Circuits 184

- 8.1 Combinational Operations—"Don't Cares" 185
- 8.2 Complete Sets of Elementary Blocks 188
- 8.3 Hardware Considerations 190
- 8.4 NOR and NAND Logic 191
- 8.5 "Interrogation" Circuit for a Four-Bit Register 200
- 8.6 NBCD to 2-Out-Of-5 Code Converter 201
- 8.7 The Design of a Half-Adder 202
- 8.8 Composition and Decomposition of Maps and Functions 206

9
Elementary Components: Switching Devices and Circuits 215

- 9.1 Engineering Specifications of Switching Devices 216
- 9.2 Switches 217
- 9.3 Electronic Diode and/or Gates 220
- 9.4 Diode Decoding/Encoding Matrices 222
- 9.5 Relays and Three-Terminal Electronic Switches 224
- 9.6 Transistor Logic 228
- 9.7 Flip-Flops 235
- 9.8 The Transistor NOR Flip-Flop 239
- 9.9 Shift Register 241
- 9.10 Binary Counters 243
- 9.11 Ring Counters 244
- 9.12 Magnetic Core Memories 245
- 9.13 All-or-None Organs in Living Organisms And in Machines 250

10
Arithmetic Operations and Machine Organization   257

10.1   Operating Rules in Arithmetic   257
10.2   Arithmetic Operations   260
10.3   Processor   262
10.4   Binary Arithmetic   262
10.5   Serial and Parallel Addition   272
10.6   Organization and Operation of a Small Computer   277

11
Sequential Machines   287

11.1   Definitions   287
11.2   A Small "Special-Event" Sequential Machine   289
11.3   States   289
11.4   State Flow Diagram   290
11.5   State Transition Table   292
11.6   Minimized State Flow Diagram and Transition Table   294
11.7   Circuit Implementation   295
11.8   Generalizations   299
11.9   Finite State Machines   302
11.10   Turing Machines   304
11.11   Unsolvability and Limitations of Computing Machinery   306

12
The Design of a Small Computer   310

12.1   Statement of the Design Problem   310
12.2   Truth Tables   312
12.3   Maps and Boolean Algebraic Expressions   312
12.4   Set-Reset Equations for the "Carry Flip-Flop"   315
12.5   Block Diagram   317
12.6   Machine Organization   318
12.7   Machine Operation—State Diagrams   319
12.8   Comments   320

Postscript 325

Appendix A 328

Appendix B 340

Appendix C 343

Appendix D 350

Index 355

# Information-Processing Machines

1

Reflections on Information-Processing Machines

Since the beginning of civilization, machines have always amazed people. This is even more true today because of the new generation of machines whose primary objectives are the collection, transmission, and manipulation of information. More specifically, high-speed electronic computers are presently taking over many activities which in the past were performed exclusively by the human mind. In the opinion of many, this age is characterized as the age of computers and automation. If this is so, the question of the relationship between man and machines, that is, the position of man in a world full of such "beastly" contrivances, is an important one.

1.1
MACHINES

It is always a difficult task to give a good definition of things that are very familiar to us. Vaguely, we know what machines are, but to define "machine" is another thing because the word has so many ramifications in everyday life. The equivalent Greek word for "machine" has meant something like a contrivance, artifact, means, expedient, and things of that sort. It is also interesting that the word "machine" has the same origin in English as the word "may," which means capability, possibility, and so on.[1]

Consequently, "machine" has meant a man-made, nonanimal contrivance, which is designed and built to possess some impressive ability for realizing a goal or achieving a certain gain. Of these features, the man-made and nonanimal aspects have become today's favored subjects of discussion and controversy in the circles of scientists, theologians, and philosophers. Machines function more and more like animals, and the differentiation between man-made devices and biologically produced animals becomes increasingly involved. In reference to the rest of the "historic" features of machines, today's automata seem to develop their own goal for existence. Then, too, the capabilities of certain machines extend to almost mystic dimensions. Astonishment and admiration for such "beasts" reach the boundaries of romantic respect among laymen. However, a careful examination of the subject of "machines" suffices to annihilate all superstitious beliefs about the domination of man by machines. Man has more to fear from himself than from the machine that he is able to create.

Whether the power of science is wielded for good or evil depends primarily not on scientific principles and theorems but on set goals and motives that humans developed directly from the domains of religion, philosophy, and human psychology. Neon light will never make men more virtuous than kerosene lamps, and a bad joke remains bad even when transmitted over the electronic wonder of television. More and more there is a growing moral demand on the scientist and the engineer to bridge the widening gap between the rapidly accelerating advances in technology and the lagging social maturity of the men that serve it. The scientist and the engineer of today cannot and should not divorce themselves from the responsibility of having to share their discoveries with others. They cannot remain oblivious to the "people problems" which the technological advancement engenders.

1.2
A CLASSIFICATION OF MACHINES

One way to classify machines may be based on the goal that the machine is built to realize. Historically, machines were invented and developed for the purpose of multiplying torque. Later, energy-converting and, more recently, information-handling machines were developed.

---

[1] Their common origin is the Greek word "mēchos."

a. Force or Torque Multiplying Machines

In these machines, force or torque is amplified or attenuated under the condition of preserving mechanical energy. Mechanical gears, levers, and like devices still in wide use today are examples of such machines. (Fig. 1.1).

Fig. 1.1 Force multiplying machine. Lever.

b. Energy-Converting Machines

These machines transform one type of energy into another. Chemical or solar batteries, steam engines, electrical motors, and generators are examples of these types of machines. Energy-converting machines operate within the law of conservation of energy and the second law of thermodynamics. The nineteenth-century steam engine played a substantial role in the industrial revolution of that time. (Fig. 1.2).

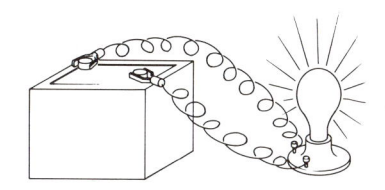

Fig. 1.2 Energy converting machine. Conversion of electricity to light.

c. Information-Handling Machines

In the twentieth century, novel types of machines appeared: the telephone, the telegraph, the gramophone, radio, television, motion pictures, and the automatic computer. Although they all consume an amount of energy and may use conversion or even torque multiplication in their operation, their primary objective is the handling of comprehensible information. These information machines collect, transmit, or process information. Three such types of information machines are identified.

1. Information-collecting machine. A microphone, a photographic camera, a radar antenna, or an orthicon television camera, devices that correspond to our sense organs (eyes, ears, and so forth), are examples of this type of machine. Energy conversion may be involved in the operation of such machines, but the overall primary goal is the collection of information.

2. Information-transmitting machine. The telephone and telegraph, analogous to the neurons that connect the sensory organs to the brain in animals, are examples of such machines.

3. Information-processing machine. A familiar example of information-processing machine is the desk calculator. When properly operated, it will perform desired processing on the input numerical data. If the human operator is also considered in conjunction with the desk calculator, then we have a man–machine information-processing system, which not only processes information, but also performs information collection, storage, and transmission. The operator may read numerical data from a ledger, feed them to the machine, or store partial results by recording them on a separate ledger for later use.

A digital computer, which accepts certain data and after a number of desired and useful manipulations delivers different data, is another example of such a machine. These machines are commonly distinguished by a high level of complexity and sophistication in design, and include in their operation, in addition to processing, the functions of information collection, storage, and transmission. Their operation in many ways resembles closely the behavior of animals.

1.3 CURRENT USES OF INFORMATION-PROCESSING MACHINES

Information processing is generally viewed as the manipulation and organization of information in a purposeful way. A large portion of data-processing equipment used in business today deals with the control of information flow through the business system. Inventory, banking, accounting, payroll and billing, and so forth are business operations involving data organization and, to a large extent, repetitive numerical computations. Traditionally, such activities, commonly referred to as "data processing," have been performed for hundreds of years by humans, aided by such machines as desk calculators, bookkeeping machines, punched-card accounting machines, and other ingenious devices listed in Table A.1 (Appendix A). However, the term "data processing" was given common currency with the widespread adaptation of automatic electronic digital computers to business problems.

Within the realm of application of information-processing machines, there exists today a commonly accepted distinction between scientific computation and business data processing. Scientific computation is mainly concerned with the solution of problems arising in scientific research, engineering design, and so forth. Business data processing, on the other hand, is primarily concerned with the organization and numerical computation of data related to business and management operations. The distinction, however, is not so strict, at least as far as the design and operation of the machine are concerned. To some extent both types of machines are used interchangeably. Organizational and "bookkeeping" operations may be required in the course of scientific computation, while numerical solution of scientific problems, such as problems of statistics, may be involved in management and business operations.

In industry, for example, information-processing machines may be used for the solution of technical and business problems interchangeably. A digital computer may be used for "industrial process control," that is, for the control and optimization of actual industrial operations, such as the production of cement, the refinement of petroleum, and so forth. It may also be used in "simulation" problems. "Simulation" of an industrial operation consists of the formulation of an appropriate mathematical model, experimentation with it by running it on a digital computer, and modification and improvement of the model so that a better understanding of the dynamic behavior of the industrial operation may be gained. Thus, the control and optimization of actual physical operations, such as the production of cement, of warehousing operations of materials and products, and of various managerial operations, may be placed under the scrutiny and supervision of an information-processing machine.

Such forms of application have today been extended to simulation, control, and optimization of far more complex systems such as space-probe operations, radar and tracking complexes, traffic and city-planning operations, hospital operations, and the analysis of regional socioeconomic processes.

More recently, information-processing machines are designed and built, which can process information and also *adapt* to changing environmental conditions imposed on them. By adaptation, the machine may alter its mode of operation and its internal "programming," that is, the structure of information flow in space and time within the machine. Thus, the problem of programming a machine, that is, of instructing the machine as to how to respond to incoming data and even how to modify its response under given conditions, is a matter of cardinal importance in the area of man–machine communication, and therefore it will be examined first in Chapter 2.

1.4
EDP AND SCIENTIFIC COMPUTERS

Several specific names are used for the particular information-processing machines that are in use or in the experimental stage. Thus, the machines used in business belong to the general class of "digital computers" and are more often referred to as "electronic data-processing" machines, abbreviated EDP. Information-processing machines, used for the processing of scientific data in the solution of engineering and scientific problems, are often referred to as "scientific computers." From the point of view of engineering design, in principle, the scientific computers and the EDP machines are the same. The former are more versatile in performing calculus by numerical approximations and are supplemented with equipment capable of generating and correlating special mathematical functions, while the latter are equipped more to handle accounting, inventory maintenance, simple managerial decision making, and other such businesslike operations.

The machines that are used for the control and optimization of industrial operations can be either scientific or EDP. The quality of manufactured products may be recorded in an EDP fashion, and statistics about the products and the manufacturing processes can be continuously updated. On-line control and optimization of industrial processes may be guided by a scientific computer. A "general-purpose" computer is capable of performing both EDP and scientific computational tasks. However, when a general-purpose computer is used to perform only very special tasks, such as airline reservations or satellite tracking, several modifications, especially on its peripheral equipment, are necessary. "Special-purpose" computers may be designed to perform such single jobs.

1.5
CYBERNETIC AND HEURISTIC MACHINES

Besides the electronic digital computers, which are today in wide use by business, industry, government, military, science, and other areas of human activity, other types of information-processing machines are also coming into being. Examples of such are the "task adapting" and several types of "cybernetic" machines. These machines receive information from their environment and may respond progressively in a different way, based on a modification of their memorized "instructions" by a process of "adaptive learning." In a way, such machines resemble the function of humans who may also modify their response because of "adaptive learning," that is, because of "experience."

It is of interest to note here that even before the first modern automatic computer was built, A. M. Turing[2] had questioned the limitations of computers. In his attempt to create a theory of what the computers can and cannot do, Turing devised a simple imaginary computer, which he proved to be theoretically capable of performing any operation that *any* other computer could do.[3] There is a straightforward mathematical answer to the question: "What *cannot* be done by computers?" An obvious case would be found in certain classes of problems where there is no algorithm—that is, no method of supplying a series of definite, unambiguous procedural steps leading to a solution. Since the digital computer operates by following a definite "program of instruction," that is, an algorithm, there are classes of problems that computers cannot solve. Several games of strategy, such as the games of chess and bridge, are examples of problems for which an *optimum* strategy that will guarantee victory cannot be known at the outset. An adaptive procedure, one which will modify the machine's response in accordance with present environmental conditions and with past "experience," is essential in such cases.

In such a case then, the machine plays a "fair" game of chess, that is, one in which the winner cannot be decided from the outset, because no one may possess a winning strategy from the beginning. An adaptive searching procedure is employed by man or machine in a game of chess, one which cannot guarantee

---

[2] Turing, A. M., "On Computable Numbers, with an Application to the Entscheidungsproblem," *Proc. London Math. Soc.*, series 2, vol. 42, pp. 230–265, 1936–1937; with a correction, *ibid.*, vol. 43, pp. 544–546, 1937.

[3] The Turing machine is discussed in Chapter 11.

optimum results (that is, victory), but only improve the chances of winning, based on the method of search employed, the breadth of search alternatives used (memory capacity of the player), and the accumulated adaptive learning, that is, the experience of the player.

In other situations, the choice of an optimum course of response may require a search through an immense number of alternatives. In strategy games like "bridge," "chess," or even "checkers," the number of alternative "paths" of action far exceeds the memory capacity of any imaginative machine. It has been estimated that there are about 10^{120} alternative paths in a game of chess and about 10^{50} possible ones in a game of checkers. A "heuristic" process, one which is based on some sort of orderly trial-and-error approach, or on a random search, with some order in it, offers a way out to the machine designer, and is of great interest in the design of artificial intelligence machines.

In this book, we use the name "information-processing machines" to mean any such machine, including the automatic digital computer. Special reference to specific machines such as to "digital computers" versus "cybernetic machines" will be made occasionally by underlining the necessary distinction. It should be understood, however, that all information-processing machines possess basic features in common. These basic common features are brought forth in the Turing machine, discussed in Chapter 11. The fact that all information-processing machines are "programmed machines," capable of manipulating information in a purposeful way and possibly of modifying their internal programming, is a common feature. The specific mode of operation may differ from machine to machine.

1.6
SPECIAL-PURPOSE AND GENERAL-PURPOSE MACHINES

Certain machines are designed and built for performing only special information-processing tasks. Such machines cannot be "instructed" to modify their sequence and types of operations to any great extent without basic redesigning of the machine. An automatic telemetering computer, for example, is designed to process certain kinds of input data and provide information about the position and velocity of moving targets. A special-purpose computer may also be used in conjunction with artillery guns, or with guided missiles and target-seeking projectiles, for fast calculation of telemetering data and control of trajectory. Smaller special-purpose information-processing units are also used within a general-purpose computer for the execution of specific information-processing routines, such as the calculation of special mathematical functions, as for example, exponentials, sinusoids, error functions, and so forth.

Frequently, special-purpose information-processing units have their "instructions" wired internally. Modification of their information-processing procedures

8   REFLECTIONS ON INFORMATION-PROCESSING MACHINES

is possible only to a limited extent by external "plugboard" wiring. The plugboard gives the machine "programmer" a number of choices in the completion of internal machine circuits. Figure 1.3 illustrates the use of a plugboard in order

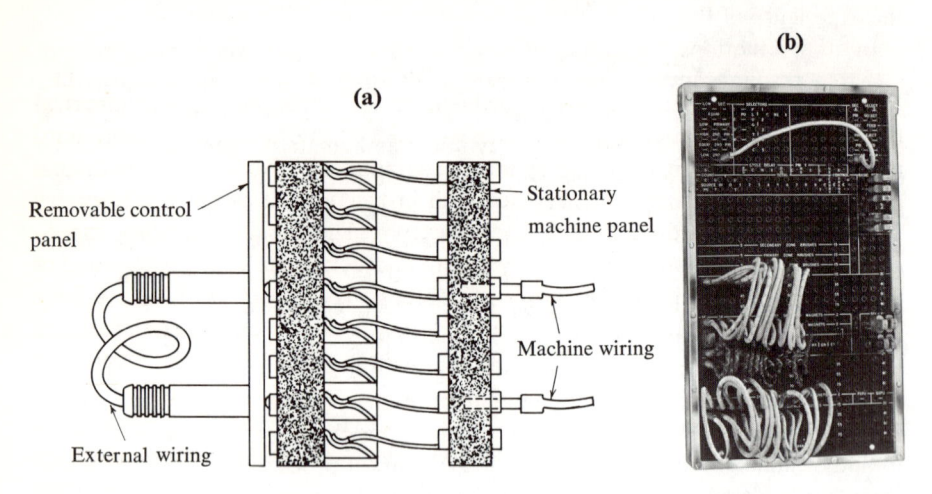

Fig. 1.3 (a) Circuit completed by external wiring on a plug board. (Reprinted by permission from IBM Reference Manual "Functional Wiring Principles," A24-1007-0, © 1958 in original notice by IBM (b) Photo of actual board (Courtesy of IBM Corp.)

to modify and complete internal machine circuits by external wiring. A removable plugboard (control panel) is fitted into place on a stationary machine panel. Wires from the machine circuits reach the receptacles on the stationary panel. For every receptacle on the stationary panel, there is a mating receptacle on the fitted plugboard. Machine circuits can be completed by appropriate external wiring of the removable plugboard.

A general-purpose information-processing machine is capable of a wide variety of information-processing tasks, and is "instructed" by a prepared program of instructions, which is translated into a special symbolic language and "fed" to the machine. Using such instruction programs, the machine user may arbitrarily modify the sequence of internal machine operations without changing the basic design of the machine.

1.7
MACHINE LOGIC, PROGRAMMING, AND TIMING

An information-processing machine is a device capable of manipulating information, when properly instructed to do so. The machine receives instructions and data from its environment, processes the data in a manner prescribed by its designer and instructed by its user, and communicates the results back to its environment.

a. Logic

For solving a problem or for processing given numerical data, several basic "instructions," such as "add," "multiply," and "store until needed" are stated in sequence by the "programmer" who is using the human procedure of reasoning. To carry out such instructions, a chain of elementary logic circuits, "building blocks," is necessary. The most elementary input–output transformations used to carry out an instruction are of the type found in the logic of propositions, discussed in Chapter 6. The spectrum of the available instruction-chains in a computer, that is, of the available chains of elementary circuits for performing basic instructions, makes up the available "logic" of the computer and decides its versatility. It is apparent that the basic instructions to be built into the computer must be selected with care by the computer systems designer and by the logic designer to make the applicability of the computer broader. A broad spectrum of available instruction-chains gives the user considerable latitude in selecting the appropriate instructions for solving his problem. The problem of selecting instructions and putting them in sequence for the solution of a data-processing problem brings us to the next concept.

b. Programming

The outstanding feature of modern digital computers is that a "plan" of instructions may be stored in the machine and modified by the machine. Constructing such a plan of instructions is commonly referred to as "programming." Programming a machine to carry out operation A, for example, means placing the appropriate instruction table in the memory of the machine so that it will do operation A. In solving problems or processing numerical data, a sequence of instructions ("commands")[4] is first prepared by the programmer, using the human process of reasoning and a common language, like English. Then, the prepared "program" of instructions is translated into a specific symbolic language that the machine can interpret (Chapter 2). After the program has been tested and perfected, it is placed in some form of permanent storage, such as "punched cards" where it is coded in the form of punched-hole patterns (Chapter 2). When needed, the program is then transferred to specified locations of the machines main storage area (general memory). From there the instructions are recalled one at a time and executed by the machine.

---

[4] The terms "instructions" and "commands" are often used interchangeably. However, IBM makes a distinction between a "command" as something used in an input–output channel and an "instruction" as something used in a main-frame processor (arithmetic unit).

Thus, a prepared program of instruction causes long chains of logic operations to occur within the machine. If a long chain of building blocks has been constructed within the machine permanently and rigidly in a specific sequence, capable of causing only one sequence of logic operation, then the resulting information-processing unit is capable of performing only one information-processing task and no other. For obvious reasons of economy, the conventional practice has been to construct short fixed chains of logic transformations that correspond to several basic "instructions." The human user (programmer) selects the sequence of such instructions. Special circuits in the computer may also permit the execution of certain built-in instruction sequences. For such sequences, special symbols representing the sequence ("subroutine") are stored in the computer, and may be called by the program to activate the sequence. This is the "stored program" feature in a computer.

A choice faces the designer at this point. He must strike a balance between more elementary instructions using very short chains of logic transformation, thereby making hardware design easier but programming harder, or he must select more complex instructions composed of long chains, thereby making the computer more complex and costly, but easier to program. The trend in the computer art has been the latter.

c. Timing

The computer responds to a sequence of instructions stated in the "program" so that the input data is properly processed. Thus, a long series of logic operations must be performed in sequence by specific units within the computer. Two different approaches have been used in the proper timing of such operations. One has been to use *master clocking*. Each operation is allotted a certain interval of time for its performance. After each such time interval, a signal pulse from a master clock is interpreted by circuits to initiate the next operation. This is the *synchronous operation*. Computers that have no master clocks are known as *asynchronous computers*. In such computers, the completion of each operation produces a "finish pulse," which then initiates the next operation. Asynchronous computers are potentially much faster than synchronous ones, but must also have more complex circuit organization.

1.8 INFORMATION STRUCTURE

a. Bits

Numbers, alphabetical characters, punctuation marks, and all the other symbols necessary for visual human communications are represented for use within the machine by coded sequences of "ones" and "zeros."

This subject will be treated in more detail in Chapter 3, and the reasons of economy and reliability for using this "binary" system of representation of information will be delineated then. Only the essential features about the structure of information will be discussed here.

A more familiar system of coding alphanumeric[5] characters may be the International Morse Code used in telegraphy. The Morse Code uses "dots," "dashes," and "spaces" in variable length sequences to represent the alphanumeric characters, and it may be implemented physically by electrical, sound, or light impulses. Samples from the International Morse Code are shown in Table 1.1.

Table 1.1. Samples from the International Morse Code. Formally, the "dot" is a pulse of light, sound, and so on, lasting approximately 1/24 sec. The "dash" and "space" between characters are pulses lasting three times as long as a dot. The "space" between words is a pulse lasting twice as long as a "dash."

Character	*International Morse Code*
A	· —
B	— · · ·
C	— · — ·
D	— · ·
E	·
F	· · — ·
G	— — ·
1	· — — — —
2	· · — — —
3	· · · — —
4	· · · · —

In the "binary" system of representation of information, two symbols are needed. These may be selected arbitrarily. We could use the symbols T ("true") and F ("false"), or Y ("yes") and N ("no"), and so forth. It is simpler to operate using the numerical characters "1" and "0". Either of these is a "*bi*nary digi*t*," or in a contracted expression a "*bit*." Thus, a bit of information is the smallest amount of information.

Bits of information may be represented physically by using devices that may

---

[5] Meaning "alphabetical" and "numerical."

possess two distinct states, such as switches that may be either "closed" or "open." The assignment of either state to represent 1 or 0 is usually arbitrary. Thus, if a sequence of binary symbols represents a given alphanumeric character, then a sequence of "bistable" (two-state) devices may be set up to physically implement the representation of this character. For example, let the numeral "5" be represented by a sequence of three bits, such as "101." Then three switches set up so that they "read" from left to right: "closed–open–closed" also represent physically the numeral "5." If we test the state of the switches by attempting to pass electric current pulses through from left to right sequentially, we will observe a "pulse-no pulse-pulse" sequence operated on the output line. If again the existence of a pulse is assigned to "1," while the absence of a pulse assigned to "0," we should read the pulse sequence as "101."

Information theory gives the amount of information gained in an experiment (observation) as

$$\triangle I = K \ln \frac{P_0}{P_1},$$

where P_0 and P_1 are the possible *combinations* of the observed system "before" and "after" the experiment, respectively, and K is a constant depending on units. If K is set equal to $1/\ln 2$, the amount of information is now $\triangle I = \log_2 P_0/P_1$, and it is measured in "bits." Thus, if three bistable devices (such as three switches) are enclosed in a "black box" and by measurement we determine the state of one of these switches, then the amount of information we gained by this experiment is

$$\triangle I = \log_2 \frac{8}{4} = 1 \text{ bit},$$

since before the observation there were $P_0 = 2^3 = 8$ possible combinations of the system, and after the observation these remained only $P_1 = 2^2 = 4$ possible combinations.

b. Characters

Since n-bits can produce 2^n in-line combinations, it seems that four-bit sequences will more than suffice to represent the ten decimal digits from 0 through 9. Indeed, four-bit sequences will produce 16 possible combinations. We may assign arbitrarily ten of them to represent the ten decimal digits. A particular assignment, called the "natural binary coded decimal" code, briefly NBCD code, is shown in Table 1.2. This code is also commonly referred to as the "8421 code" as is discussed in Chapter 3 (Section 3.11). Using this code we may, for example, represent the decimal number 582 by the grouped sequence: 0101 1000 0010.

Table 1.2. Four-bit coding of the decimal digits—the NBCD or 8421 Code.

Decimal Digit	NBCD Code	Decimal Digit	NBCD Code
0	0000	5	0101
1	0001	6	0110
2	0010	7	0111
3	0011	8	1000
4	0100	9	1001

In order to represent both alphabetical and numerical characters, we need at least 36 combinations. We find then that sequences of five-bits will not suffice since they only provide us with 32 possible combinations. Therefore, it is customary to use six-bits in order to encode both letters and digits. The 64 possible combinations provided by the six-bit sequences are sufficient to supply many additional symbols, such as punctuation marks and other special symbols found on a typewriter. Such six-bit combinations may then be referred to as "characters."

c. Words

Information that appears to the machine is of two different kinds: "program information," that is, instructions which tell the machine the tasks to perform, and "data information," that is, numerical and alphabetical information to be operated upon. Either kind of information is represented to the machine in the form of "words."[6]

Thus, if the machine is to add the decimal numbers 187 and 92, it must receive a "word" representing the number 187, another "word" representing the number 92, and another "word" that will represent to the machine the instruction to "ADD."

If we use the NBCD code to represent these numbers, we need a 12-bit word for the decimal number 187 (0001 1000 0111) and only an eight-bit word for the number 92 (1001 0010). Since the machine handles continuous streams of bits, and it must have some means of determining when one number leaves off and another one starts, it was found expedient to represent information by fixed-length bit sequences, that is, by fixed-length "words." This has been generally the policy in most digital computers. Several modern computers, however, use variable-length bit sequences to represent information.

---

[6] Although most computers use one word per instruction, an occasional computer may pack two instructions in one word, or have one instruction occupy two words.

The length of the word may be fixed by the manufacturer of the machine and determines the largest number that can be handled by the machine. Thus in the NBCD code, which uses four-bit characters to represent decimal digits, a 60-bit word can represent up to 15 decimal digit numbers. The largest number represented by a 60-bit word in the NBCD code is $10^{15} - 1$. When six-bit characters are used to represent alphanumeric information, a 60-bit word can hold ten such characters.

d. Instructions and Data Words

There is no way to tell whether a given word is supposed to be an "instruction" or a "datum" by examining it, since all words are of the same length. If the machine receives a piece of data and it tries to interpret it as though it were an instruction, its ensuing actions would be unpredictable. Several safeguards are employed to make sure that this would not happen.

The programmer sees that specific areas of the machine's memory are assigned to hold the instruction words. Then the machine will interpret a word coming from that area as an instruction, while a word coming from the area of the memory reserved for data as datum. Also, the internal structure of an instruction word is different from that of a datum word. A portion of the word is devoted to "task description," another portion to "addresses of data referred to," that is, where to find the data to be operated upon, and so forth.

Thus, the machine handles continuously two different streams of information, "instructions" and "data," and can tell which is which by the origin, the internal structure of the words, and what it is asked to perform. If the machine has been asked to add a word to another, it assumes that these words represent decimal numbers and reads the word bits in groups of four. It is the responsibility of the programmer to see that information is in the proper form and in its proper location in the machine's memory.

1.9
GENERAL MACHINE ORGANIZATION

The initial ideas of logic design and organization, which the pioneers of information-processing machines developed during the 1940s, were so powerful that present work is still being conducted within the framework of what was already feasible over twenty years ago. Machine organization and language remain basically the same, and a generation of machines that will encompass new, broad, and informative ideas does not yet appear as a certainty.

A complete information-processing system, which may be a combination of mechanical, electrical, or other components, includes five kinds of basic units: *memory* components which accept information, store it for short or long periods of time, and then deliver it unchanged on demand; *arithmetic* components (processors) which process (transform) information in accordance with prescribed ways; *control* components which command and administer operations within the machine; *input* components which serve as sensors for receiving information from the environment and transferring it to the machine's main memory; *output* components which serve for the dissemination and the display of information to the environment.

Input and output components consist of punched-card readers, magnetic tape units, typewriters, high-speed printers, and so forth. They are often integrated in separate units referred to as peripheral equipment. (See Section 1.10.) The main arithmetic and control components also form separate integrated units, although smaller processing and control components may also be scattered throughout the machine for the execution of special tasks. Similarly, there exists a main memory unit (general store) for the storage of instructions and data, although information is also stored to and retrieved from smaller temporary storage components within the arithmetic unit and elsewhere called *registers*. Usually registers have one "word" capacity and are used to store arithmetic *operands*, instruction *commands*, memory *addresses*, or computation *results* only for limited time between operations. Besides the above-mentioned basic kinds of components there exists a multitude of other auxiliary components, such as electronic circuits for pulse forming, clocking, amplifying, and so on.

Only the essential features of the information-processing machine organization are discussed here. A more specific and detailed discussion is given in Chapter 10, Section 10.6. Meanwhile, the student must have a conceptual picture of the general organization and of the principal units and functions involved. A block diagram is given in Fig. 1.4.

The machine shown in Fig. 1.4 includes five basic units: the *input*, the *general store* (memory), the *processor* (arithmetic), the *control*, and the *output*. Two buffer units are also shown as parts of the input–output equipment. *Registers*, which are temporary storage units of limited capacity, are used throughout the main body of the machine (memory-processor-control units), in order that a "word" of information, such as an "operand" or an instruction, may be temporarily stored while awaiting action. Off-line devices, such as card punches, paper tape punches, and other auxiliary equipment, are not shown here. A list of the main functions for each one of the five units of the block diagram shown in Fig. 1.4 follows:

1. Storage. Short-term and long-term storage of data and of programmed instructions is essential for the automatic operation of a data-processing machine. Storage of large amounts of information is performed by the main memory unit of the machine. Information is constantly exchanged between the memory and the processor unit during the operation of the machine. In large data-processing systems, communication between the memory and the processor or between the memory and the input–output equipment is done through the premises of intermediate units. These units act as temporary storages and are called "buffer" units. By their intervention, they permit the units of both sides to operate at different information flow rates. This is desirable because

16 REFLECTIONS ON INFORMATION-PROCESSING MACHINES

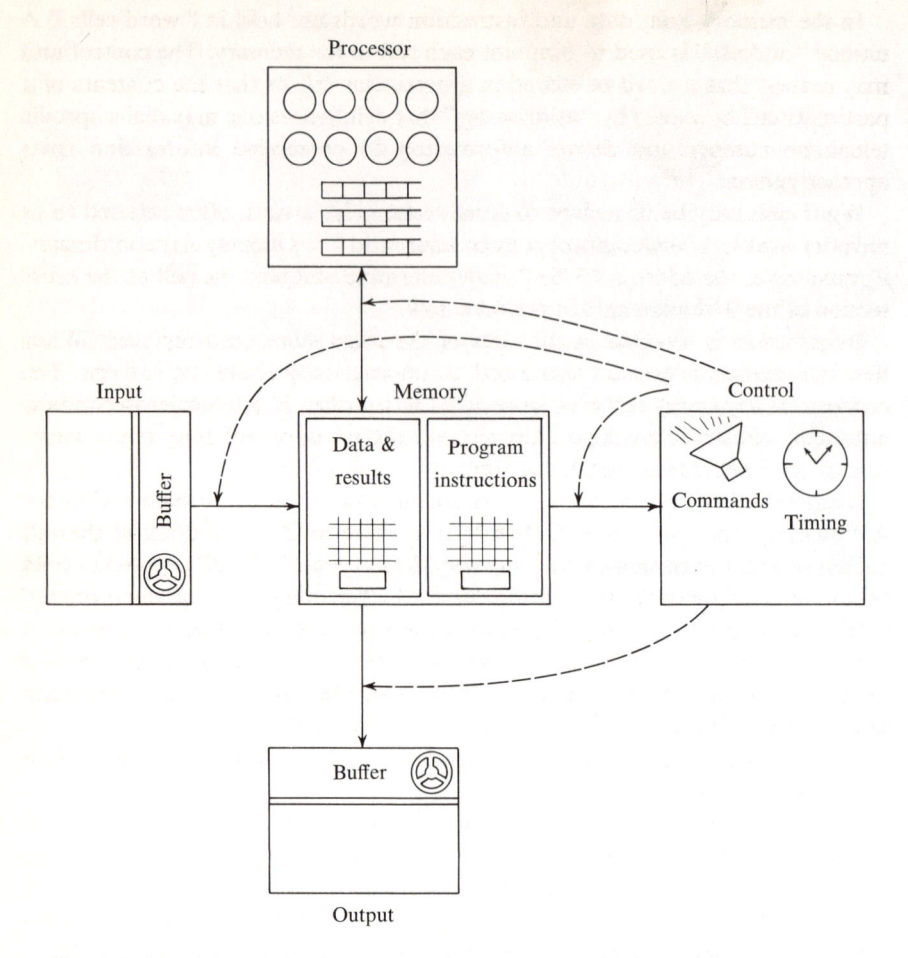

[Figure: General organization of an information processing machine, showing Processor at top, Input with Buffer on left, Memory (Data & results, Program instructions) in center, Control (Commands, Timing) on right, and Output with Buffer at bottom.]

Fig. 1.4 General organization of an information processing machine.

the input–output equipment operates slower than the memory unit and even slower than the processor.

Shorter-term storage devices are used throughout the computer for temporary storage of small amounts of information. Such devices as "registers," "shift registers," and "delay devices"[7] are in wide use throughout the main memory, the processor, and the control unit.

---

[7] Shift registers are temporary memory cells, like registers, whose capacity may exceed that of one word. A single word may enter the cell on one side and be shifted with respect to itself, to the right or to the left, along the register. This shifting operation is essential for multiplication of two numbers as we shall see later.

"Delay devices," which can receive a piece of information such as a word, circulate it in a temporary storage loop, and then deliver it, are essential in arithmetic circuits (see Chapter 10). Delay lines or circular shift registers may be used for this purpose.

In the memory unit, data and instruction words are held in "word cells." A unique "address" is used to pinpoint each cell in the memory. The control unit may request that a word be stored in a particular cell or that the contents of a particular cell be copied by "addressing" that cell, just as one may dial a specific telephone number and deliver information to or receive information from another person.

Word cells may be conceived to form rectangular arrays, often referred to as memory banks. A particular cell may be addressed by its rectangular coordinates. For example, the address "3753" may determine uniquely the cell at the cross section of the 37th column and the 53rd row.

Information is retained in the cells of the main store until replaced. When new information is entered into a cell, it automatically erases the old one. The contents of a memory cell may be recalled in a matter of a few microseconds in automatic electronic machines. On request, the memory will copy the contents of any given cell into a specific register.

Memory units encompass other components in addition to the memory banks. A "memory address register" (MAR) is used to hold the address of the cell currently under examination. A "memory data register" (MDR) is used to hold temporarily incoming or outgoing data. A "memory control component" (MCC) is a small control unit, which supervises the execution of operations within the memory. It initiates memory cycles, locates cells, takes required action, keeps track of the operations, and issues the "done" signal to the main control unit (Chapter 10, Section 10.6).

2. Processing. Instructions and data enter the memory unit through the input equipment. Under the command of the programmed instructions that are read, timed, and coordinated by the control unit, the processor unit performs addition, subtraction, multiplication, division, as well as various data editing operations on binary coded numerical data. As we shall see in Chapter 10, Section 10.2, the four arithmetic operations may be reduced to performing the three simpler operations of addition, complementation, and shifting.

Editing refers to the proper grooming of data before they are presented to the human user of the machine. For example, the price of an inventory item may be stored in the machine's memory as 0000063950. Before it appears on the printed list of items correctly as $639.50, the initial zeros must be *deleted*, the dollar sign *appended*, and the decimal point *introduced*. These operations comprise the editing functions of the processor.

3. Control. All operations within the machine take place under the administration of the control unit. The operations occur in sequence of alternating "fetch" and "execute" cycles. A complete fetch and execute sequence should involve the following procedure: *Fetch* the program instruction from the memory (one at a time); an "instruction register" is used to keep the instruction word stored while the instruction is being executed; recognize the instruction command and issue the *execute* signals for the appropriate operations; time and monitor the execution of the command; receive the "done" signal. The "instruction counter" keeps track of the address sequence for the instructions. Once an instruction has been executed, an appropriate signal to the instruction counter causes it to advance one position in address to the next instruction step. This procedure continues until the machine comes to the END instruction and turns itself off.

4. Input–Output. All information enters the machine through the input equipment and is stored in the general store (main memory unit). All information from the machine passes through the output equipment. Buffer devices are used, which serve to gear the information flow rate. Thus input buffer equipment, such as punched-card readers, are used so that information punched into cards at a slow rate may be transmitted to the machine's memory at a much higher rate. Often, two intermediate steps may be used. Information may be punched first manually into punched cards. Then the information is transferred from the punched cards into high-density magnetic tape. The machine then reads the tape at extremely high speeds, hundreds of thousands of characters per second, much faster than punched cards. Similarly, printing, display, or other output devices are fed through buffer magnetic tape units in order that they may operate at a lower rate than the machine does internally.

Because of the tremendous speed mismatch between the computer and the human and the high cost of computer time, on-line "conversation" between many users and one machine is possible by properly multiplexing the input channels. Today, researchers are working on more efficient schemes that allow many users to "converse" with a single machine concurrently. During the time it takes one user to read information supplied by the machine or prepare information for it, the machine is kept busy by solving problems of other users.

1.10
PERIPHERAL HARDWARE

A number of equipment is essential for the preparation and conveyance of information to the machine and for the dissemination of information from the machine. The most common "input" and "output" equipment is discussed briefly.

a. Input-Output Media

The most commonly used media in transferring information to and from the machine are punched cards, punched paper tapes, and magnetic tapes.

b. Punched Cards

The punched card is a rectangular paper card of about 3×8 in. and a few thousandths of an inch in dimensions. It uses rectangular holes in accordance to the IBM or Hollerith coding system[8] to provide for the representation of alphanumeric and other special characters by the relative positions of the holes in the card.

The card that is shown in Fig. 1.5(a) has 80 columns numbered from left to

(a)

(b)

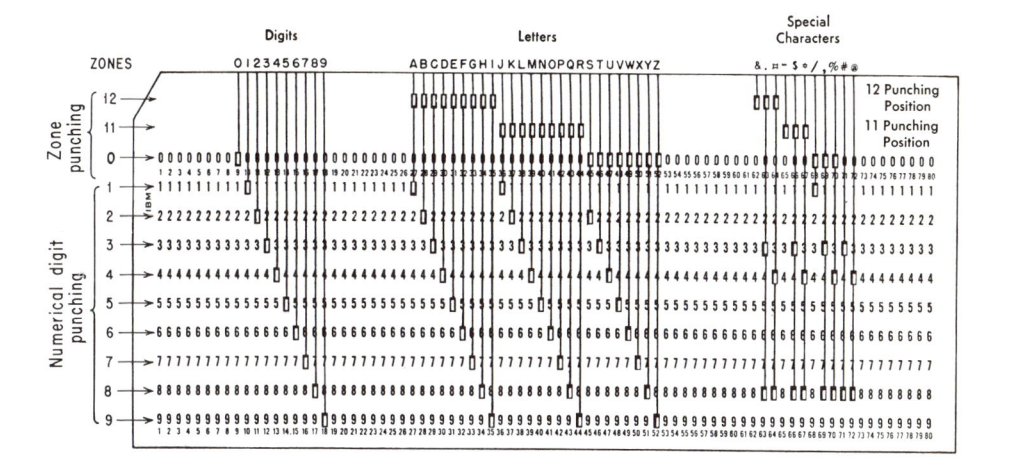

Numerical Zone	12-Zone	11-Zone	0-Zone
1	12 and 1 = A	11 and 1 = J	
2	12 and 2 = B	11 and 2 = K	0 and 2 = S
3	12 and 3 = C	11 and 3 = L	0 and 3 = T
4	12 and 4 = D	11 and 4 = M	0 and 4 = U
5	12 and 5 = E	11 and 5 = N	0 and 5 = V
6	12 and 6 = F	11 and 6 = O	0 and 6 = W
7	12 and 7 = G	11 and 7 = P	0 and 7 = X
8	12 and 8 = H	11 and 8 = Q	0 and 8 = Y
9	12 and 9 = I	11 and 9 = R	0 and 9 = Z

Fig. 1.5 (a) IBM card, and character coding. (Reprinted by permission from IBM Reference Manual "Functional Wiring Principles," A24–1007–0, © 1958 in original notice by IBM Corp. (b) Table of column-coded combinations of punches for the IBM card.

[8] Dr. Herman Hollerith developed a punched-card system for tabulating census data during the decade 1880 to 1890. The IBM code is based on that system.

right, and 12 horizontal rows. Each column may be used to represent uniquely any one character provided for by the code. Therefore, the maximum capacity of each card is 80 characters. The characters provided for by the code are shown at the top of Fig. 1.5(a), and they correspond to the specific combinations of punches in the columns as shown in Fig. 1.5(b).

The two rows above zero row are without printing and are referred to as twelve (top row) and eleven. Notice that the alphabetical characters require two punches per each, one of which is either a twelve, an eleven, or a zero-row punch. A twelve, eleven, or zero punch is known as a "zone punch," while a punch in the rows from one through nine is known as a "digit punch."

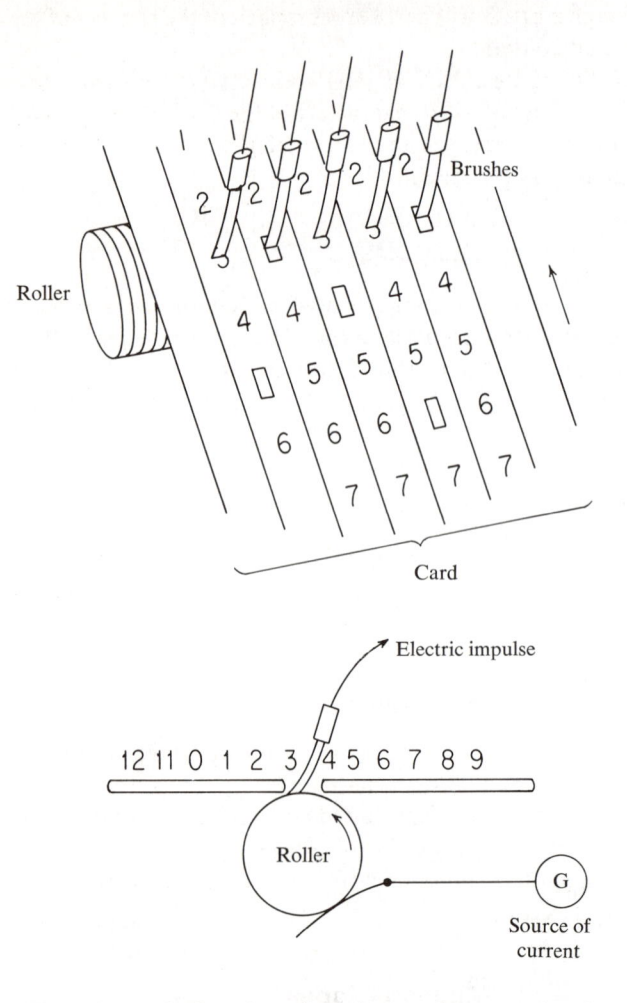

Fig. 1.6 Punched-card reading. (Lower figure reprinted from IBM Reference Manual "Functional Wiring Principles," A24–1007–0, © 1958.)

c. Card Readers

A card-reading machine can interpret the Hollerith code on the basis of the relative times at which electric impulses are emitted. The technique used for sensing the coordinates of the punches on a given card is illustrated in Fig. 1.6. The card is moving with the twelve-edge leading, over a metallic roller which is connected to an electric current source. Metallic brushes, aligned along the 80 columns, sweep over the card. When a brush makes contact with the roller through a hole, as with the "three"-hole in Fig. 1.6, an electric circuit is completed and an electric impulse emitted through the brush lead. If the card is pushed along with a constant velocity, the time relation of impulses emitted through a brush will be an exact replica of the spacing of the corresponding holes in that column.

A row of 80 brushes, each aligned with one card column, provides for the simultaneous sensing of all 80 characters in the card. A card reader can feed the machine with the contents at an average of 1000 cards per minute, that is, well over 1000 characters per second.

d. Keypunches

The keypunch is used so that an operator may punch alphanumeric and other characters in cards. The keypunch resembles a typewriter, as shown in Fig. 1.7. The characters are punched one at a time in column by column from left to right. As a key is pressed, the corresponding hole for the character on that key is punched into the column presently before the punching hammers. The card is then automatically advanced to the next column.

Card positioning, column skipping, and upper and lower case shifting are provided. The main components of the keypunch are a card hopper, a card feed, a punch-and-print station, a read station, a stacker, a keyboard, and a program drum. The physical location of these components on the IBM 29 Printing Card Punch is illustrated in Fig. 1.7.

e. Card Punches

The machine can copy any information stored in its memory in the form of punched cards using its own card-punch facility. For example, after a certain program has been tested and perfected, it can be copied from the machine's memory into punched cards for permanent storage. Present machines can enter information into punched cards at an average rate of about 250 characters per second.

f. Magnetic Tapes

A commonly used medium for supplementary external storage and for "buffering" (see Section 1.9) is the magnetic tape. It possesses

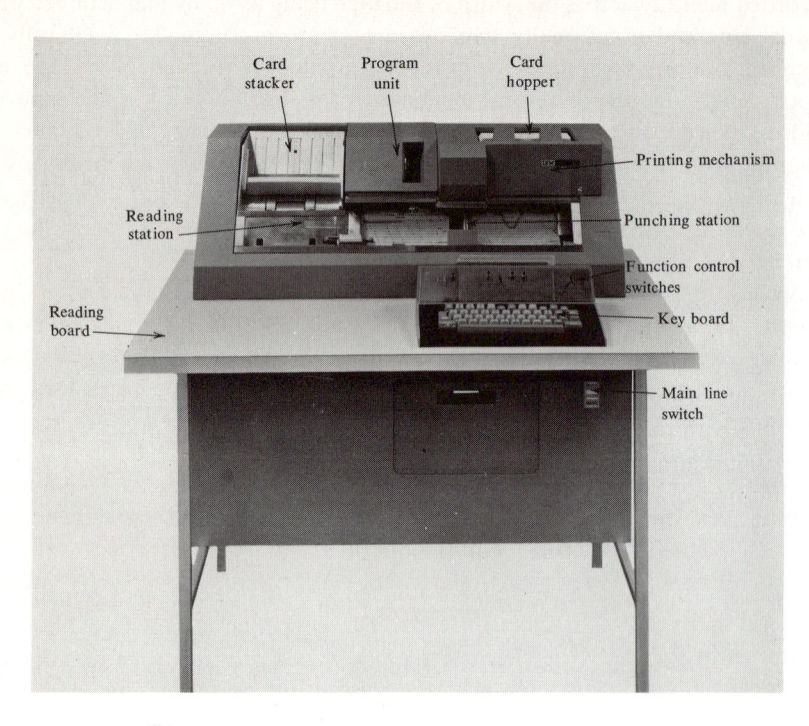

Fig. 1.7 (a) IBM 29 Card Punch; (b) Combination keyboard. (Reprinted by permission from IBM Reference Manual "IBM 29 Card Punch," A24–3332–2.)

inherently large storage capacity of the order of many million characters per reel of tape. The tape itself is usually made of mylar, about 1 to $1\frac{1}{2}$ mil thick, $\frac{3}{4}$ in. wide and 2000 to 4000 ft long wound on $10\frac{1}{2}$ in. diam reels. Bits of information are stored in the form of minute magnetized spots. The magnetization of each spot is complete ("saturation" magnetization), and the direction of magnetization is allowed to align in one of two possible ways, one to denote "1" and the other to denote "0." Eight to twenty separate tracks or "channels" of bits are recorded laterally across the width of the tape using as many magnetic recording heads. One and occasionally two characters are accommodated laterally for each stack of bits. Characters are grouped into blocks the length of the tape, each block containing one or several machine words. Each block has an identifying address placed on a separate channel, and a "parity-checking" channel (see Chapter 3, Section 3.11) wherein a digit is carried along with each machine word as a check on the accuracy of the data. In "even-parity-check," the parity digit is 1 when the total number of 1's in a machine word is odd, and 0 when the total number of 1's in a machine word is even.

Because of the sequential access to any section of the tape and its great length, the average access time for information stored on magnetic tapes is comparatively long, in relation to other media (ferrite cores, magnetic disks or drums, and so on). The tapes, however, are mechanized to run either forward or backward and with great speeds for faster access. The average access time to any particular block of information on the tape is on the order of 10 sec.

The adjustable speed of receiving or delivering information makes the magnetic tape an ideal device for buffer uses, whereby information-flow-rate mismatches between the machine and different peripheral equipment may be accommodated. Through magnetic tape units, the machine may receive at high speeds information accumulated from slow-rate sources, such as typewriters, and also deliver information at high speeds to slower printing or displaying output equipment.

g. Printers

Visible records of the machine's output may be obtained at relatively high speeds using high-speed printing devices. Two types of high-speed printers are commonly used: the mechanical (or electro-mechanical) and the electrostatic printers.

The *mechanical printer* employs physical contact between paper and a hard-type surface. The so-called *line printers*, which are in wide use today, employ stacks of print wheels containing 120 alphanumeric and other special character print positions. Each print wheel is rotated and positioned by electric impulses to the desired character. Special hammers activated again by electric impulses strike the corresponding print wheels. Ink ribbons are used for the typing, and paper is moved and guided between the print wheels and the hammers. Average speeds of several hundred lines per minute, with up to 120 characters per line, are attained by line-at-a-time printing.

Even higher speeds are achieved with the *on-the-fly* electromechanical printers. The print wheels or solid drums with rows of characters are in continuous motion during operation. When precisely in a desired position, the wheels or the drum are struck by well synchronized hammers. High-speed printing of over a 1000 in./min is feasible with the on-the-fly printer.

In the *electrostatic printer*, character formation on paper is accomplished in the following way: High voltages are applied to a selected configuration of pins and bars on the face of the print head. This causes arcing between the pins and the bars. The strong electrostatic fields of the arcs cause charges to deposit on the plastic coating of the special paper as the paper passes in the close vicinity of the print head. In this manner, the patterns of arcs on the print head are transferred as similar configurations of charged areas onto the paper. Then the paper is advanced to an inking station, where dry powdered ink is attracted, and subsequently fixed thermally only to those areas that are charged, thus producing permanently printed characters.

An overview of several peripheral equipment is provided in Fig. 1.8, where the IBM System/360 data-processing system is shown. This machine is available in several models.

1.11 HISTORY AND CLASSIFICATION OF DIGITAL COMPUTERS

Digital computers may be classified with reference to their internal organization, in plugged or wired program computers versus externally programmed computers. With respect to data-processing features and capabilities, computers may also be classified as: scientific and commercial, small, medium, and large systems, as well as special-purpose computers.

The term "data" should refer more aptly to input information that is to be operated upon. The following types of data sources characterize the classes of application for digital computers:

1. data generated as a result of measurements on physical quantities,
2. alphabetical or numerical information encountered in statistical and accounting work, and
3. numbers involved in the solution of mathematical problems.

These numbers may have some relationship to the previous two data sources, or they may have no physical significance.

In the following discussion, our historical references are limited to those that deal directly with the development of modern high-speed electronic data-processing machines, which have come into use since about 1945. A great variety of such machines exists now. They allow computations to be carried out up to several million times faster than was possible with even the most elaborate machines used as recently as 1940. This advance opens the door to the possibility of successfully attacking significant and complicated problems never before attempted.

The large variety of past and present data-handling machines may be classified as mechanical, electromechanical, or electronic. Perhaps the simplest mechanical calculator is the addition slide rule, the ancestor of the present-day multiplication slide rule. Both use lengths in computations to represent quantities, the

Fig. 1.8 IBM System/360, Model 65. The IBM System/360 is a general purpose machine available in several models, which are identical in design concepts and compatible in programming but varying in size, speed, cost, and other features. The wide range in complexity and capabilities is provided in order to offer the comprehensive range of computing versatility demanded today in industry, business, and science.

numbers being spaced in a linear scale on the addition slide rule and in a logarithmic scale on the multiplication slide rule.

The next very important mechanical calculator used rotation in the processing of information, employing various combinations of gears and digit-carrying cylinders. Its great importance lies in its large capacity of information processing. Pascal (France, 1642) and more completely Leibnitz (Germany, 1671) used such ideas and constructed some of the earliest devices of this type. They served as the basis for several later developments. The idea of Leibnitz's stepped cylinder was widely used later by many. Hahhn (Germany, 1770) developed the first dependable four-process calculator, using Leibnitz's stepped cylinder. A crank-possessing Leibnitz calculator was perfected fifty years later by Thomas (France, 1820). This device is considered to be the grandfather of all present-day desk calculators. It was copied widely in Europe, and after its appearance in the United States, stimulated further developments. Such important developments in the United States include the first key-driven four-process calculator by Hill (1857), Baldwin's (1872) first reversible four-process calculator, and Verea's (1878) first model that could do direct multiplication (not through successive additions as was done before). The first simple key-driven adding machine, known today in somewhat improved form as the Comptometer, was developed by Felt (1884). Hopkins (1901) made the first ten-key adding machine, and Baldwin and Monroe (1920) constructed the first fully automatic Monroe calculator. This machine uses electric power and belongs to the next class of electromechanical calculators.

Developments of electromechanical calculators involved two basic ideas, punched cards and difference machines. Punched holes in cards are used to represent information and also to control the course of calculations. The idea first came from Joseph Marie Jacquard (France, 1801). Punched cards were used in sequence to control the selection of threads employed in weaving so that the loom could weave designs into the fabric. The idea of a difference machine, first developed by Muller (Germany, 1786), was to compute mathematical tables, by adding together in a sequential manner a set of successive differences between numbers. The idea was more advanced than the state of the technology at the time of Muller and never materialized as hardware at that time. Babbage (England, 1830) borrowed Muller's and Jacquard's ideas to make models of a difference machine that was card controlled. He used the holes in the cards to control the course of computations. During the decade before the 1890 census in the United States, Hollerith and Powers (U.S.) working for the U.S. Census Bureau constructed the first key punch, sorter, and tabulator, all using large punched cards. The cards were used to carry information, not to control the calculation process. Electrical communication was used within the machine, whereby information in the form of holes on cards was transformed into electric impulses to activate relays. Their developments, further improved, are used today in punched-card equipment of such companies as International Business Machines, Sperry-Rand, International Computers and Tabulators, Ltd., and Compagnie des Machines Bull of France.

The next major development in electromechnical calculators came in 1944 when Howard Aiken, in cooperation with IBM engineers and some graduate students at Harvard University, completed the Mark I relay computer. Binary logic was developed, and information was represented by patterns of closed and open relays. The machine, in spite of the shortcoming of mechanical operation and low speed, had much success.

For greater speed and higher reliability of operation, electronic devices were sought for use in the electromechanical machines. Electron tubes were first used as grid controlled two-state devices. The transition from relays to electron tubes involves no change in fundamentals of computer designs, only substitution of faster and more reliable components.

The development of automatic electronic computers received great help from something that is seemingly unrelated—radar technology. In radar designs during World War II, the techniques of pulse generation, shaping, transmission, detection, and interpretation were developed and perfected. Such pulse techniques formed the foundation of electronic circuits in digital computers.

In spite of considerable mistrust of the notoriously unreliable vacuum tubes, Eniac (Electronic Numerical Integrator and Automatic Computer) was completed in midsummer 1947 by a group of designers under the direction of John W. Mauchly and J. P. Eckert at the Moore School of Engineering, University of Pennsylvania. Codes of pulse trains were used to represent decimal digits. The great success of Eniac stimulated many other projects. An explosive development of electronic computers followed during the decade of the fifties. A new era, that of solid-state super computers, started with the introduction of transistor and other semiconductor devices and is still under development at present. The tables in Appendix A review the chronological developments discussed.

BIBLIOGRAPHY

Burroughs Corp. Staff, *Digital Computer Principles*. New York: McGraw-Hill, Inc., 1962.
Chapin, Ned, *An Introduction to Automatic Computers*. Princeton, N.J.: D. Van Nostrand, Inc., 1957.
Irwin, W. C., *Digital Computer Principles*. Princeton, N.J.: D. Van Nostrand, Inc., 1960.
Ledley, R. S., *Digital Computer and Control Engineering*. New York: McGraw-Hill, Inc., 1960.
Nett, R. and Hetzler, S. A., *An Introduction to Electronic Data Processing*. Glencoe, Ill.: The Free Press, 1959.
Oakford, R. W., *Electronic Data Processing Equipment*. New York: McGraw-Hill, Inc., 1962.
Von Neumann, J., *The Computer and the Brain*. New Haven, Conn.: Yale University Press, 1958.
Williams, S. B., *Digital Computing Systems*. New York: McGraw-Hill, Inc., 1959.

2
Programming the Machine

Information-processing machines can do many operations that up to now have been done only by humans. After reading Chapter 1, one might be tempted to describe an automatic information-processing machine as a "giant electronic brain": something like a magic box, into which one could casually feed raw information and, after a slight pause, receive all sorts of answers, reports, analyses, and so forth.

This rather naive attitude toward present information-processing machines is a result of exaggerated ideas as to the powers of these machines. In fact, the presently available electronic computer is a room filled with equipment that cannot do any data processing without having been "instructed" by humans. Machines cannot think in the creative and higher associative sense of the word. Their ability to devise methods for solving problems without having specific directions given to them is practically nil. For a machine to solve a problem, it must be given a definite "plan" of instructions that tells it what to do. This devising of a complete set of instructions and writing it in a shorthand language that the machine can interpret is referred to as "programming" the machine.

2.1 INFORMATION CONVEYANCE AND COMMUNICATION

Information is conveyed by means that are perceptible by special sensory organs in the "receiver." For example, the house thermostat works by receiving information through a thermocouple, a device that senses temperature variations in the environment. Sounds, written symbols, sensing of "hot" or "cold," "sweet" or "bitter," "fresh" or "stuffy," and all such perceptions properly interpreted by conditioning and "learning" in animals and in machines are used to convey information.

A man receives information from his environment through his eyes, his ears, and his other sensory organs. Recognition and interpretation of written or verbal words are done in humans mostly by association with memorized facts, aided also by the exercise of reasoning. For practical purposes, communication of ideas and opinions is effected among humans by exercising reasoning, inquiry, and redundancy, all used to defeat the abundant ambiguity present in all human languages. By reasoning with the context of a sentence, one may decide, for example, about the proper interpretation of identically sounding words, such as "peace" and "piece."

Machines, however, cannot *reason*. In spite of obvious superiority over man in speed and accuracy of information processing, machines lack the wide variety of human faculties that are derived from the ability to reason. Therefore, machines require a *simple* and *precise* language for purposes of communication.

Since machines cannot reason, they cannot "solve" problems passed to them either (at least not problems that require deductive reasoning). A precisely stated algorithm[1] must be provided to the machine for each task. This refers to a formalized procedure whereby requested objectives are attained by a definite and unambiguous chain of operations, each operation requiring the results of preceding ones. The algorithm must be expressed in discrete procedural steps, each step involving a single "instruction command." Each instruction command must be communicated to—and interpreted by—the machine by way of an unambiguous and simple symbolic language. The methods and the means used to accomplish communication with the machine are discussed in this chapter.

2.2 PLUGBOARD AND EXTERNAL PROGRAMMING

The idea of storing instructions, which direct the internal operations, in the machine's memory was first implemented in an information-

---

[1] The word "algorithm" is a corruption of al Khuwarizmi, an Arab mathematician of the ninth century, who was instrumental through his writings in bringing to the West the present method of numeration. The use of Hindu-Arabic numerals during the middle ages was referred to by the word algorithm.

processing machine in the late 1940s. The facility for providing an information-processing machine with a "plan" of instructions and especially the ability of the machine to modify this "plan" during its execution are features of great importance that warrant some special consideration. In this chapter, we discuss the essential task of programming and the basic methods used.

a. Plugboard Programming

Instructions may be fed to the machine in different ways. According to one method, instructions may be "wired" on a plugboard and attached to the machine when needed. In this fashion, the machine receives all the necessary instructions simultaneously in one step. The plan of instructions wired on the plugboard is susceptible to a certain degree of alteration, if one uses short connecting wires that are plugged and interchanged accordingly.

The plugboard is simply a removable board, also known as the "control panel," which is fitted into place on a stationary machine panel. Both boards bear receptacles (holes or hubs). For every receptacle on the stationary panel of the machine there is a mating receptacle on the removable plugboard connected to it by feed-through electrical contact (Fig. 2.1). Wires from the control circuits of the machine, that is, those circuits which control and supervise such arithmetic operations as ADD or MULTIPLY, reach the receptacles on the fixed panel of the machine. Machine circuits can be coupled by appropriate external wiring of the replaceable plugboard. For a given machine, several plugboards may be prewired separately, each for a different information-processing task, and then they may be used as needed for the execution of the different "programs."

In order to illustrate the wiring of a plugboard, consider that the board bears a sequence of receptacles, numbered 1, 2, 3, and so on, connected through a mechanical stepping switch to a battery within the machine. These receptacles will be referred to as "program-step receptacles"; they are illustrated in Fig. 2.2. Different groups of receptacles are also connected to such control circuits within the machine as the circuits which control arithmetic and other operations of the machine, also illustrated in Fig. 2.2. These receptacles will be referred to as "instruction receptacles." The program-step receptacles are emitters of voltage signals, which may be received by the instruction receptacles if appropriate electrical connections have been effected by proper wiring of the removable plugboard.

We wire the plugboard in order to carry out the following computation:

$$\frac{(a+b)c}{d}.$$

The program steps may be described as follows:

Program Step 1: Prepare the machine by CLEARING, that is, by removing any spurious information that may have been left in it.

PLUGBOARD AND EXTERNAL PROGRAMMING   31

Fig. 2.1 Examples of plugboards.

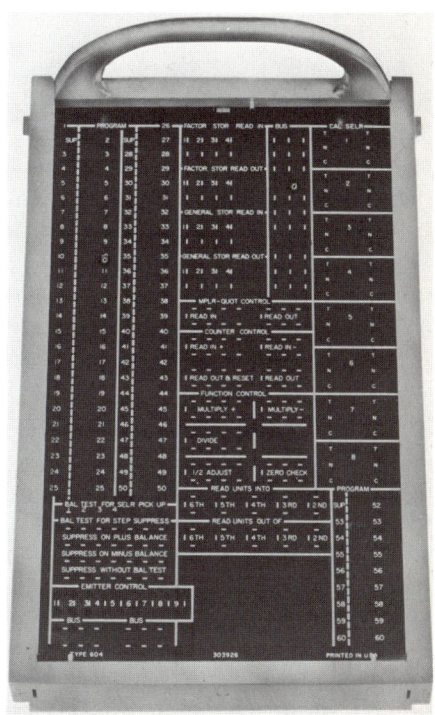

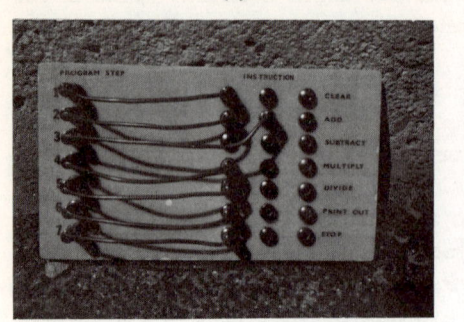

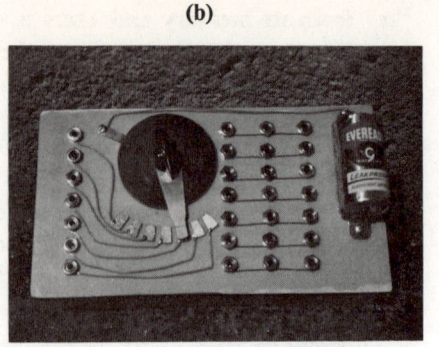

Fig. 2.2 (a) Front view of wired plug board for carrying out calculation of $x = (a + b)c/d$; (b) rear view.

Program Step 2: ADD the number "a" to the number that is already in the machine, in this case, "zero," (that is, $a + 0 = a$), and "store" the result.

Program Step 3: ADD the number "b" to the number that is already in the machine, in this case "a," (that is, $a + b$), and store the result.

Program Step 4: MULTIPLY the number "c" to the one that is already in the machine [that is, $(a + b)c$], and store the result.

Program Step 5: DIVIDE the number that is already in the machine by "d," {that is, $[(a + b)c/d]$}, and store the result.

Program Step 6: PRINT OUT the result.

Program Step 7: STOP (or GO BACK to program Step 1 and repeat the process over).

Each one of the numbers a, b, c, and d must be available to the machine's arithmetic unit at the beginning of the program steps 2, 3, 4, and 5 correspondingly. This may be accomplished by having those numbers (data) transferred beforehand from punched cards or tape into the machine's memory and then have them recalled by the control circuits of the machine at the appropriate times. Such actions must also be programmed and wired on the plugboard; they are, however, omitted in this illustration in order to simplify the program.

In accordance with the program described above, the plugboard is wired as shown in Fig. 2.2. The Program Step 1 receptacle is connected with the instruction receptable CLEAR. The Program Step 2 receptacle is connected to the instruction receptacle ADD. Similarly the Program Step 3 to ADD, the Program Step 4 to MULTIPLY, the Program Step 5 to DIVIDE, the Program Step 6 to PRINT OUT, and the Program Step 7 to STOP.

The stepping switch may now be operated. As it makes contact consecutively with the program step receptacles, it causes the activation of the connected control circuits in the above described time sequence. Thus, during "time 1," that is, while the stepping switch is in contact with the Program Step 1 receptacle, the CLEAR control circuit is activated by the voltage signal of the battery and the machine is "cleared." During "time 2" the machine receives a number

"a" from its memory and ADDs it to the one which is already stored at the location where the machine keeps the result of a computation; it then stores the result, that is, $a + 0 = a$. During "time 3" the machine receives a number "b" from its memory and ADDs it to the number "a," and stores the result, and so on. After the Program Step 6 has been completed, the machine may be instructed to either STOP, or to repeat the program anew with a new set of data numbers until some condition is satisfied, such as counting a hundred repetitions of the above program, then STOP. The latter course of procedure is known as "looping," a very important function of the machine, which allows a portion of a program to be repeated any number of times desired. If, for example, in the previous illustration we were asked to program and wire the processing of one hundred sets of numbers a, b, c, and d, the use of "looping" in programming would save us a lot of programming and wiring time.

The size of a plugboard depends on the size of the machine, that is, on the variety of the available control circuits. The complexity of the wiring depends on the extent of the program at hand. The mechanical stepping switch is replaced by an electronic one in high-speed computers.

Plugboard programming is used today in smaller machines which handle problems of rather limited complexity, often requiring information processing of largely iterative nature. The versatility of such machines is also limited to the variety of programmed plugboards available.

b. External Programming

Larger machines are programmed by instructions that the human programmer compiles in a "shorthand" language of selected special symbol combinations. The machine is capable of interpreting these coded instructions and of further translating into its own internal binary language. For example, the programmer may write: CCF 123 to mean "Copy the Contents From memory cell with address '123' into the register A." This instruction command is first entered in a punched card in the form of coded combinations of holes, as discussed in Section 1.10. Then, the computer, under the guidance of a special "translator" program referred to as an "assembler" or "compiler" program (Section 2.9), interprets each command read off a punched card and replaces each alphanumeric character in the command by a sequence of binary digits, so that at the end the whole instruction command is translated and stored in the machine's memory as a binary "word" occupying one cell.[2] While the program is being encoded in binary language and stored in the machine's memory, it is also being tested for errors. The checked and corrected program, which is now stored in the machine's memory in the form of binary coded words, is ready to be transferred as a permanent record on some physical medium. The

---

[2] Rather rarely a single command may be spread over two words or occupy only part of a single word. Two commands per word were used in IBM 701 for more efficient use of the available memory space.

instruction words may be recorded as coded combinations of holes in punched cards or paper tape or they may be stored on magnetic tape as sequences of magnetized spots packed very closely.

The advantages of this method of programming are obvious, since the same machine may be run over and over with different programs to solve various different problems, by simply changing to a different tape or a different stack of punched cards each time. The program of instructions that is assembled and recorded on a tape or in punched cards is, each time, "read" by the machine and stored in its memory, from which it is recalled to instruct the solution of the specific problem at hand. When the desired task is executed, a new program may be fed to the machine for the solution of another problem. In the rest of this chapter, we discuss further the operations involved in the preparation of such programs.

2.3 PREPARATION OF AN ALGORITHM

Writing programs for an information-processing machine consists of determining the nature and the logical sequence of operations that are to be performed by the machine, in order to obtain solutions to problems of one sort or another. For that, the problem at hand must first be thoroughly analyzed using the human way of reasoning. Eventually a solution must be found, or a definite method leading to a solution. Then an algorithm must be formed.

It should be noted here that an algorithm is not just the solution or a mathematical formula providing the solution to a given problem. The formula that provides the roots of a quadratic equation is not by itself an algorithm for solving the quadratic equation. The formula must be broken down to a finite number of small, well-defined, procedural steps, each of them requiring the results of preceding ones.

Imagine a person who is not allowed to exercise any sort of judgment on his own, unless he is provided with explicit directions for making a decision. If this person is given a desk calculator, pencil, and paper to carry out the solution of a problem, he must also be given explicit instructions of what he has to do. In a similar way, a machine operates without exercising common sense or any kind of judgment on its own. If a set of room-temperature readings is entered, the machine will not be able to recognize them as Fahrenheit or Centigrade readings and will carry the designated calculations blindly, even though the results may be meaningless. If the temperature readings come from an egg-hatching chamber, you cannot ask the machine to let you know if anything unusual happens, unless you can specify in advance what constitutes "anything unusual."

Another consideration in instructing the machine is that if you want the machine to take a specific action when it meets with certain circumstances, you must so signify. You must also foresee in advance any reasonable circumstances that may arise and instruct the machine accordingly. You should not expect the machine to recognize the end of a problem and stop, unless it has been given the means and instructions for doing so.

All these considerations must be taken into account when preparing an algorithm; they are also very important in improving an algorithm.

2.4 PREPARATION OF A PROGRAM

In its very essence, a program with its set of instructions enables the machine to follow the proper sequence of operations for the solution of a problem. Each operation is decomposed into its mathematical and logical elements, which are then carried out by the machine in accordance with its basic circuit designs and chains of elementary operations that have been built into the machine by the system designer.

The economics of the proper method of decomposition of a procedure for solution and of the choice of the most suitable sequence of instructions are, at present, based largely on intuition and depend to some appreciable extent on the skill of the author of the program and on the depth of his understanding of the possibilities latent in the machine. One can see that this is a consequence of the natural growth of the relatively young art of programming. It should be expected, however, that in the future, by way of some radical breakthrough, programming should possess its own mathematical methods and become another well-developed mathematical discipline. New mathematical methods are expected to give an accurate and exhaustive description of the internal processes occurring in a machine and, at the same time, to afford the possibility of forming self-consistent programs to deal with problems of one sort or another. An understanding of the internal processes and of the programming task through this special mathematics will also aid in the design of new machines especially suited to given problems, by clarifying and determining the character of the class of elementary and logical operations that are best adaptable to those problems the machine is called upon to solve.

The preparation of a program involves a number of steps. The information-processing task in hand must first be analyzed, and then an algorithm must be prepared. A *flowchart* (also referred to as a *flow diagram*), which provides a pictorial representation of the procedural steps and actions to be taken, should be prepared next for the complete algorithm. These two steps are helpful to the programmer so that he may clarify what he wants the computer to do before he writes the actual program. The machine is not capable as yet of understanding either a verbal algorithm or a flow diagram. Next, the prepared algorithm must be written in some form of shorthand *symbolic language*, which the machine can interpret. This constitutes the actual "program." ALGOL, FORTRAN, or

PL/I are such symbolic languages. Each program step is written in symbolic notation, and an appropriate memory location may be assigned for its storage. Finally, the program is *loaded* on the machine, tested, corrected, perfected, and punched out in cards.

To begin with, the programmer must know what the computer can do. For that he must know the machine's capacity and capabilities in general, and he must also refer constantly to the manual of listed "instructions,"[3] that is, the elementary operations such as ADD, SUBTRACT, COMPARE, and so on, that the machine is capable of carrying out. He should compose his algorithm with a series of "instructions," which can be carried out by the machine and are not too complex, but will cover the essential aspects of the task at hand.

Certain arithmetic operations and other specific short chains of action are internally "wired" in the machine and may be triggered on command by external programming. For example, the multiplication of two numbers, which are available at *proper* locations within the machine, will take place automatically if the instruction "multiply" is properly relayed to the machine. An "instruction word," such as "MUL020" may be used. If the machine recognizes the instruction, it "reads" the number stored in the A-register,[4] multiplies it with the number stored in the memory location 020, and stores the result back into the A-register. Similarly, the instruction word "CCF186" may trigger the operation of transferring and copying the contents from the memory location designated by the address "186" into the A-register (accumulator).

In following, we use two examples in order to illustrate the procedure of forming an algorithm and a flowchart for a given computational task. One can always form an algorithm and a flowchart for a given computation problem with no reference to any specific programming symbolic language. However, knowledge of the programming language to be used should dictate how detailed the algorithmic statements should be.

EXAMPLE ONE: The numbers a, b, c, and d are measurements of some kind. Given one hundred sets of such measurements, compute and tabulate x using a computer, such that

$$x = \frac{(a+b)c}{d}.$$

Depending on the versatility of the available control circuits of the machine, that is, on the structure of the programming language to be used, the algorithm may be stated in a more or less detailed sequence of statements. A detailed logarithm for the above problem may be the following one:

---

[3] Also sometimes referred to as "commands."
[4] A-register is a temporary storage device (see Section 1.9). The A-register, commonly known as the "accumulator," is used for temporary storage of the operands and results of arithmetic operations.

1. Clear the machine and start.
2. Read N, where N is the number of computations to be performed ($N = 100$).
3. Prepare to count the number of computations performed by setting a count-index I equal to zero. The index I will be increased by one after each computation and compared to N to determine when the desired number of computations has been performed.
4. Read the first number a from its stored location into the arithmetic unit, so that it becomes available for computations.
5. Add the number a to the number that is already available in the "accumulator." The accumulator is a special register that holds the result of the most recent computations.[5] Since the machine has just been cleared, the accumulator holds a "zero." Thus, this operation amounts to $a + 0 = a$. At the end of this step, the accumulator holds the number "a."

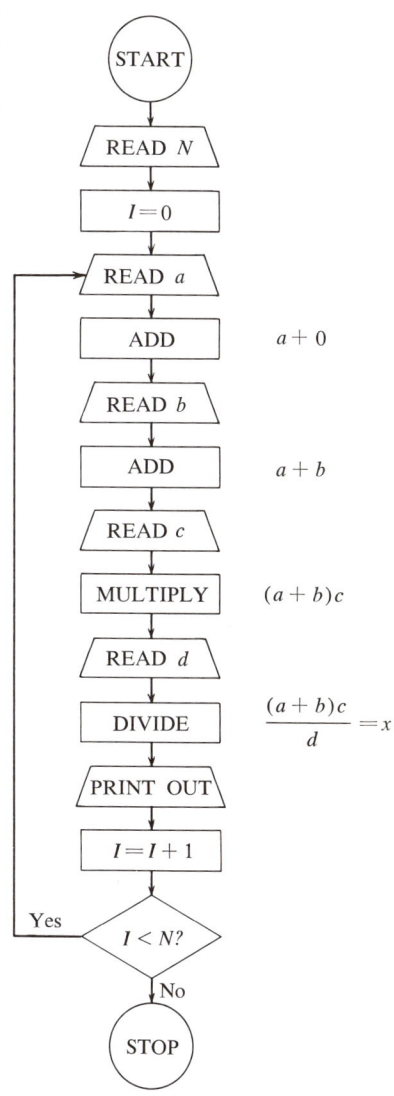

Fig. 2.3 Flowchart for Example One.

Flowchart steps (right column):
- START
- READ N
- $I = 0$
- READ a
- ADD → $a + 0$
- READ b
- ADD → $a + b$
- READ c
- MULTIPLY → $(a + b)c$
- READ d
- DIVIDE → $\dfrac{(a + b)c}{d} = x$
- PRINT OUT
- $I = I + 1$
- $I < N?$ — Yes (loop back to READ a), No → STOP

---

[5] *Ibid.*

6. Read b (similar as step 4).
7. Add the number b to the number that is already available in the accumulator, that is, $a + b$.
8. Read c.
9. Multiply the number c to the number that is already available in the accumulator, that is, $(a + b)c$.
10. Read d.
11. Divide the number that is already available in the accumulator by the number d, that is, $[(a + b)c/d]$.
12. Print out the number that is available in the accumulator, that is, "x," and clear the accumulator.
13. Accent the count by one, that is, $I = I + 1$.
14. Make the decision: "Is I less than N?" If "yes," go back to step 4. If "no," end the program and stop the machine.

Figure 2.3 shows the flowchart that corresponds to the above algorithm. In drawing flowcharts, it is a standard practice to use circles for the START and STOP operations and diamond-shaped boxes for decision operations. The direction of the flow is indicated with arrows.

EXAMPLE TWO: We are given a matrix of numbers like the one shown in Fig. 2.4, composed of M rows and N columns. We are asked

$$\begin{bmatrix} a_{11} & a_{12} & a_{13} & a_{14} & \cdots & a_{1N} \\ a_{21} & \cdot & \cdot & \cdot & & \cdot \\ a_{31} & \cdot & \cdot & \cdot & & \cdot \\ a_{41} & \cdot & \cdot & \cdot & & \cdot \\ \cdot & \cdot & \cdot & \cdot & & \cdot \\ \cdot & \cdot & \cdot & \cdot & & \cdot \\ \cdot & \cdot & \cdot & \cdot & & \cdot \\ a_{M1} & \cdot & \cdot & \cdot & & a_{MN} \end{bmatrix}$$

Fig. 2.4 Matrix representation for Example Two.

to program a machine that will add these numbers in consecutive rows and print out the accumulated sum at the end of each row, and the grand total. The computational task may be expressed in mathematical symbolic language as follows:

$$\sum_{J=1}^{M} \sum_{I=1}^{N} a_{JI}$$

The algorithm may be written as follows:
1. START the machine.
2. READ N, M, and the elements a_{JI} of the matrix.
3. SET the J-count equal to zero.
4. AUGMENT J by one, that is, $J = J + 1$.
5. SET the I-count equal to zero.
6. AUGMENT I by one, that is, $I = I + 1$.
7. FORM the sum: SUM = SUM + a_{JI}, that is, ADD the number a_{JI} (defined by the values of the J and I indices above) to the number already in the accumulator and store the result back into the accumulator.
8. Is $I \geq N$? If "no," go back to step 6. If "yes," continue on to the next step.
9. Print out the SUM, that is, the contents of the accumulator. (The accumulator, however, retains the information that is simply copied in the print out.)
10. Is $J \geq M$? If "no," go back to step 4. If "yes," continue on to the next step.
11. STOP.

The flowchart corresponding to the above algorithm is shown in Fig. 2.5.

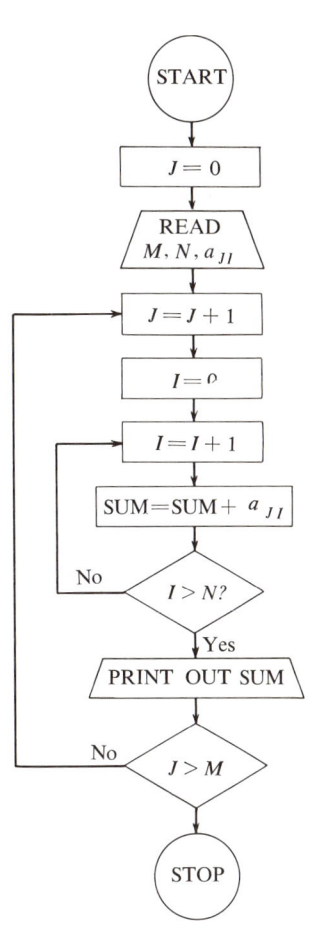

Fig. 2.5 Flowchart for Example Two.

Notice that we have formed the algorithms and flowcharts for the above two examples without reference to a specific programming language. It is recommended that the student practice preparing algorithms and flowcharts for several computational tasks of his own making. A few problems are also provided at the end of this chapter for this purpose.

A very important property of computers is illustrated with the above two examples. This is the ability of the machine to "loop." In the first of the above examples, the machine is requested to loop 100 times until all measurements have been processed. The second example illustrates two "nested" loops, that is, one loop operating inside another. The machine operates on the inside loop processing data until a specific condition ($I \geq N$ in this case) is satisfied. When this occurs, the machine moves to the outside loop, performs a specific task (in this case it augments J by one), and returns to the inside loop after I is set back to 1. This cycle is repeated until another specific condition is satisfied ($J \geq M$ in this case).

After the preparation of a flowchart, the next step is to prepare the actual program using one of the available symbolic languages. The program is then punched into cards, each card bearing short portions of the program in the form of coded combinations of holes. (See Section 1.10.) The cards are "read" by the machine and interpreted into its own internal "machine language," which commonly is a binary language like those discussed in Chapter 3.

Preparing a program directly in machine language would be a tedious and highly error-prone job. Use of the English language unaltered would make programming very easy. However, computers are not capable, as yet, of interpreting statements written in plain English. In addition, the English language, like any other human language, is replete with redundancies and ambiguities (see Section 2.1).

A direct, shorthand, unambiguous symbolic language, somewhere between the machine language and English in complexity, had to be devised for purposes of programming. Today, many such languages are in existence. They are classified in two categories: the "assembly" languages and the "procedure-oriented" languages.

The assembly languages resemble more closely the machine languages, except that "mnemonic" codes are used in place of the machine's numerical codes. Mnemonics are combinations of alphabetical and numerical symbols, each representing a program "instruction," such as "transfer," "add," "compare," and so forth.

The procedure oriented languages resemble more closely the grammar and structure of the English language. They use "statements" and "sentences." The widely used FORTRAN language is an example of a procedure oriented language.

In the rest of this chapter we discuss the most basic ideas which are essential in understanding assembly and procedure oriented programming techniques. It is not intended to teach the student any particular programming language. However, the uninitiated student should acquire an understanding of the fundamentals of programming techniques, which should prove very useful if afterwards he were to attend a specialized course on computer programming.

2.5
ASSEMBLY LANGUAGES—MNEMONICS

The programmer must complete the algorithm by writing a "statement" for each operation that the machine is to execute. Then he must use a special symbolic language to translate the program "statements" into "instruction words."[6] The symbolic language used must be simple, direct, and unambiguous. An "assembly language" employing "mnemonics" is often used to convey the task description of a given "statement." A mnemonic is a set of letters, which represent an action that the machine can take. The mnemonic instruction word is entered into the machine like data, using punched cards, paper tape, or magnetic tape as a carrier. In smaller machines, data and instructions may also be entered using a typewriter to type-in the machine "words." The machine reads and translates the mnemonic instruction word into a machine word composed exclusively of 1 and 0 bits. The instruction word consists of the same number of 1's and 0's as any other machine word in a "fixed-word-length" machine. There is no way to tell if a given word is a datum or an instruction word by just looking at it. The machine can tell the difference by its place of origin in the memory or elsewhere, and by its intended purpose. This word is then stored in the machine's memory to be recalled by the control, recognized, and processed.

An example of a mnemonic word was given in the previous section with the transfer instruction: "CCF186." Arrow notation is also often used in instruction words. The above instruction in arrow notation would be

$$(186) \rightarrow A$$

meaning: "The contents of memory location 186 to be transferred into register-A." The parentheses are used to signify "contents of."

A programmer should have a dictionary of mnemonics for any given machine in order to write a program using this method. The method of mnemonics is used widely, and there are many such mnemonic languages available today. Such languages are related to corresponding "assembling programs," which are special programs allowing the machine to translate mnemonic instructions into its internal binary code. The machine's internal binary codes are discussed in Chapter 3.

A section of a program, stored in memory locations 034, 035, and 036, may look in mnemonics and in arrow notation, like the following:

034	CCF123	$(123) \rightarrow A$
035	ADD124	$(A) + (124) \rightarrow A$
036	CCI124	$(A) \rightarrow 124$

The control unit, at regular time intervals dictated by a counter, will fetch from the memory location 034 (reserved for storage of program instructions) and execute the first instruction, which reads: "Transfer contents of memory location 123 into A-register (accumulator)." Next it will fetch and execute the instruction word stored in memory location 035, which reads: "Add the contents of the A-register to those of memory location 124 and replace the result back into the

---

[6] "Instruction" and "command" are sometimes used interchangeably. (See page 9, Chapter 1, footnote.)

A-register." Next, the control unit fetches and executes the instruction word stored in memory location 036, which reads: "Transfer the contents of the A-register into the memory location 124."

For purposes of illustration, we provide in Table 2.1 a list of 15 mnemonics.

Table 2.1. Table of Mnemonics (RPC 4000–PINT)

Transfers

INA000	INput to Accumulator
CCF—	Copy Contents From memory location—into the accumulator
CCI—	Copy Contents of accumulator Into the memory location—
CZI—	Copy Zero Into memory location—

Arithmetic

ADD—	ADD: $(A) + (—) \rightarrow A$
SUB—	SUBtract: $(A) - (—) \rightarrow A$
MUL—	MULtiply: $(A) \times (—) \rightarrow A$
DIV—	DIvide: $(A) \div (—) \rightarrow A$

Jumps

JIN—	Jump If Negative to instruction in memory location—
JIP—	Jump If Positive to instruction in memory location—
JMP—	Jump to instruction in memory location—

Miscellaneous

CAR000	typewriter CARriage return
PFA000	Print From Accumulator
TAB000	TABulate
HLT—	HaLT the program. Then when machine is started again, go to command in memory location—

These mnemonics are used later to program two illustrative examples, in order to illustrate the principles of assembly language programming. Although they belong to a lesser used programming language,[7] these mnemonics have a tutorial advantage in their simple composition. Programming examples using the IBM System/360 mnemonics are also illustrated later in Appendix C.

A few comments are in order at this point in order to supplement the information provided by the table of mnemonics. Notice that all the mnemonic instructions here are "words" of fixed length,[8] with three letters each in sequence with

---

[7] Known as PINT—Purdue INTerpretive system, devised for the General Precision RPC 4000 machine.

[8] The RPC is a "fixed-word-length" machine.

three numerical digits. The letters refer to a specific "instruction" and the digits give the memory location referred to in the instruction. For example, CCI101 should be interpreted as: "Transfer and copy the contents of the A-register into the memory location with address 101."

If the operation called for by the instruction does not involve a memory location, then a 000 is appended to the corresponding mnemonic. In this case the zeros, which are used only to maintain the proper length of the instruction words, are *ignored* by the machine. Thus, INA000 means simply: "Transfer the next input word into the accumulator."

Arithmetic operation instructions always imply that one of the two operands is stored in the accumulator and the other is stored in the memory location addressed in the instruction; the result of the operation is to be stored back into the accumulator.

The assembly language illustrated by the mnemonics of Table 2.1 uses a *one-address system*. This means that each mnemonic is followed by a single address. Thus, the command ADD101 relays the operation: (A) + (101)→A, where the parentheses are used to signify "contents off." *Two-address* and *three-address systems* are also used in modern computers. In the two-address system the mnemonic is followed by two addresses. For example, the instruction word ADD101102 would relay the operation: (101) + (102)→101. A two-address system eliminates the need for the accumulator register, at the expense of additional circuit design and programming.

The RPC 4000 is a rather small machine, which has no printer. It uses a typewriter for the output. As a result, the machine must be instructed to "return the carriage" of the typewriter each time new output is to be "printed out," in order to avoid having everything typed out on one line.

The assembly language, illustrated in Table 2.1, is now used in programming two illustrative examples.

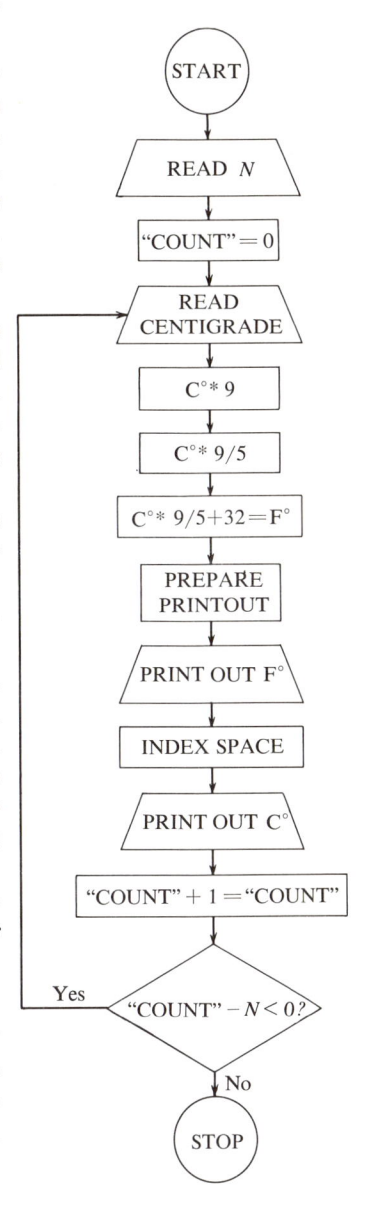

Fig. 2.6 Flowchart for temperature conversion example.

2.6 TEMPERATURE CONVERSION EXAMPLE

Let us now assume that we wish to use the machine to convert N temperature readings from degrees Centigrade into degrees Fahrenheit. The conversion formula is

$$°F = \frac{9}{5}°C + 32.$$

We must first analyze the task and prepare an algorithm. Figures 2.6 and 2.7 show the flowchart and the prepared assembly language program using the mnemonics provided in Table 2.1. The temperature readings in this example are fed to the machine one at a time through an input typewriter. Instead, all the

Storage Location of Instruction	Instruction & Operand	Comments (Not part of the program)
000	INA000	READ N INTO ACCUMULATOR[9]
	CCI100	STORE N INTO 100
	CZI101	PUT ZERO IN 101 ("COUNT")
	→INA000	READ CENTIGRADE INTO ACCUMULATOR
	CCI102	STORE CENTIGRADE IN 102
005	MUL019	$°C \times 9$
	DIV020	$°C \times 9/5$
	ADD021	$°C \times 9/5 + 32 = °F$
	CAR000	TYPEWRITER CARRIAGE RETURN
	PFA000	PRINT FAHRENHEIT FROM ACCUMULATOR
010	CCF102	BRING CENTIGRADE INTO ACCUMULATOR
	TAB000	TYPEWRITER SPACE
	PFA000	PRINT CENTIGRADE FROM ACCUMULATOR
	CCF101	BRING "COUNT" INTO ACCUMULATOR
	ADD022	"COUNT" $+ 1 =$ "COUNT"
015	CCI101	COPY "COUNT" FROM ACCUMULATOR IN 101
	SUB100	"COUNT" $- N$
	└─JIN003	IF "COUNT" $- N < 0$, JUMP IN 004 (REPEAT PROGRAM)
	HLT000	END
	9	⎫
020	5	⎬ Constants
	32	⎪
022	1	⎭

[9] $N =$ the number of temperature readings to be converted; may be entered into the machine manually using the typewriter, or it may be read automatically off the input paper tape.

Fig. 2.7 Temperature Conversion Example Program (RPC 4000-PINT).

readings could have been stored in the machine's memory and the computer asked to perform all the temperature conversions automatically, if the machine had an "index register" available. The operation of an index register is explained in Chapter 10, Section 10.6.

2.7
INCOME TAX EXAMPLE

A company executive decides to automate Federal tax withholdings for his employees. By consulting the Federal tax table which is provided by the Internal Revenue Service, he finds that the tax schedule for single persons with one exemption is as follows:

If $\$\ 000 \leq$ gross $< \$\ 900 \ldots$ Tax $= 0$
If $\$\ 900 \leq$ gross $< \$1400 \ldots$ Tax $= 0.14 \times$ (gross-900)
If $\$1400 \leq$ gross $< \$1900 \ldots$ Tax $= 0.15 \times$ (gross-1400) $+ 70$
If $\$1900 \leq$ gross $< \$2400 \ldots$ Tax $= 0.16 \times$ (gross-1900) $+ 145$

This is a more involved programming example than the temperature conversion example discussed previously. A corresponding flowchart is shown in Fig. 2.8 and the complete program in the assembly RPC 4000-PINT language is shown in Fig. 2.9.

2.8
PROCEDURE-ORIENTED LANGUAGES

The programming of information-processing machines for the solution of computational problems in science, engineering, or business is greatly simplified by the use of "procedure-oriented languages" such as FORTRAN, PL/I, or ALGOL. Procedure languages simplify programming greatly because one is not required to learn the details of machine operation. Above all, they are intended to provide the programmer with the convenience of expressing briefly, and in a form familiar to him, formulas and arithmetic processes so as to make programming as independent of any particular machine as possible. As a language it may be used for communication of algorithms either among people or between people and machines.

A program written in a procedure language is in itself an expression of the algorithm for solving a problem in a form easily recognizable by people; it employs a language more or less clearly resembling English or algebraic expressions and avoids the often obscure mnemonics. This program, which is referred to as the "source program," is usable with any machine that possesses the appropriate "compiler," sometimes also referred to as a "processor." The compiler is a special program for translating the source program into an "object program," which is now expressed in the machine's own special internal binary language and consists of elementary instructions for the machine. The object

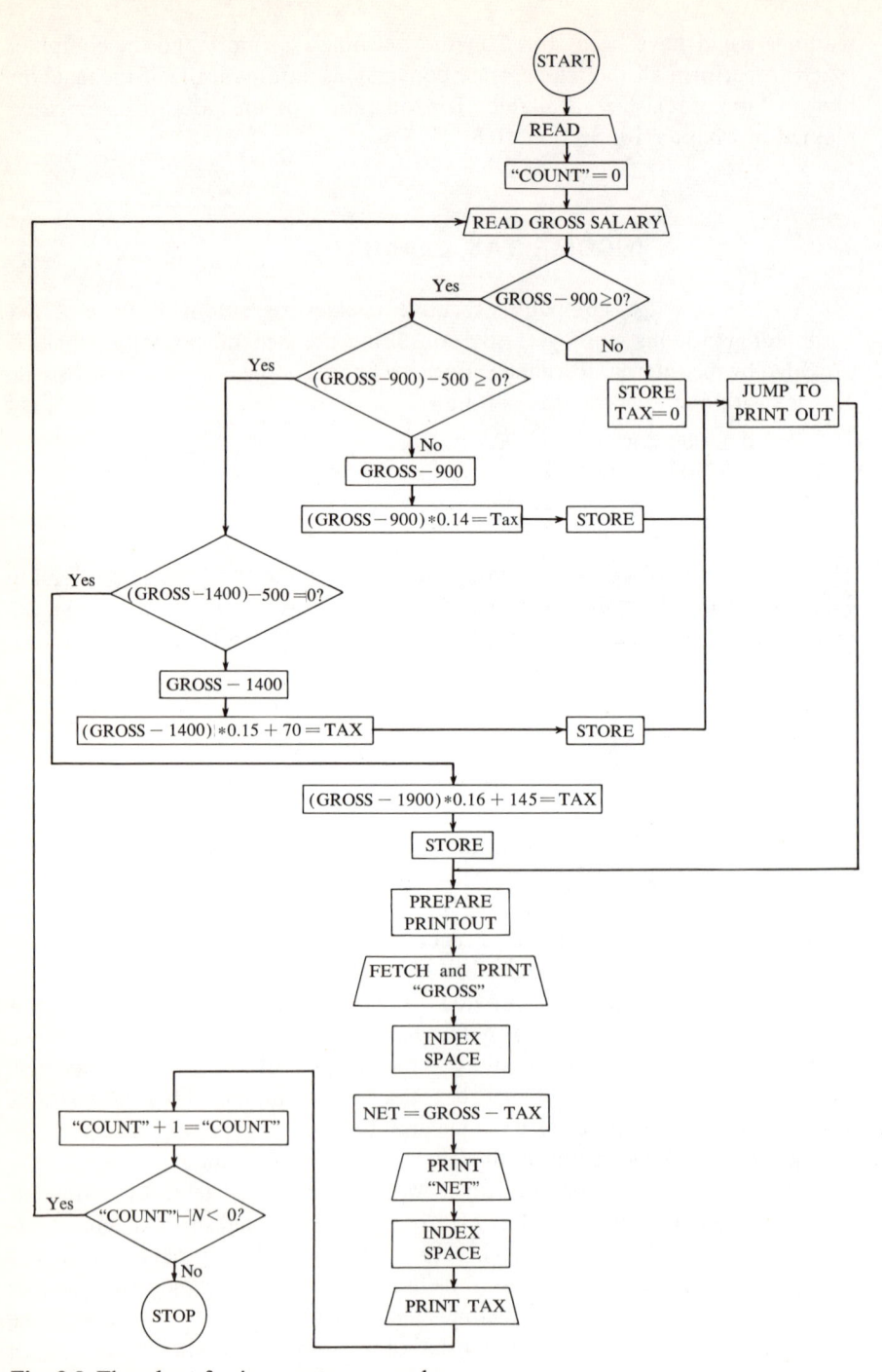

Fig. 2.8 Flowchart for income tax example.

Storage Location of Instruction	Instruction & Operand	Comments (Not part of the program)
000	INA000	READ N INTO ACCUMULATOR
	CCI100	STORE N IN MEMORY 100
	CZI101	PUT ZERO IN 101 (COUNT)
	→INA000	READ GROSS SALARY INTO ACCUMULATOR
	CCI102	STORE GROSS IN 102
005	SUB041	GROSS − $900
	JIP009 ──	JUMP TO NEXT BRACKET IF GROSS − 900 ≥ 0
	CZI103	TAX = 0
	JMP025 ──	JUMP TO PRINTOUT
	SUB042 ←	(GROSS − 900) − 500
010	JIP015 ──	JUMP TO NEXT BRACKET IF GROSS − 1400 ≥ 0
	ADD042	GROSS − 900
	MUL045	TAX = 0.14 (GROSS − 900)
	CCI103	STORE TAX IN 103
	JMP025 ──	JUMP TO PRINTOUT
015	SUB042 ←	(GROSS − 1400) − 500
	JIP022 ──	JUMP TO NEXT BRACKET IF GROSS − 1900 ≥ 0
	ADD042	GROSS − 1400
	MUL046	0.15 (GROSS − 1400)
	ADD043	TAX = 0.15 (GROSS − 1400) + 70
020	CCI103	STORE TAX IN 103
	JMP025 ──	JUMP TO PRINTOUT
	MUL047 ←	0.16 (GROSS − 1900)
	ADD044	TAX = 0.16 (GROSS − 1900) + 145
	CCI103	STORE TAX IN 103
025	CAR000 ←	TYPEWRITER CARRIAGE RETURN
	CCF102	BRING GROSS INTO ACCUMULATOR
	PFA000	PRINT GROSS SALARY
	TAB000	TYPEWRITER SPACE
	CCF102	BRING GROSS INTO ACCUMULATOR
030	SUB103	NET = GROSS − TAX
	PFA000	PRINT NET SALARY
	TAB000	TYPEWRITER SPACE
	CCF103	BRING TAX INTO ACCUMULATOR
	PFA000	PRINT TAX
035	CCF101	BRING COUNT INTO ACCUMULATOR
	ADD048	COUNT = COUNT + 1
	CCI101	STORE IN 101
	SUB100	COUNT − N
	JIN003	REPEAT PROGRAM IF COUNT − N < 0
040	HLT000	END
	900	
	500	
	70	
	140	
045	0.14	Constants
	0.15	
	0.16	
048	1	

(JUMP BACK TO 003)

Fig. 2.9 Income tax example program (RPC 4000-PINT).

program which is thus prepared is the one executed. It is only then that data is read into the machine and information processing really takes place. The compiler, which is in itself a program, does not participate in the execution phase. For example, in order to instruct a machine using FORTRAN programming language, the appropriate FORTRAN compiler must be available for the machine. Most modern machines possess FORTRAN compilers, thus making FORTRAN a widely used programming language.

A program written in a procedure language consists of a sequence of programming "statements." Each programming "statement,' is a declarative or imperative sentence in an abbreviated language resembling somewhat the English or algebraic language; it corresponds to an elementary operation such as the reading of information off punched cards (an "input operation"), the transferring of data from one location inside the machine to another, the performing of arithmetic operations, or the printing of computational results (an "output operation"). A programming statement may also be used to alter the course of machine operations on the basis of some well-defined criterion: If $I > N$, follow procedure A; if not, follow procedure B.

The rules governing the use of statements and the format of a program written in one of the procedure languages available today, differ from one language to another. The most essential points will be illustrated separately for ALGOL and for FORTRAN in the following sections. ALGOL is used first for tutorial purposes, because it is an easy language to explain and use.

Writing a program in ALGOL and in FORTRAN language will be exemplified later on by programming and executing the temperature conversion example and the income tax example discussed previously. We shall use the same examples as before to illustrate the simplification afforded by procedure-oriented language programming. Since it is not possible to discuss here in detail the ALGOL or FORTRAN programming, we will concentrate on the main features of these languages. It is best to illustrate these features in conjunction with the two examples.

2.9
COMPILERS AND ASSEMBLERS

Except in rare circumstances, machines are programmed in a language which is somewhere halfway between the internal basic machine language and spoken English. If the programmer would have to write the instruction words as they appear in the machine's memory, that is, as long sequences of 1's and 0's, programming a machine would have been impractical. Only in some special-purpose computers, usually used by the military, programming in basic machine language is done in order to avoid the high cost of a language translator for the computer. Thus, generally, the programmer will use a symbolic language which is much easier than the internal machine language but not as ambiguous and uneconomical as the human language.

We have seen the use of mnemonics or arrow notation in forming a language. PINT, made especially for the RPC 4000 machine, is one example of such a language. This machine must then be provided with the means of translating programs written in PINT (source program), into its own binary internal language (object program). This is accomplished by providing the machine with the necessary hardware to do the simulation, and with a special "assembler," which is a program written to do the translation.

The machine performs the assembling by itself. A program written in a mnemonic language, most often referred to as an "assembly language," is punched in cards and fed to the machine together with the assembler. The machine follows the assembler's instructions, performs the translation of the program from assembly language into its internal binary language, and on command, punches out on cards the finished product.

An important characteristic of assembly languages, such as the RPC 4000-PINT, is that the sequence of instruction statements is duplicated in the translation. Thus, an assembly language will resemble the language structure found in the object program produced from it; there is a kind of one-to-one correspondence between the source program and the object program instruction words.

ALGOL and FORTRAN are languages formulated to resemble English or algebraic language statements. They are "procedure-oriented" languages, written to be executed on any machine. They follow the human way of reasoning, the natural manner of writing mathematical expressions with only minor restrictions on the format.

A translator program and the necessary hardware means are again essential if a machine is to accept ALGOL or FORTRAN programming. The translator program in this case is referred to as an ALGOL or FORTRAN "compiler." The machine is said to perform "compiling" when a source program written in ALGOL or FORTRAN or any other procedure-oriented language is translated into an object program formed in internal machine language.

In the translation of a procedure language (also referred to as a "compiler language") such as ALGOL or FORTRAN, the sequence of instruction statements is not necessarily preserved in the translation. Often the compiler will translate the source program into the machine's assembly language as an intermediate step to the final object program.

2.10
EXAMPLE OF ALGOL PROGRAMMING

ALGOL, which stands for ALGOrithmic Language, was initially developed between 1955 and 1960 by an international group of programmers.[10] Today the basic ALGOL language is widely used with various improvements and modifications as ALGOL-60, BALGOL, and so forth.

[10] "Report on the Algorithmic Language ALGOL." *Communication of the ACM*, vol. 3, no. 5, May 1960.

50   PROGRAMMING THE MACHINE

The examples selected to illustrate ALGOL programming do not involve sophisticated mathematics. They have been chosen so that the reader will have no difficulty in understanding the problem to be solved, and he can therefore concentrate on the features of the ALGOL description of the computational process.

Certain features of the language are discussed in a rather sporadic way, only to provide the reader with a few pointers. These comments are not intended to be a substitute for a course on ALGOL programming, and the interested student is urged to refer to more specialized textbooks on this subject. It is, however, hoped that this brief analysis will help identify and describe certain important elements of computer programming. Special attention is paid to procedure-oriented languages because these algorithmic languages (ALGOL and FORTRAN being two of them) introduce an important alternative approach to the problem of communication between man and machine. ALGOL in particular is rather simple to understand because it resembles more closely the grammar and syntax of the English language.

It is recommended that a *flowchart* be always constructed before an attempt is made to write the actual program. The flowchart has proven to be a very useful tool for inspecting and improving the algorithm which has resulted from the analysis of the problem. It helps the programmer the way a block diagram helps the engineer.

The flowchart indicates the logical sequence of the data flow and of the machine operations to be performed and thus assists in exposing possible flaws in the algorithm. The details of a flowchart are a matter which varies with problems to be solved and with individual programmers. The notation and the format used also varies among individuals and texts.

The flowchart for the temperature conversion

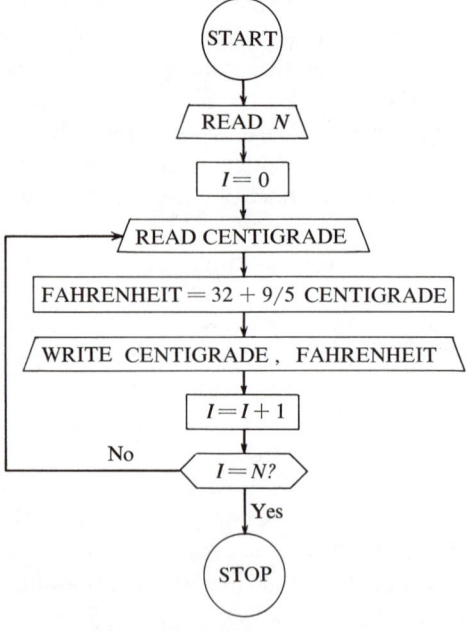

Fig. 2.10 Flowchart for temperature conversion example (ALGOL).

$$°F = \frac{9}{5} °C + 32$$

is shown in Fig. 2.10. When compared with the flowchart prepared for the same problem for use with an assembly language (see Fig. 2.6), we find that the arithmetic portion of the algorithm is represented by a single box. This is so because in ALGOL, as in almost any other procedure language, a single algebraic statement, namely: °F = (9/5)°C + 32, is all that is needed to program the actual temperature conversion.

In keeping with previously employed symbolism, we depict "start" and "stop" statements on the flowchart with circles. A diamond-shaped block is used for "decision," that is, points at which the program *branches* due to a decision. At such instances the program may *loop* back to some previous instruction step, or branch forward to some other part of the program. The particular symbols used here are not universal or standardized and may differ from those used in other texts.

The ALGOL program for the temperature conversion example is shown in Fig. 2.11. After the program is prepared,[11] it is punched on cards in the indicated sequence. These cards constitute the source program. Properly assembled,

---

Temperature Conversion Example—ALGOL

BEGIN COMMENT TEMPERATURE CONVERSION EXAMPLE;
INTEGER I, N;
REAL CENTIGRADE, FAHRENHEIT;
READ (N);
FOR $I \leftarrow 1$ STEP 1 UNTIL N DO
  BEGIN
  READ (CENTIGRADE);
  FAHRENHEIT $\leftarrow 32 + 9 \times$ CENTIGRADE/5;
  WRITE (CENTIGRADE, FAHRENHEIT);
  END;
END.

Fig. 2.11 ALGOL program.

the cards are read into the machine under the control of the compiler program. The compiler then translates the program into internal machine-language instruction-words, and the resulting program (object program) is stored in the memory and if desired also punched on a new deck of cards for future use.

The compiled program is printed out as shown by Fig. 2.12. Notice its absolute agreement with the initial program of Fig. 2.11. Also notice the printout at the bottom, giving the compilation[12] time elapsed (004 seconds) and the statement that: NO ERRORS DETECTED. This matter of error detection capabilities of the compiler will be discussed in Section 2.12.

---

[11] It is ordinarily written on the ALGOL Coding Form, a specially printed paper form. The specific version used in both examples is the "Extended ALGOL" of Burroughs Corp. The programs were executed on the B5500 Burroughs computer.

[12] Additional printout information during this procedure is of no particular importance at present.

```
BEGIN COMMENT TEMPERATURE CONVERSION EXAMPLE;
INTEGER N, I;
REAL CENTIGRADE, FAHRENHEIT;
READ (N);

FOR I←1 STEP 1 UNTIL N DO
    BEGIN
    READ (CENTIGRADE);
    FAHRENHEIT ← 32 + 9×CENTIGRADE/5;
    WRITE (CENTIGRADE, FAHRENHEIT);
    END;
END.
```

```
     WRITE   IS SEGMENT 004,  PRT 048        BLK CTR IS SEGMENT 005,  PRT 005
     FILEOUT IS SEGMENT 007,  PRT 012        FILE IN IS SEGMENT 008,  PRT 013
 PRT SIZE=0049;  NO. SEGS.=010;  TOTAL SEGMENT SIZE=00117;  DISK STORAGE REQ.=00117;
NO ERRORS DETECTED.  ELAPSED COMPILATION TIME = 004 SECONDS.
```

Fig. 2.12 Temperature conversion program compiled.

For purposes of illustration the prepared program was executed by performing a few temperature conversions. The results, as printed by the machine, are shown in Fig. 2.13.

°C →	°F
0.0000000000	32.0000000000
17.7777777777	64.0000000000
37.0000000000	98.6000000000
37.7777777777	100.0000000000
100.0000000000	212.0000000000
-17.7777777777	0.0000000000

Fig. 2.13 Temperature conversion printout (ALGOL program-Fig. 2.12).

2.11 COMMENTS ON ALGOL PROGRAMMING

With reference to the ALGOL program shown in Fig. 2.11 the following delineating comments are appropriate. Notice that a program must begin with the word "BEGIN" and end with the word "END." Also within the program, a sequence of statements, called a "sentence," may be "bracketed" between the words "BEGIN" and "END;". For this reason these words are referred to as *statement brackets*. The complete sequence of statements which covers the whole problem-solving procedure, that is, the source program, is also bracketed between the words "BEGIN" and "END."

Each statement in the program, that is, each machine instruction, must be terminated with a semicolon. For example, we may write: REAL CENTIGRADE, FAHRENHEIT; READ (N); and so on.

The word "comment" may be used to make some pertinent comment which would clarify the contents of the program. The machine will ignore everything between the word "comment" and the following semicolon. For example, the statement: COMMENT TEMPERATURE CONVERSION EXAMPLE; will be ignored by the machine. It is there only for the information of the user of the program.

The program uses a number of *identifiers*. These are special names, chosen by the programmer, to identify various entities in the program, for example, variables, input-data sets, editorial formats, output-data sets, and so on. Each identifier must be unique; it must not conflict with any other words used in the program. It must also start with a letter; otherwise it can be an almost arbitrary combination of letters and digits. Thus, the integers "I," "N," and the real numbers "CENTIGRADE," "FAHRENHEIT" are identifiers in this program.

No *labels* have been used in this program. A label is a special name or identifier which is used to identify a program statement for later reference purposes.

Notice that the symbol " × " is used for multiplication in order to avoid misunderstanding which may result from the use of the symbol "·"

2.12
THE DO-STATEMENT

An important feature of ALGOL language is illustrated with the above program (Fig. 2.11), in the use of the "DO-statement." As shown in the flowchart of Fig. 2.10, the temperature conversion is to be repeated N times. The machine keeps track of the number of repetitions by using an integer "index" variable number I, which is set initially equal to zero and is augmented by one between repetitions of the portion of the program having to do with the actual temperature conversion. After each conversion, the machine compares I with N and loops back to the statement "READ CENTIGRADE" until I becomes equal to N, in which case it moves forward to "STOP."

Programming of such repetitive computations, where a certain portion of the program is to be repeated over and over, is greatly simplified by the use of the DO-statement. Properly used, the DO-statement makes it possible to execute a portion of the program repeatedly, by automatically changing the value of an integer "index" variable I between repetitions, until a certain condition, such as $I = N$, is satisfied. For the programming example above (Fig. 2.11) the DO-statement reads,

FOR $I \leftarrow 1$ STEP 1 UNTIL N DO,

which instructs the machine to set the index I, located at some specified memory register, equal to 1, to augment it between repetitions by steps of 1, and continue repeatedly until I becomes equal to N. In this manner the machine repeats the "sentence" of the program which follows the DO-statement; that is, that portion of the program immediately following the DO-statement, which is enclosed between the statement brackets BEGIN and END;. In this example the portion of the program to be repeated (Fig. 2.11) is

BEGIN
READ (CENTIGRADE);
FAHRENHEIT $\leftarrow 32 + 9 \times$ CENTIGRADE$/5$;
WRITE (CENTIGRADE, FAHRENHEIT);
END;

The use of the DO-statement will be illustrated again later in this chapter with FORTRAN programming. It constitutes the most powerful and widely used feature of FORTRAN, ALGOL, and other procedure languages.

2.13 ERROR DETECTION BY THE COMPILER

A great number of possible *syntactical errors* (briefly "syntax errors") which can be made by the programmer are detected by the compiler during compilation. Any given compiler provides us with a list of usually over one hundred possible detectable syntax errors and the appropriate error messages. This feature is of considerable aid in removing errors from the program. The process of removing errors is commonly referred to as program "debugging."

Of course, *logical errors*, such as wrong mathematical formulas, cannot be detected by the compiler. Also not all syntax errors are necessarily detected during compilation. Therefore, *testing* of the compiled program and *debugging* are essential parts of the program preparation procedure.

To illustrate error debugging by the compiler, the temperature conversion program was rewritten, with some intentional errors included. The compiled printout is shown in Fig. 2.14.

```
                          BEGIN COMMENT TEMPERATURE CONVERSION EXAMPLE;
                          INTEGER N, I;
                          REAL CENTIGRADE, FAHRENHEIT;
                          READ (N);
                          FOR I←1 STEP 1 UNTIL N DO
                             BEGIN
                             READ (TEMPERATURE);
         ***** ERROR 100              +                           DID NOT
   UNDECLARED                                                     CATCH
   IDENTIFIER      FAHRENHEIT ← 32 + 5×CENTIGRADE/9;              LOGICAL
         ***** ERROR 104   WRITE (CENTIGRADE, FAHRENHEIT;         ERROR
   MISSING RIGHT                      +
   PARENTHESIS    END;
                          END.

           WRITE    IS SEGMENT 004, PRT 048      BLK CTR IS SEGMENT 005, PRT 005
           FILEOUT  IS SEGMENT 007, PRT 012      FILE IN IS SEGMENT 008, PRT 013
       PRT SIZE=004;  NO. SEGS.=010;  TOTAL SEGMENT SIZE=00117;  DISK STORAGE REQ.=00117;
       002 ERRORS DETECTED.  ELAPSED COMPILATION TIME = 004 SECONDS.  LAST ERROR IS ON CARD
```

Fig. 2.14 Compiled printout of ALGOL program indicating errors detected.

Observe that at the bottom of the printout it is stated that "002 ERRORS DETECTED." At the left side of the printout the errors' location is marked by asterisks and by an error code. From the manual of the compiler it is found that "ERROR 100" means "UNDECLARED IDENTIFIER." Indeed, in the program statement: "READ (TEMPERATURE);" the identifier TEMPERATURE has not been previously declared. Next, it is found, again from the manual that "ERROR 104" means "MISSING RIGHT PARENTHESIS," as is actually the case in the statement: "WRITE (CENTIGRADE, FAHRENHEIT);".

Notice that in the conversion formula

$$\text{FAHRENHEIT} \leftarrow 32 + 5 \times \text{CENTIGRADE}/9;$$

there is a logical error, which the compiler did not catch. The correct formula would have been

$$\text{FAHRENHEIT} \leftarrow 32 + 9 \times \text{CENTIGRADE}/5;$$

There are over 100 error checks in a compiler, each one assigned a special code number, such as ERROR 101, and so forth. The corresponding list in the compiler manual provides the necessary explanation of the detected error.

An experienced programmer knows that he is bound to make mistakes, and that his program must be checked for errors before it is run with actual data in the machine. Therefore, he depends on the error detection capabilities of the compiler. In addition, he may provide his own program with periodic checks in order that he may easily detect possible errors.

Since the detection of programming mistakes is a serious business (it may cost valuable computer time if not detected and corrected), there exist special debugging programs for error detection. Such programs provide a means for reproducing portions of the program and for understanding what is going on at different times during the run of the program.

2.14
A SECOND ILLUSTRATIVE EXAMPLE

The income tax problem, as described in Section 2.7, is now programmed in ALGOL. The flowchart is shown in Fig. 2.15. The written ALGOL program is shown in Fig. 2.16. One should observe again the use of the DO-statement

FOR $I \leftarrow 1$ STEP 1 UNTIL N DO

The compiled program is shown in Fig. 2.17. Notice that it is stated at the bottom that "NO ERRORS DETECTED." This simply means that the compiler was not able to find any errors. If any exist they are beyond the compiler's ability to detect them.

The program was run using arbitrarily chosen data and the results were printed out, as shown in Fig. 2.18.

56 PROGRAMMING THE MACHINE

```
        START
          │
        READ N
          │
         I = 0
          │
    ┌──►READ GROSS
    │     │
    │   GROSS < 900?  ──Yes──►  TAX = 0  ──┐
    │     │ No                              │
    │   GROSS < 1400? ──Yes──►  TAX = 0.14 × (GROSS − 900)  ──┤
    │     │ No                              │
    │   GROSS < 1900? ──Yes──►  TAX = 0.15 × (GROSS − 1400) + 70  ──┤
    │     │ No                              │
    │        ──────►  TAX = 0.16 × (GROSS − 1900) + 145  ──┤
    │     │                                 │
    │   NET = GROSS − TAX ◄─────────────────┘
    │     │
    │   WRITE GROSS, NET, TAX
    │     │
    │   I = I + 1
    │     │
    └──No─ I = N?
          │ Yes
        STOP
```

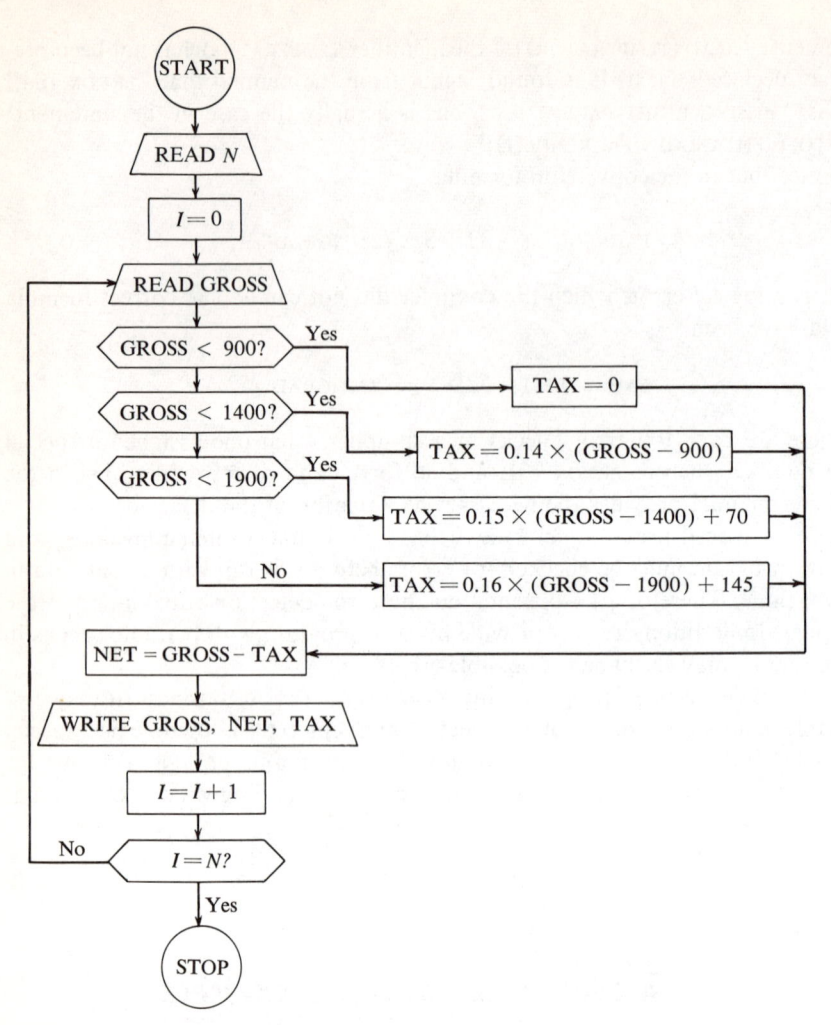

Fig. 2.15 Flowchart of the income tax example.

2.15
FORTRAN PROGRAMMING

Like ALGOL, FORTRAN—which stands for FORmula TRANslation—is a procedure oriented language. It is structurally similar to ALGOL. A FORTRAN program consists of statements, which are classified into the following four categories:
(a) arithmetic statements,
(b) input–output statements,
(c) Those statements which cause a change in the sequence of the program execution (such as the DO-statement), and

```
BEGIN COMMENT INCOME TAX EXAMPLE;
INTEGER N, I;
REAL GROSS, NET, TAX;

READ (N);
FOR I←1 STEP 1 UNTIL N DO
   BEGIN
   READ (GROSS);
   IF GROSS < 900 THEN
      TAX←0
   ELSE IF GROSS < 1400 THEN
      TAX←0.14 × (GROSS − 900)
   ELSE IF GROSS < 1900 THEN
      TAX←0.15 × (GROSS − 1400) + 70
   ELSE
      TAX←0.16 × (GROSS − 1900) + 145;
   NET←GROSS − TAX;
   WRITE (GROSS, NET, TAX);
   END;
END.
```

Fig. 2.16 Income Tax Example–ALGOL.

```
BEGIN COMMENT INCOME TAX EXAMPLE;
INTEGER N, I;
REAL GROSS, NET, TAX;

READ (N);
FOR I←1 STEP 1 UNTIL N DO
   BEGIN
   READ (GROSS);
   IF GROSS < 900 THEN
      TAX ← 0
   ELSE IF GROSS < 1400 THEN
      TAX ← 0.14×(GROSS − 900) + 0
   ELSE IF GROSS < 1900 THEN
      TAX ← 0.15×(GROSS − 1400) + 70
   ELSE
      TAX ← 0.16×(GROSS − 1900) + 145;
   NET ← GROSS − TAX;
   WRITE (GROSS, NET, TAX);
   END;
END.
     WRITE    IS SEGMENT 004, PRT 049       BLK CTR IS SEGMENT 005, PRT 005
     FILEOUT  IS SEGMENT 007, PRT 012       FILE IN IS SEGMENT 008, PRT 013
  PRT SIZE=0050;   NO. SEGS.=010;  TOTAL SEGMENT SIZE=00138;  DISK STORAGE REQ.=00138;
NO ERRORS DETECTED.   ELAPSED COMPILATION TIME = 004 SECONDS.
```

Fig. 2.17 Compiled income tax ALGOL program.

Gross Income	Net Income	Tax
281.4000000000	281.4000000000	0.0000000000
900.0000000000	900.0000000000	0.0000000000
1200.0000000000	1158.0000000000	41.9999999999
1889.0000000000	1745.6500000000	143.3500000000
2399.9900000000	2174.9916000000	224.9984000000

Fig. 2.18 Income tax runs using the program of Fig. 2.16.

(d) the "Comment-statement" which provides information about the procedure and in general about the program without itself requiring any computation.

FORTRAN differs basically from ALGOL in the organization of programs. Some such differences are the following.

Unlike ALGOL, FORTRAN does not follow very closely the grammar and the syntax of the English language. Hence FORTRAN requires that each statement start on a new line. When punched in a punch card no more than one statement is allowed per punch card, and it is always restricted to the 7th through the 72nd columns. If the statement is longer than the space available in one punch card, it may occupy more than one card, but is always restricted to the above limits on each card.

Every name used to identify a variable, such as CENTIGRADE, must have six letters or less and if the variable is an integer, it must begin with an I, J, K, L, M, or N.

The FORTRAN programmer has to obey a greater number of syntactical and grammatical rules in the use of symbols and of statements. For example, he is not allowed to mix decimal numbers with integers in arithmetic operations. If he writes: $a + b$, where a has been predefined as an integer and b has been predefined as a decimal, the program will not work. Many such syntactical errors are detectable during the compilation of the program.

FORTRAN provides considerably more flexibility in the way the input and output operations are performed, than does ALGOL. The price one must pay for this is some increase in programming complexity.

The DO-statement, written in a different format than the one we saw previously in connection with the ALGOL programming examples, is widely used in FORTRAN programming. Its use is illustrated with programming examples in the following section.

The "Comment-statement" is used to provide information about the procedure, without itself participating in the computations. A Comment-statement is recognized by the machine by the letter "C" which is punched in the first column of the punch card, and then ignored as far as computations are concerned.

An essential advantage of FORTRAN over ALGOL programming is that FORTRAN does not require statement brackets like the words BEGIN and END. Although this facility prevents the programmer from displaying plainly the logic chain of thought which is used in the algorithm, it makes the organization of statements considerably simpler for the programmer. In order to facilitate the interpretation of a previously written program by another user quite often a Comment-statement (also referred to as a *C*-statement) is used by the programmer to denote the beginning and the ending of sentences or portions of the program.

With these differences between FORTRAN and ALGOL in mind, we exemplify FORTRAN programming in the following section, using the temperature conversion and the income tax examples. It may be added here that the best features of both these languages, namely the relative simplicity of ALGOL programming and the great flexibility of FORTRAN programming, have been incorporated in PL/I, another procedure-oriented language.[13]

Our discussion of procedure-oriented languages was not intended to substitute for a course on FORTRAN or ALGOL programming. Rather we intended to point out the essential principles of programming. With an understanding of these principles, coupled with practice on the techniques of algorithm and flow-chart forming, the student can learn easily any programming language.

2.16 EXAMPLES OF FORTRAN PROGRAMMING

Several versions of the FORTRAN procedure-oriented language have been devised. Of those, FORTRAN II and FORTRAN IV have been the most popular. Today, FORTRAN IV is used most widely in several slightly different dialects, which may vary from installation to installation, or even at the same installation from time to time.

In order to illustrate FORTRAN programming, we used FORTRAN IV to program the temperature conversion example and the income tax example, previously programmed in assembly language and in ALGOL. The FORTRAN program for the temperature conversion example, following the flow chart of Fig. 2.10, is shown in Fig. 2.19.

A few explanatory notes are in order at this point. Notice that each statement alone occupies a single line. The top statement, preceded by the letter "C" is a "Comment-statement," informing the user about the nature of the computational procedure that follows. The number column to the far left keeps a count of the statements that follow. The numbers 1, 5, 2, and 3 immediately to the left of the program statements are reference numbers used in order to refer to these statements throughout the program. For example, the statement

IF $(I.LT.N)$ GO TO 5

instructs the machine to go back to the program step $I = I + 1$ if $I < N$.

A few additional notes of explanation are included with the program in

---

[13] See Bates and Douglas in the Bibliography at the end of this chapter.

60 PROGRAMMING THE MACHINE

```
FORTRAN IV C LEVEL 1, MOD 1        MAIN
              C       TEMPERATURE CONVERSION EXAMPLE
0001                  INTEGER I, N
0002                  REAL CENTI, FAHREN
0003                  READ ( 5,1 ) N
0004              1   FORMAT(15)
0005                  I = 0
0006              5   I = I+1
0007                  READ ( 5,2 ) CENTI
0008              2   FORMAT ( F20.10 )
0009                  FAHREN = 32.0 + 9.0*CENTI/5.0
0010                  WRITE (6,3) CENTI, FAHREN
0011              3   FORMAT ( 2F30.10 )
0012                  IF ( I .LT. N ) GO TO 5
0013                  STOP
0014                  END
```

Explanatory Notes

C denotes a "Comment-statement"
Specify variables I and N as "integers"
"Centigrade" and "Fahrenheit" are to be decimal numbers
{Variable N is to be read from a card in accordance
 with a certain machine procedure (Format 15)
The variable I takes the value "zero"
The variable I is augmented by 1
{The variable "CENTI" is to be read from a card in accord-
 ance to a certain machine procedure (Format F20.10)
Calculate FAHREN in accordance with the provided formula
{Print output variables CENTI and FAHREN in accordance to
 certain machine procedure (Format 2F30.10)
If variable "I" is "Less Than" the variable "N," then GO
 back TO program statement #5, otherwise move to next
 statement
STOP
END (end of program)

Fig. 2.19 FORTRAN programming for the temperature conversion example. (The program was compiled with the IBM System/360, model 75 computer.)

Fig. 2.19. Notice that the word FORMAT is used in the program to signify certain input and output procedures, such as "read from cards" or "print-out," that the machine is instructed to execute.

The FORTRAN program shown in Fig. 2.19 is now repeated in Fig. 2.20, modified somewhat, in order to illustrate the use of the DO-statement. (See Section 2.12.) The DO-statement in this program appears as follows:

DO 4 I = 1, N.

This is a typical form in which the DO-statement appears in FORTRAN languages. It is interpreted to mean: "repeat up to statement 4, beginning with $I = 1$, each time increasing I in steps of 1, until $I = N$." The statement 4 in the program is: WRITE (6,3) CENTI, FAHREN.

The temperature conversion program (Fig. 2.19) was modified slightly by deliberately introducing two syntax and one logical error. The errors intro-

```
              C       TEMPERATURE CONVERSION EXAMPLE
                      INTEGER I, N
                      REAL CENTI, FAHREN
                      READ ( 5,1 ) N
                   1  FORMAT(15)
                      DO 4 I = 1,N
                      READ ( 5,2 ) CENTI
                   2  FORMAT ( F20.10 )
                      FAHREN = 32.0 + 9.0*CENTI/5.0
                   4  WRITE (6,3) CENTI, FAHREN
                   3  FORMAT ( 2F30.10 )
                      STOP
                      END
```

Fig. 2.20 FORTRAN Program for the temperature conversion example, making use of the DO-statement. (The program was compiled by a Sigma-7 BCM computer.)

duced are the same as those which were used in Section 2.13 to demonstrate the syntax-error detection capabilities of the ALGOL compiler (see Fig. 2.14). The

FORTRAN program illustrating the deliberately introduced errors is shown in Fig. 2.21. Just as in the case of the ALGOL program (Fig. 2.14), the machine was incapable of diagnosing the logical error in the conversion formula, which was presented to the machine as FAHREN = 32.0 + 5.0 ∗ CENTI/9.0, instead of the correct one: FAHREN = 32.0 + 9.0 ∗ CENTI/5.0. However, the FORTRAN compiler will detect most of the common syntax errors and will not compile the source program into an object program. It will instead signal an ERROR,

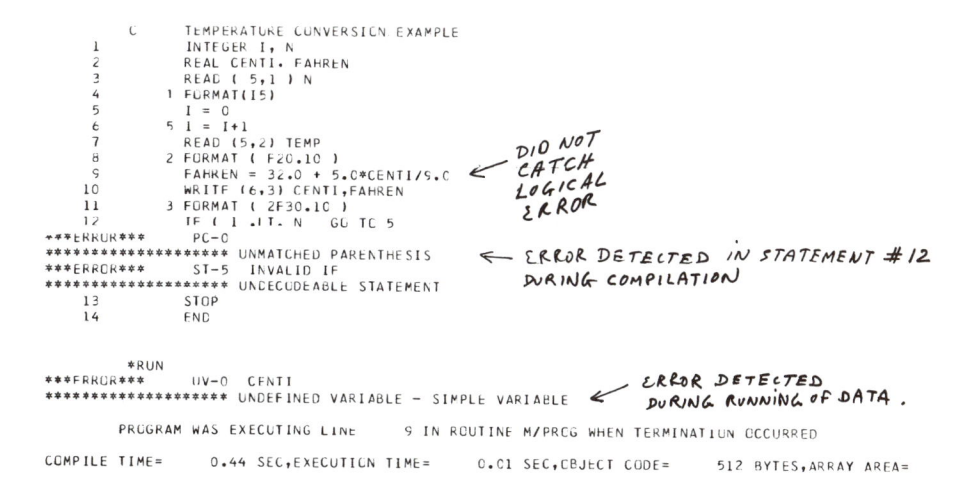

```
              C       TEMPERATURE CONVERSION EXAMPLE
       1              INTEGER I, N
       2              REAL CENTI, FAHREN
       3              READ ( 5,1 ) N
       4         1    FORMAT(I5)
       5              I = 0
       6         5    I = I+1
       7              READ (5,2) TEMP
       8         2    FORMAT ( F20.10 )              ← DID NOT
       9              FAHREN = 32.0 + 5.0*CENTI/9.0      CATCH
      10              WRITE (6,3) CENTI,FAHREN          LOGICAL
      11         3    FORMAT ( 2F30.10 )                ERROR
      12              IF ( I .LT. N ) GO TO 5
 ***ERROR***          PC-0
 ******************   UNMATCHED PARENTHESIS    ← ERROR DETECTED IN STATEMENT #12
 ***ERROR***          ST-5   INVALID IF            DURING COMPILATION
 ******************   UNDECODEABLE STATEMENT
      13              STOP
      14              END

        *RUN
 ***ERROR***          UV-0  CENTI                ← ERROR DETECTED
 ******************   UNDEFINED VARIABLE - SIMPLE VARIABLE    DURING RUNNING OF DATA.
        PROGRAM WAS EXECUTING LINE    9 IN ROUTINE M/PROG WHEN TERMINATION OCCURRED
 COMPILE TIME=      0.44 SEC,EXECUTION TIME=   0.01 SEC,OBJECT CODE=    512 BYTES,ARRAY AREA=
```

Fig. 2.21 FORTRAN program for the temperature conversion example with intentional syntax and logical errors included. (Compiled with the IBM System/360, model 75 computer.)

just as it did when it detected an UNMATCHED PARENTHESIS in the statement IF (I.LT.N) GO TO 5.

It is of interest to notice a difference between the error detection of an UNDECLARED IDENTIFIER in the compilation of the ALGOL program (Fig. 2.14) and the detection of the same error, here referred to as an UNDEFINED VARIABLE, in the compilation of the FORTRAN program (Fig. 2.21). The ALGOL compiler was able to detect the error, because in accordance with the rules of ALGOL programming every variable used must be declared (that is, defined) at the beginning of the program. However, the FORTRAN compiler places no significance on names beyond inspecting the proper form in which they appear, such as the number of letters used (must be six or less) and the first letter (if it is I, J, K, L, M, or N, then the machine will expect an integer). Therefore, an UNDEFINED VARIABLE is not detected by the FORTRAN compiler; one may use such variables indiscriminately throughout a FORTRAN program.

However, such variables are detected during the running of the program with actual data, since no computations are possible that involve undefined variables.

It may be noted that there is a limit to the amount of error checking that can economically be designed into a compiler. The syntax-error checking capabilities vary widely among the compilers of the various FORTRAN versions. In any case there are certain errors that cannot be detected during the compilation, and the program user is advised to test the correctness of his program with trial-runs on data properly selected for such tests.

The income tax example, also used earlier in this chapter to illustrate assembly language programming and ALGOL programming, is programmed again in FORTRAN IV following the flowchart shown in Fig. 2.15. The compiled FORTRAN program is shown in Fig. 2.22. The program was rewritten making

```
        FORTRAN IV G LEVEL 1, MOD 1          MAIN

              C         INCOME TAX EXAMPLE
    0001              INTEGER I, N
    0002              REAL GROSS, NET, TAX
    0003              READ ( 5,1 ) N
    0004            1 FORMAT ( I5 )
    0005              I = 0
    0006           12 I = I + 1
    0007              READ ( 5,2 ) GROSS
    0008            2 FORMAT (F10.2)
    0009              IF ( GROSS .LT. 900.0 ) GO TO 4
    0010              IF ( GROSS .LT. 1400.0 ) GO TO 6
    0011              IF ( GROSS .LT. 1900.0 ) GO TO 7
    0012              GO TO 8
    0013            4 TAX = 0.0
    0014              GO TO 10
    0015            6 TAX = 0.14*( GROSS - 900.0 )
    0016              GO TO 10
    0017            7 TAX = 0.15*( GROSS - 1400.0 ) + 70.0
    0018              GO TO 10
    0019            8 TAX = 0.16*( GROSS - 1900.0 ) + 145.0
    0020           10 NET = GROSS - TAX
    0021            5 WRITE ( 6,11 ) GROSS, NET, TAX
    0022           11 FORMAT( 3F30.10 )
    0023              IF ( I .LT. N ) GO TO 12
    0024              STOP
    0025              END
```

Fig. 2.22 FORTRAN program for the income tax example (Compiled with the IBM System/360, model 75 computer.)

use of the DO-statement. The new program is illustrated in Fig. 2.23. Based on explanations provided previously with the programs of the temperature conversion example, the student can easily identify the statements involved in this program.

The FORTRAN programs for both examples were run on an IBM System/360, model 75, machine using the same data as that shown in the ALGOL printouts of Figs. 2.13 and 2.18.

```
C       INCOME TAX EXAMPLE
        INTEGER I, N
        REAL GROSS, NET, TAX
        READ ( 5,1 ) N
   1    FORMAT ( I5 )
        DO 5 I = 1, N
        READ ( 5,2 ) GROSS
   2    FORMAT (F10.2)
        IF ( GROSS .LT. 900.0 ) GO TO 4
        IF ( GROSS .LT. 1400.0 ) GO TO 6
        IF ( GROSS .LT. 1900.0 ) GO TO 7
        GO TO 8
   4    TAX = 0.0
        GO TO 10
   6    TAX = 0.14*( GROSS - 900.0 )
        GO TO 10
   7    TAX = 0.15*( GROSS - 1400.0 ) + 70.0
        GO TO 10
   8    TAX = 0.16*( GROSS - 1900.0 ) + 145.0
  10    NET = GROSS - TAX
   5    WRITE ( 6,11 ) GROSS, NET, TAX
  11    FORMAT( 3F30.10 )
        STOP
        END
```

Fig. 2.23 FORTRAN program for the income tax example using the DO-statement. (Compiled with the Sigma-7 BCM computer.)

2.17
ADDITIONAL COMMENTS ON PROGRAMMING

The flexibility offered by external programming has revolutionized the field of information processing by machine. Storing instructions in the form of instruction words, enables the machine to perform operations on its stored instructions. This gives the programmer the facility to alter the instructions automatically, under specific conditions, during the execution of the program. Instructions modification has become a routine part of the programming art.

In its very essence, a program enables the machine to carry out the proper sequence of operations in the execution of an algorithm for the solution of a problem. During the follow-through of the program the control unit initiates an alternating sequence of "fetch" and "execute" steps. It examines the contents of the instruction counter to determine the address of the next instruction; it fetches the instruction from the address specified into its own instruction register, and breaks it down into its component parts before executing.

Each instruction contains the description of the task to be performed, the address(es) of the operand(s) and supplementary information. In *one-address* systems, the instruction contains the task command and only one address—that of the operand. The other operand has been already placed in the A-register. For example, in the RPC 4000-PINT language, the instruction ADD101 contains the task command "ADD" and the address "101" of the operand; the instruction is interpreted as $(A)+(101)\to A$.

When an instruction, like ADD101, is assembled in machine language it occupies a machine word. In a ten-digit word the instruction may look like: CCDDDDIIII, where CC is a two-digit "operation code" substituting for the command "ADD," DDDD is the address of the operand ("D-address"), and IIII is the address of the next instruction ("I-address"). If two instructions are packed in one word for more efficient use of the available memory space, then each instruction is treated independently.

In *two-address* systems (used in the IBM-1620 machine) the instruction carries the addresses of both operands, thus eliminating the need for the A-register. The instruction ADD101102 should be interpreted as: $(101)+(102)\to 101$. In *three-address* systems, the address of the memory location for returning the result of the operation is also included in the instruction.

The sequence of steps to be followed in the preparation of a program are the following: Thorough analysis of the information-processing task; preparation of an algorithm; flowcharting and improvement of the algorithm; actual programming in the selected symbolic (source program) language; program assembling or compiling; test and debugging; production of the object program in punched cards, magnetic tape, or other permanent storage medium. The programmer may perform all these chores without ever coming in physical contact with the machine. This is especially so in large machines, where the machine's operator is responsible for managing the machine.

When a program is debugged and ready to run, data and other essential information is prepared on punched cards or the like. Then a "loader" is used to transfer all the necessary information into the machine's memory. The loader is a special program which is designed to "load" the object program and the data from the input media (cards, tapes, and so forth) into properly designated locations in the main memory. Many programs use "subroutines," that is, preprepared programs for specific information-processing tasks. These subroutines are stored in tapes and are available in program "libraries." The loader looks up the subroutine on the tape, allocates space in the memory and makes sure that the subroutine is stored properly. Also, if needed, the loader will establish linkages between the subroutines and the main program. Only when all this is done, the loader will turn control over to the object program for the execution of the actual data-processing task.

It is of interest to note the manner in which the actual time the machine is ON and operating is divided among different functions. In a scientific computer this ON time is divided about equally among compiling, debugging, and actual information processing. In EDP machines the actual information-processing time may be up to 80 percent.

We should point out again that the outstanding properties of externally programmed information processing machines are: the ability to *loop* and the ability to *branch conditionally*. These properties were demonstrated in the illustrative examples of this chapter.

Looping refers to the ability of the machine to step back in its program and, on command, repeat a portion of it over and over. Conditional branching refers to the ability to branch once the machine has reached a certain step in the execution of its program and as long as a certain condition is satisfied.

One of the principal goals of programming research is to make computer programs universally applicable for the different species of computers in existence. Another is to make the program simple and "mechanical" so that errors in programming are easily checked. Finally, there is the desire to reduce the amount of information fed into the machine to the absolute essential for solving the problem at hand.

BIBLIOGRAPHY

Bates, F., and Douglas, M. L., *Programming Language/One*. Englewood Cliffs, N.J.: Prentice-Hall, Inc., 1967.

McCracken, D. D., *A Guide to ALGOL Programming*. New York: John Wiley & Sons, Inc., 1962.

McCracken, D. D., *A Guide to FORTRAN IV Programming*. New York: John Wiley & Sons, Inc., 1965.

McCracken, D. D., *Digital Computer Programming*. New York: John Wiley & Sons, Inc., 1957.

McCracken, D. D., and Dorn, W. S., *Numerical Methods and FORTRAN Programming*. New York: John Wiley & Sons, Inc., 1964.

Problems

2.1 You are given two sets of N numbers each. They may represent two N-dimensional vectors

$$A = (a_1, a_2, \ldots, a_N),$$
$$B = (b_1, b_2, \ldots, b_N).$$

You are requested to program a machine which will perform the "dot product" of the two vectors, that is, $A \cdot B$. The dot product, signified by DP is defined as:

$$DP = a_1 b_1 + a_2 b_2 + \ldots + a_N b_N.$$

(a) Form an algorithm which will describe concisely and unambiguously the necessary steps that the machine should follow in order to provide us with the dot product of the two given sets of numbers.

(b) Draw a flowchart for the above algorithm.

(c) Using a symbolic language of your own invention write a complete program. You may use any kind of symbolic "shorthand" to write your program, as long as you define *clearly* and *consistently* the symbols and rules followed.

2.2 Given two sets of numbers $(a_1,a_2,\ldots,a_N)$ and $(b_1,b_2,\ldots,b_M)$, program a machine which will determine the maximum number $(a_I)_{max}$ of the first set and the minimum number $(b_J)_{min}$ of the second set, and form the product $(a_I)_{max}(b_J)_{min}$. Follow steps (a), (b), and (c), as in Problem 2.1.

2.3 Given a set of numbers $a_1,a_2,\ldots,a_N$, program a machine which will sort out the numbers in a descending order, that is, place the largest number first, the next-to-the-largest second, and so on. Follow steps (a), (b), and (c), as in Problem 2.1.

3
Number Systems and Coding

In order to understand the functioning of computers, which depend so heavily on numbers and counting, let us begin by examining the definitions of some fundamental concepts in arithmetic. Many of these concepts are not defined on the basis of mathematical justification, because they were developed long before any mathematical justification originated. Nevertheless, these mathematical foundations will be of interest in our discussion of designing a machine that uses numbers and performs arithmetic operations. If some of the definitions given here seem trivial, try stating your own before you read the text. It may prove to be a revealing experience.

3.1 SETS

A collection of things specified by some *defining property*, such as "all natural numbers" or "a bunch of bananas," constitutes a "set" or "class."[1] The bridge-playing group of Smith, Brown, Jones, and Clark, segregated from the rest of the world by *enumeration*, also constitutes a class.

---

[1] The two terms are synonymous. Other terms used in mathematics are "aggregate" or "manifold."

A banana, a black cat, or Mr. Jones are units which cannot be subdivided without losing their identity; thus they are characterized as *primary* or *elementary* units. A collection of black cats or a bunch of bananas are characterized as *secondary units, because they can be subdivided* to form two similar bunches or sets. A tertiary unit or a higher-order unit consists of classes whose members are lower-order units, such as in a "company," a "platoon," or a "squad" composed of individual soldiers. Counting with natural numbers in tens, hundreds, thousands, and so forth is a further example of the use of the idea of secondary and higher-order sets. To delimit a secondary unit so that it cannot be further partitioned without losing its identity, some delineating property such as "the fourth platoon of B Company" must be used.

3.2 NUMBER

Notions of quantity are associated with all classes of objects and are manifested by numbers. Thus, "number" is a property of "class," but not a property of the objects in the class themselves. The different numbers are instances of "number" exactly as different men are instances of "man." Thus, the number three which is the common property of all "trios," is also an instance of number.

A few additional comments should help to delineate further the meaning of "number."[2] Just as a certain object has a definite size, color, and other such properties, a group of objects may be characterized by the property of "number," such as "five," for example. Similarly, just as the size of an object is changeable when the object is subjected to physical stresses or breakage, so the number property of a group of objects is also changeable; it is born with the formation of the group and it "dies" when the group undergoes change. This, more specifically, is the property of the "physical" number. Thus, the number of a group of objects or a group of animals is changeable and in continuous motion, always depending on the physical entity it represents.

Contrary to the notion of the physical number is the notion of the "mathematical" number, that is, the number of a mathematician, that is eternal and free from change. Each mathematical number is abstracted from any direct association with a particular group of objects or animals; it does not necessarily represent a group of some kind of objects or animals. Thus, the mathematical number "four" represents the *common property* of all groups with "four" individual members: four cats, four collars, four men, or the group of John,

[2] First definitions of "number" may be found in Aristotle and Plato. A. N. Whitehead and B. Russell, in *Principia Mathematica*, Cambridge, 1910, produced the meaning of "numbers" and of their laws from purely logical axioms. D. Hilbert and P. Bernay, in *Grundlager der Mathematik*, Berlin, 1934, did the same based on logical and mathematical axioms.

James, Jack, and Paul. This notion of "number" is the one used in computer arithmetic and it is the most common in mathematical operations.

In addition to these two concepts, there is also the notion of the "idea" of a number, such as the idea of "two" or the idea of "three." The difference between the *mathematical* number and the *idea* of a number is that the idea of a number, like any other idea, is unique, while a mathematical number may be repeated at will and be different each time. To demonstrate this difference, consider the arithmetic formula $2 \times 2 = 4$. The two's in this formula cannot be the idea of "two," because in this formula we have more than one "two," while the idea of two is necessarily unique and therefore cannot be multiplied. In a similar fashion in geometry, a square may be divided by its diagonals into four equal isosceles triangles, none of which can be the idea of the isosceles triangle.

Of the three notions of a "number," that of the "physical" number is the fundamental one. To understand this, one must recognize that all other concepts of a "number," including the concept of "counting," begin from the concept of the "physical" number. This truth may be demonstrated by the philosophical question of Plato: "Could a pure scientist of numbers count or generate the idea of numbers without ever having had the assistance of any outside objects or animals, that is, outside himself? Can anybody 'read' *five* or *seven* 'inside' himself, without placing five or seven objects, men or other units, in front of him? Can anybody understand 'five' itself?" The essence of this question may also point to the difference between "pure" mathematics and "applied" mathematics—the numbers of a mathematician or of a scientist and the *use* of numbers by a merchant or a farmer to count objects or animals.

3.3 NUMERICAL SYMBOLS

For reasons of efficiency, numbers are represented by symbols. Thus, a numerical symbol represents information bearing the notion of quantity. To achieve this and the efficient manipulation of quantitative (numerical) information, a coding system of symbols with set operation rules is necessary. For example, the arabic numerical symbols may be used in the decimal coding system (with appropriate rules and marks to take care of decimal points, plus and minus signs, exponents, and so on) to satisfy efficiently the needs of ordinary pencil and paper arithmetic. In this number system, the notion of "fourness" is represented by the numerical symbol "4" which is properly used to refer to classes such as a bunch of four bananas, the four Gospels of the New Testament, or the sides of a square. It also refers to the quantitative notion of membership, common in the following delineating statements about such classes (or sets) as "the four bananas I just bought," "the four Gospels of the New Testament," "the sides of this square."

3.4 EQUAL NUMBERS

Two collections of objects for which a one-to-one correspondence may be set between their respective members are said to have an equal number of members, even if we do not know what that number is. Thus, in a piece of intrinsic semiconductor material, the number of the thermally generated electrons is equal to the number of generated holes, independent of the temperature and of the number of pairs generated and recombined per second, because of the existing one-to-one correspondence between the members of these two classes. Also, the numbers of husbands and wives are equal in a monogamic society, independent of the number of marriages and marriage resolutions per year.

3.5 COUNTING

The notion of "oneness" is obtained by subdividing a secondary class of objects to the point that it can no longer be subdivided without its members losing their integral identity. The resulting class is then a collection of one primary unit, a "unit class." The notion of oneness, associated with such a group, is represented symbolically in the decimal system of numbers by the numerical symbol "1." The concept of "two" and the concept of counting may be derived and also visualized if the operational process of "adding one" is established and applied to the unit class successively thereafter.

"Adding one" consists of augmenting the membership of a class by implanting a new valid member within the class. The concept of counting is implicit here. Successive integral sets can now be formed, and by their quantitative notions of membership, the concept of all "natural numbers" can be created.[3]

Additional operational rules may be formed to establish the concepts of rational, negative, and irrational numbers, or those of exponential, imaginary, and complex numbers.

3.6 SYSTEMS OF NUMERATION

The fundamental need for numerical symbols to represent numbers is apparent from our previous discussions. If, in an unimaginative way, a different symbol is used for each number, a "baseless" system of symbols results. Writing numbers in a baseless system is like writing spoken words

---

[3] When an operational rule like addition and a property like the "associative property" are established for the members of the set of natural numbers, the set is called a "group."

without using an alphabet, as for example in the writing of Chinese. The inefficiency of such a system of symbols is obvious and it becomes even more striking when we attempt to use such symbols in manipulating numerical information. Endless tables of addition, subtraction, multiplication, and division would be required, because no operational rules of any general value can be established. Considerable retardation in the development of mathematics, due to the awkwardness of the systems of numeration in the civilizations of the ancient Greeks and Romans, testifies to that effect.

An unimaginably large vocabulary of numerical symbols for representing numbers may be avoided by using a system that uses a "base" or "radix." In the banana business, for example, counting in large quantities is done in terms of "stems" and "bunches" or "hands."

The commonly used decimal system of numeration was initiated by our early ancestors who used their fingers as a natural aid to counting. In this system, a number is represented as a sum of powers of ten, where each power of ten is weighted by a digit between zero and nine, inclusive. The number ten is then used as a radix, and the system depends on the notion of place value (positional system) and on the use of the "zero" concept; that is, in writing a number, we write only the weights to be attached to the various powers and a decimal point (radix point) which tells what powers of ten are to be weighted. A number N written in the radix 10 system as $a_n a_{n-1} \ldots a_2 a_1 a_0 . a_{-1} a_{-2} \ldots a_m$ can be expanded as follows:

$$N = a_n \times 10^n + \ldots + a_2 \times 10^2 + a_1 \times 10^1 + a_0 \times 10^0 + a_{-1} \times 10^{-1} + a_{-2} \times 10^{-2} + \ldots + a_{-m} \times 10^{-m}$$

where the weighting coefficients

$$a_n, a_{n-1}, \ldots, a_2, a_1, a_0, a_{-1}, a_{-2}, \ldots, a_{-m}$$

are all positive integers less than ten. According to this, the decimal number 341.75 means

$$341.75 = 3 \times 10^2 + 4 \times 10^1 + 1 \times 10^0 + 7 \times 10^{-1} + 5 \times 10^{-2}.$$

If our ancestors had not used their thumbs in counting, we would probably be counting in the octonary system, which in some ways is more convenient. Fractions in the octonary system are formed by successive bisecting, and multiplication is easier than in the decimal system of numeration.

Generally, for any other radix r, a number is written as $a_n a_{n-1} \ldots a_2 a_1 a_0 . a_{-1} a_{-2} \ldots a_{-m}$ [the reference dot (radix point) denotes the beginning of negative powers], and it may be expanded in a series of weighted powers of the radix r:

$$N = a_n \times r^n + a_{n-1} \times r^{n-1} + \ldots + a_2 \times r^2 + a_1 \times r^1 + a_0 \times r^0 + a_{-1} \times r^{-1} + a_{-2} \times r^{-2} + \ldots + a_m \times r^{-m}.$$

The weighting coefficients are positive integers less than r (including "zero"). Note that the number of necessary symbols is r and that numbers are represented

NUMBER SYSTEMS AND CODING

by their own weighting coefficients, according to the place-value notion. As an example, let us see how the decimal number 53 can be represented in several systems of numeration using the arabic numerical symbols.

$$
\begin{aligned}
(53.0)_{10} &= 5 \times 10^1 + 3 \times 10^0 \\
&= 6 \times 8^1 + 5 \times 8^0 &&= (65.0)_8 \\
&= 2 \times 5^2 + 0 \times 5^1 + 3 \times 5^0 &&= (203.0)_5 \\
&= 1 \times 3^3 + 2 \times 3^2 + 2 \times 3^1 + 2 \times 3^0 &&= (1222.0)_3 \\
&= 1 \times 2^5 + 1 \times 2^4 + 0 \times 2^3 + 1 \times 2^2 + 0 \times 2^1 + 1 \times 2^0 &&= (110101.0)_2
\end{aligned}
$$

The system with radix $r = 2$ uses weights consisting entirely of ones and zeros. That system of numeration is of particular importance in computer arithmetic and will be studied further in subsequent paragraphs.

3.7 COUNTING WITH NATURAL NUMBERS

Sets of objects may be formed to provide a visualization of numbers by the quantitative notion of their membership. Counting the next higher number by sets means to open the set and to implant a new valid member within the set. The concept of counting, which is thus established in terms of sets, may now be restated as a general rule with reference to counting with numbers presented in any numerical system using a radix. In such a system, there are r accepted numerical symbols (marks) from zero to $r - 1$, inclusive.

To record the next number higher than a given natural integer number, we first search, starting from the right-hand side of the number toward the left-hand side, to find the first digit that is not at its highest admissible symbol value (that is, less than $r - 1$). Then we raise that symbol to the next higher admissible symbol and lower all the others on its right to the lowest admissible symbol. For example, in the decimal system, the next number after 378 is 379 (raise 8 to 9), then 380 (raise 7 to 8 and lower 9 to 0), and so on. This rule is applicable to decimal, binary, or any other system of numeration processing a radix. Table 3.1 illustrates counting from zero to twenty-six in the decimal, binary, and octonary systems.

3.8 RADIX CONVERSION

Transforming a number from one radix to another is done by expanding and regrouping the number in accordance with the expansion formula

$$N = \sum_{n=-\infty}^{+\infty} a_n r^n$$

applied for the accepted symbols a_n each time.

Table 3.1. Counting in decimal, binary, and octonary systems.

Decimal radix-ten	Binary radix-two	Octonary radix-eight	Decimal radix-ten	Binary radix-two	Octonary radix-eight
0.125	0.001	0.1	13	1101	15
0.250	0.010	0.2	14	1110	16
0.500	0.100	0.4	15	1111	17
0	0	0	16	10000	20
1	1	1	17	10001	21
2	10	2	18	10010	22
3	11	3	19	10011	23
4	100	4	20	10100	24
5	101	5	21	10101	25
6	110	6	22	10110	26
7	111	7	23	10111	27
8	1000	10	24	11000	30
9	1001	11	25	11001	31
10	1010	12	26	11010	32
11	1011	13			
12	1100	14			

Since $8 = 2^3$, a binary number is easily written in the octonary system by dividing it in groups of three digits starting from the right-hand side. Then each group is transformed to an octonary digit separately. For example, the binary number 11010111 is transformed as follows:

$$(11010111)_2 = (011)(010)(111) = (327)_8.$$

Because of such a relation between the two systems, the octonary system acquires a particular importance in computer arithmetic because it occasionally suggests efficient "shorthand" ways of writing binary numbers.

More generally, the radix conversion can be done efficiently and simply by hand, using the method of "division by radix." In so doing, a given decimal number can be converted to a number with base r, by successively dividing the given decimal number by the new radix r, in the conventional manner of performing a division on the decimal system of numeration. In each division, the remainder represents a new digit of the converted number, the process beginning from the right-hand end, that is, from the power r^0-position, while the quotient of each division is then divided successively until it becomes smaller than r. When this happens, the next division gives a quotient equal to zero and a remainder which then becomes the highest-power digit (left-most digit) of the number written in the new basis of radix r. To demonstrate, let us convert the decimal number $(53)_{10}$ to base 3:

$$(53)_{10} = 17 \times 3 + 2 \longrightarrow \times 3^0$$
$$17 = 5 \times 3 + 2 \longrightarrow \times 3^1$$
$$5 = 1 \times 3 + 2 \longrightarrow \times 3^2$$
$$1 = 0 \times 3 + 1 \longrightarrow \times 3^3$$

(less than 3) "STOP" thus: $(53)_{10} = (1222)_3$

In another example, let us convert the decimal number $(839)_{10}$ to base 2:

$$(839)_{10} = 419 \times 2 + 1 \longrightarrow \times 2^0$$
$$419 = 209 \times 2 + 1 \longrightarrow \times 2^1$$
$$209 = 104 \times 2 + 1 \longrightarrow \times 2^2$$
$$104 = 52 \times 2 + 0 \longrightarrow \times 2^3$$
$$52 = 26 \times 2 + 0 \longrightarrow \times 2^4$$
$$26 = 13 \times 2 + 0 \longrightarrow \times 2^5$$
$$13 = 6 \times 2 + 1 \longrightarrow \times 2^6$$
$$6 = 3 \times 2 + 0 \longrightarrow \times 2^7$$
$$3 = 1 \times 2 + 1 \longrightarrow \times 2^8$$
$$1 = 0 \times 2 + 1 \longrightarrow \times 2^9$$

(less than 2) "STOP"

Thus, $(839)_{10} = (1101000111)_2$.

3.9
APPROPRIATE NUMERICAL SYSTEM FOR MECHANIZING ARITHMETIC

The electronic engineer who is faced with the problem of designing the arithmetic circuits for the processor unit must start first by devising some method of substantiating numerical information within the computer. Some physical entity has to be selected, such as the position angle of a wheel or the voltage between a pair of wires, and its measured values must be coded properly to represent the numerical information.

Selection of the proper means of representing numbers and alphabetic characters within the machine is determined by considerations of economy, speed, and reliability, not necessarily in this order. Electrical means of implementing coded information offer greater speed and economy and are almost exclusively used in the arithmetic sections of large-scale computers. For example, 100 different numbers and symbols can be represented by 100 different voltage levels above reference on a single wire, or by combinations of 10 levels on one wire and 10 on another, or by $5 \times 5 \times 4$ on three wires, and so on. The use of more than one wire in addition to the reference wire, however, is to be avoided for obvious reasons of economy and circuit design complications.

If the decimal system of numeration is used, ten different voltage levels may be chosen between a pair of wires to represent the ten numerical digits. Because it is not possible to maintain an exact voltage with electronic circuits, ten voltage ranges are selected instead (see Table 3.2). The circuits and the devices that are used to make up the decision and memory elements within the computer must be able to recognize the ten states and also reproduce them at their outputs.

Table 3.2. Voltage-range coding.

Voltage greater than or equal to…(volts)	… but less than (volts)	Represents digit
$-\infty$	10	0
10	20	1
20	30	2
30	40	3
40	50	4
50	60	5
60	70	6
70	80	7
80	90	8
90	$+\infty$	9

A matter of utmost importance is raised now: that of reliability. An error of six per cent on the nominal value of 85 V (which represents the digit 8) would make it look like 79.9 V or like 90.1 V and be read as the digit 7 or the digit 9 instead of the correct 8. The maximum allowable error varies for the different voltage ranges defined in Table 3.2, but an error in establishing or detecting the nominal voltage values of less than about five percent must be secured for reliable operation. If not impossible, it is very difficult to accomplish this over long periods of time, using electronic devices like vacuum tubes or transistors whose parameter values drift with time. The complexity of electronic circuits that will establish, keep and detect ten nominal voltage values with a small allowable error, causes additional problems of reliability.

Things are quite different, though, if only two voltage levels are required for representing numerical information. This is the case when binary symbols are used in representing numerical information. Only two nominal voltage levels, such as zero volts (off-condition) to represent the digit "0" and ten volts (on-condition) to represent the digit "1" are required. The allowable error for reliable operation is much greater now, up to fifty percent for example, if any voltage below 5 V is considered as "0" and any voltage above 5 V (with a nominal value of 10 V) is considered as "1." This leaves much more room than

76   NUMBER SYSTEMS AND CODING

before for parameter variations from aging or from replacing components in the electronic circuits. Thus, binary implementation is more reliable.

There is nothing inherently binary in many conventional electronic devices, such as the vacuum tube or the transistor. However, they operate much more reliably in a binary fashion, as in a digital computer. There is also another reason for using the binary system in representing numerical information. Many reliable, inherently binary devices are available, such as the electromagnetic relay, the switching diode, a hole in a card, an electronic flip-flop circuit, the ferrite core, and many others, all of which possess two discrete and clearly distinguishable and attainable stable states.

The superiority of the binary system of numeration for use in digital computer design has been demonstrated above by reasons of reliability and simplicity in device and arithmetical circuit design. Before discussing the ramifications of this system in computer arithmetic, we briefly review some basic facts.

3.10
THE PURE BINARY SYSTEM

A number written in the binary system consists exclusively of the binary digits 0 and 1 which, by contraction of the words "binary digits," are called "bits." For instance, 1001.11 is a six-bit number. Operations in the binary system are performed according to the same general rules of decimal arithmetic and carry operations. To minimize the size of the carry, computers add only two numbers at a time. In this way, the carry is never larger than the second admissible symbol in the system, a 1 in this case. The table for binary addition has 2 × 2 entries as shown in Table 3.3.

Table 3.3. Binary addition table.

		Augend	
		0	1
Addend	0	0	1
	1	1	10
			carry ↗

Information written in binary form must be translated into the decimal system and back to the binary for the purpose of presenting the numerical information in a form legible to the operator of the computer. To translate an N-digit decimal number into the pure binary code, M-bits are required according to the relationship $2^{M-1} < N \leq 2^M$. For example, since $2^8 < 467 < 2^9$, nine bits are required to encode 467 into pure binary code: 111010011.

To facilitate man–machine communication, several binary coded systems have been devised. The most popular of these still use the decimal system as a basis for representing numbers, but in a binary coded fashion.

3.11
BINARY CODED NUMERICAL SYSTEMS— MACHINE LANGUAGES

The exclusive practice in all digital computers today is to represent numbers by coded combinations of binary symbols. This practice will continue in the foreseeable future, because there is no prospect for an early development of "decimal" devices that are comparable to the simplicity, cost, and reliability of existing binary devices.

Digital machines, when incorporated into large systems, may communicate with each other in any selected appropriate *language* of numeration. However, for communication with human operators, the conventional decimal system has more advantages. For this reason, numerous methods (codes) that make use of binary symbols to represent decimal digits have been formulated. The method by which binary digits are combined to represent characters like the decimal digits or the letters of the alphabet forms a *code* which then constitutes the *machine language*. A code is judged by its success in keeping a continuous check for errors on the transmitted and manipulated information, by its ability in checking the correctness of arithmetic operations, and by its ability in making the operations and the translation to and from the human language as simple as possible. A good code should possess several of these qualities.

Perhaps the most obvious way to represent a decimal number is by grouping binary digits in decades and counting an equivalent number of ones for each decimal digit. Thus, the number 36 will correspond to (0000000111) (0000111111). A counter and a few built-in rules for the carry operation would accomplish mechanization of addition and the other arithmetical operations.[4] However, this code is obviously exceedingly inefficient in terms of hardware equipment and bit storage space. Only 100 numbers may be represented with 20 binary digit places, while in the pure binary system the same number of bit places may represent over a million numbers, 2^{20} to be exact. The pure binary system itself is highly efficient in calculations and arithmetical manipulations, but the human language of decimal digits and alphabetical characters cannot be translated simply into the pure binary code. Versions of sixteen-bit pure binary code are used in the IBM 1130 and 1800 computer systems.

---

[4] A modified form of this system was used in the Eniac (Table A.2).

a. The 8421 or NBCD Code

In sacrificing part of the arithmetical efficiency of the pure binary code, we gain ease of translation in using binary coded decimal codes. The 8421 code, otherwise called the "natural binary coded decimal" (NBCD system), translates decimal numbers into a binary code which replaces each decimal digit with a combination of four coded binary symbols. A correspondence is thus established between each decimal digit and a set of four bits.

There are sixteen ways by which four binary digits may be arranged in a row. Ten of these combinations may be selected to represent the ten decimal numerical digits. Thus, $16!/6! = 29,059,430,400$ codes are possible this way! Even after eliminating codes that may be derived from each other by complementation or rearrangement of the bits, there are $16!/384 \times 6!$ or about seventy million possible codes left. Quite a wide choice!

Table 3.4 illustrates a natural choice which corresponds to counting in binary,

Table 3.4. Four-bit coding of the decimal digits: the 8421 code.

Decimal digits	Code 8421—weights	Decimal digits	Code 8421—weights
0	0000	5	0101
1	0001	6	0110
2	0010	7	0111
3	0011	8	1000
4	0100	9	1001

using four bits from zero to nine inclusive. The code is commonly referred to as the 8421 code. It is a weighted code, that is, a decimal digit represented in this code is equal to the sum of the products of the bits in the coded message times the successive powers of two (weights), starting from the right end progressing toward the left of the coded message. Such a code is additive, because the sum of the binary coded decimal digits, in accordance with the rules defined in Table 3.3, is the code of their sum.

An N-digit decimal number is translated into N four-bit sets of machine language. Such encoding is illustrated by an example in Table 3.5.

The remaining unused six combinations of four bits (sixteen combinations are possible with four bits, but only ten were used in Table 3.4) may be used for representing other arithmetic symbols, such as plus and minus signs, percent, dollar, and so forth. The system is called "8421" because of the place significance of the four bits allowed per decimal digit. It is a weighted positional system; the weighting factors are shown in Table 3.4.

Table 3.5. Binary coded decimal system exemplified. No space is necessary between the four-bit groups when transmitting. Spaces have been used here for convenience in reading.

```
Place significance (weighting factors):
         800  400  200  100      80  40  20  10      8  4  2  1
         ---( -    -    -    -)  ( -  -   -   -)    (-  -  -  -)
         ---        Hundreds          Tens              Units
Example: Decimal 3                     7                  6
                   Coded 0011          0111               0110
```

Addition in the 8421 system is performed according to the rules of binary addition defined in Table 3.3. When addition is performed in this code, bit combinations not belonging to the code may result if the sum of two decimal digits is more than nine. This situation does not occur in the pure binary code, but it is common among the binary coded decimal codes. An example of "out-of-code" addition in the 8421 system is illustrated in Table 3.6.

Table 3.6. In-code and out-of-code addition in the 8421 system.

```
   4   0100              4   0100
  +3   0011             +8   1000
   7   0111 ← in-code   12   1100 ← out-of-code (forbidden
                                                  combination)
                            (0001  0010) ← desired
```

The sum in the out-of-code addition in the 8421 system is a pure binary number, which when converted to decimal is just ten greater than the proper number. Correction of the out-of-code addition is an arithmetical problem. Correction rules are stated and incorporated within the computer.

The 8421 system makes arithmetic more complicated than the pure binary system does. It presents difficulties in the carry operation of out-of-code addition and also makes somewhat inefficient use of the available equipment. However, the 8421 system possesses a number of advantages. First, it is easy to understand. Second, it offers the possibility of a code check because the even decimal digits are coded with a 0 at the end, and odd decimal digits are coded with a 1 at the end. For more efficient error checking purposes an additional bit may be added, bringing the total number of bits per decimal digit to five. The additional bit is called a "parity" bit and is a 0 or 1, as the case may be, to maintain the total number of 1's per coded character always odd (or even, as may be chosen). The parity check is used for protection against corruption of the code by erroneous

80 NUMBER SYSTEMS AND CODING

transmission of a single bit. However, two-bit errors in the same transmission may go undetected. The odd-parity 8421 code is shown in Table 3.7. This code is used in the IBM 1401 and 1620 computer systems. The odd- (or even-) parity check is commonly used in connection with several popular binary coded decimal codes.

With the addition of two more bits (BA) to the 8421 case, a six-bit code may be created which will accommodate 64 different characters. The decimal digits,

Table 3.7. Odd-parity 8421 code.

Decimal Digits	Code P8421—weights	Decimal Digits	Code P8421—weights
0	10000	5	10101
1	00001	6	10110
2	00010	7	00111
3	10011	8	01000
4	00100	9	11001

the letters of the alphabet, and many additional symbols may be assigned binary codes in this way. Adding the parity bit C and the BA bits to the 8421 code results in a seven-bit code CBA8421, including parity. An example of such a six-bit code with a parity bit added is shown in Table 3.8. The seven bit positions in this code are divided into three groups: one parity check position (C), two zone positions (BA), and four numerical positions (8421). This code is used as a standard interchange code among various computer systems, to provide compatibility of data interchangeability.

b. The "Excess-Three" Code

This code, commonly referred to as the XS3 code, is formed by adding the binary three (0011) to the 8421 code, as shown in Table 3.9.[5]

The decimal number 376 is coded in XS3 as 0110 1010 1001. A pleasant feature of this coding system is that the nine's complements of decimal numbers are coded by the one's complements of their binary codes.[6] For example, the nine's complement of the number 376 which is the number 623, is coded as 1001 0101 0110 (623), which is the one's complement of 0110 1010 1001 (376). This property is useful in performing arithmetic.

[5] Univac II—Sperry Rand Corp. uses XS3, seven-bit, odd-parity code.
[6] The nine-complement of a decimal number is obtained by subtracting each digit from nine. The one-complement of a binary number is obtained by substituting zero for each one and visa versa.

BINARY CODED NUMERICAL SYSTEMS 81

Table 3.8. The CBA8421 odd-parity code used as standard interchange code on the IBM 1401, 1410, 7010, 7040, and 7044 Data Processing Systems.

Character	Odd-parity bit	Zone bits	Numeric bits	Character	Odd-parity bit	Zone bits	Numeric bits	Character	Odd-parity bit	Zone bits	Numeric bits
0	1	00	1010	L	0	10	0011	&	1	11	0000
1	0	00	0001	M	1	10	0100	$	1	10	1011
2	0	00	0010	N	0	10	0101	*	0	10	1100
3	1	00	0011	O	0	10	0110	]	1	10	1101
4	0	00	0100	P	1	10	0111	;	1	10	1110
5	1	00	0101	Q	1	10	1000	△	0	10	1111
6	1	00	0110	R	0	10	1001	—	0	10	0000
7	0	00	0111	S	1	01	0010	/	1	01	0001
8	0	00	1000	T	0	01	0011	,	1	01	1011
9	1	00	1001	U	1	01	0100	%	0	01	1100
A	0	11	0001	V	0	01	0101	⋎	1	01	1101
B	0	11	0010	W	0	01	0110	\	1	01	1110
C	1	11	0011	X	1	01	0111	⧣	0	01	1111
D	0	11	0100	Y	1	01	1000	ƀ	0	01	0000
E	1	11	0101	Z	0	01	1001	#	0	00	1011
F	1	11	0110	BLANK	1	00	0000	@	1	00	1100
G	0	11	0111	.	0	11	1011	:	0	00	1101
H	0	11	1000	⧠	1	11	1100	>	0	00	1110
I	1	11	1001	[	0	11	1101	√	1	00	1111
J	1	10	0001	<	0	11	1110	?	1	11	1010
K	1	10	0010	⧧	1	11	1111	!	0	10	1010
								⧧	0	01	1010

Table 3.9. XS3 coding.

Decimal digit	Code	Decimal digit	Code
0	0011	5	1000
1	0100	6	1001
2	0101	7	1010
3	0110	8	1011
4	0111	9	1100

Other advantages of this code include ease of understanding the coding techniques and ease of doing arithmetical operations. Addition of two XS3 numbers is brought "in code" easily. The addition is performed for every four-bit coded decimal digit separately as with binary numbers. If the two digits add to less than ten, the XS3 codes add to less than sixteen, and no carry is created in the fifth place, then simply subtract three from the binary sum, which means subtract 0011 or add 1101 and neglect the carry. If the sum is greater than nine, then add 0011 and transfer the carry. This addition is illustrated in Table 3.10.

Table 3.10. Addition in the XS3 code.

$$
\begin{array}{cc}
3 + 6 & 7 + 8 \\
\hline
0110 & 1010 \\
1001 & 1011 \\
\hline
1111 & 10101 \\
+1101 & +\ 0011 \\
\hline
\text{drop} \to 1)\ 1100 \to 9 & \text{carry} \to 1)\ 1000 \to 5
\end{array}
$$

3.12
FIXED- AND FLOATING-POINT NUMBER REPRESENTATION

When writing a number, we can place the radix point in any location we wish and multiply the number by an appropriate power of the base. This method of writing a number is called the "floating-point" representation, in contrast to the "fixed-point" representation in which the radix point is placed in its true location and no powers of the base are used.

It is possible that errors may be caused as a matter of course during the processing of numbers in machines which employ fixed-length words. The errors are introduced by truncation or round-off, which may be employed on the least significant digits of a number, when the length of the number exceeds the length of the machine word. For example, a forty-bit word can accommodate up to ten-digit numbers in the NBCD code. If the size of the number is above ten digits long, then overflow will occur in the machine registers and in the memory cells.

Errors which occur and accumulate because of truncation and round-off may be quite sizable when fixed-point number representation and arithmetic is used. However, by using floating-point representation, the consequences of the restrictions of fixed-word size can be greatly alleviated.

In a ten-digit word, a floating-point number is represented with its eight most significant digits preceded by two special digits that specify the relative location of the decimal point and followed by the sign. In the "excess-fifty" floating-point convention, the number

$$\pm .abcdefgh \times 10^{xx-50}$$

is represented by

$$xxabcdefgh \pm$$

The following are examples of representation of numbers in the excess-fifty floating-point convention:

Number	Floating-Point Convention
—1234.567890	5412345678—
0.0000123000	4612300000+
−0.1234567890	5012345678—
0	0000000000+

It should be noted that a number F is assigned the value 0 when $0 < |F| < 0.10000000 \times 10^{-50}$; it is not defined if $|F| \geq 10^{50}$. Also, the number is truncated to the eight most significant digit positions, as the first and third numbers in the above examples. (Truncation and rounding-off are discussed in Chapter 4.)

Scientific computers, which face demands of great accuracy in computations, commonly use floating-point representation. Many modern machines have a built-in floating-point. That is, they have circuits which perform arithmetical operations automatically in floating-point. To facilitate programming the appropriate instructions are included in the machine language.

Machines with fixed-point arithmetic can be programmed to perform floating-point arithmetical operations, if the necessary instructions are available. Then the programmer must write special subroutines for floating-point arithmetic operations. This implies added cost in storage space for the special subroutines and in computing time because of added subroutine instructions.

However, fixed-point arithmetic is faster and therefore is preferred over floating-point arithmetic in many business machines or whenever high accuracy is not demanded and very large numbers are not likely.

A machine which is designed to perform arithmetic operations directly with floating-point numbers keeps track of the radix-point location automatically. In fixed-point machines, the programmer must do this.

BIBLIOGRAPHY

Flores, I., *Computer Design*. Englewood Cliffs, N.J.: Prentice-Hall, Inc., 1967.
Maley, G. A. and Earle, J., *The Logic Design of Transistor Digital Computers*. Englewood Cliffs, N.J.: Prentice-Hall, Inc., 1963.
Stibitz, C. R. and Larrivee, J. A., *Mathematics and Computers*. New York: McGraw-Hill, Inc., 1957.
Uspensky, J. V. and Heaslet, M. A., *Elementary Number Theory*. New York: McGraw-Hill, Inc., 1939.

Problems

3.1 Express $(357.36)_{10}$ in expanded form
$$a_n 10^n + \ldots + a_0 10^0 + a_{-1} 10^{-1} + \ldots + a_{-m} 10^{-m}.$$
Also express it in binary form.

3.2 Express $(212011.02)_3$ in binary and then in octal form.

3.3 Express $(347)_{10}$ and $(512)_{10}$ in binary form.

3.4 The hexadecimal number system has a base (or radix) of 16. A common symbol set for this system in ascending order is:
0, 1, 2, 3, 4, 5, 6, 7, 8, 9, A, B, C, D, E, F.
(a) Convert $(8A.C)_{16}$ into a decimal number.
(b) Convert $(8A.C)_{16}$ into a binary number.
(c) Convert $(11011011.0111)_2$ into a hexadecimal number.

3.5 Write addition and multiplication tables in the quinary and the octonary (also referred to as "octal") number systems.

3.6 Write the 9's and 10's complements of the following decimal numbers: (a) 9109, (b) 099.129, (c) 628.39

3.7 Write the 1's and 2's complements of the following binary numbers: (a) 101101, (b) 1011.011, (c) 0000001.1

3.8 Convert each of the following octal numbers into a decimal and a binary number: (a) $(125)_8$, (b) $(101)_8$, (c) $(765)_8$, (d) $(35.24)_8$, (e) $(0.34)_8$

3.9 Convert the following numbers:
(a) decimal 574 into bases 16, 7, and 2.
(b) hexadecimal 2EA into bases 10, 8, 4, and 2.
(c) decimal π into octal and binary to five significant figures.

3.10 Convert $(0.110101)_2$ and $(1101.0110)_2$ into decimal numbers. Are there exact decimal equivalents? Under what conditions does a binary number have an exact decimal equivalent?

3.11 Convert $(33.47)_{10}$ and $(12.53125)_{10}$ into binary numbers. Are there exact binary equivalents? Under what conditions does a decimal number have an exact binary equivalent?

3.12 Express the decimal number 453 in (a) straight binary code, (b) in NBCD, and (c) in Excess-3 code.

4
Reliability

Fault-free operation of a machine for an indefinite length of time is inconceivable. The "probability of survival" of the machine over some finite length of time, may serve as a measure of the *reliability* of the machine. In large physical systems, such as information-processing machines, the task of building high reliability into the machine is a most serious and often difficult problem which is shared by the systems engineer, the programmer, the circuit and device designer. The reliability of an information-processing machine may be raised by building proper safeguards into the hardware design as well as in the machine programming. In this chapter, we discuss the principles of error-detecting and error-correcting codes, which are used for programming computers so that more reliable operation may be secured. We also discuss a few very interesting and instructive features of reliability problems of a different nature which are met in the design of computer hardware.

4.1
RELIABILITY DEMANDS

It is instructive to look first into the causes of high reliability demands in the design of information-processing machines. Reliability demands in the design of arithmetic and other units of a computer often exceed one part in a billion. No computation—scientific, engineering, or business—would require such accuracy. Most delicate scientific problems do not require accuracy higher than one part in a thousand. However, for an information-processing machine to produce the outcome of a long computation with an accuracy of one part in a thousand, the machine must perform the millions of intermediate arithmetic operations with accuracies exceeding one part in a billion. This surprising demand is imposed because of the accumulative arithmetic errors, which are generated in the course of normal operation of the machine and which pile up in successive operations.

4.2 SYSTEM RELIABILITY

With such stringent reliability demands in the operation of computers, the hardware and systems engineers have been searching very seriously for reliability in their design work. To their dismay, however, they soon discovered one of the most fundamental truths about reliability of systems, namely, that it is not directly calculable if only the reliabilities of the components are known unless the manner of their interconnections is also stated clearly and their interactions are understood and are measurable. Thus the effects from the interactions between the components of a given system, in most cases, grossly determine the reliability of the system itself.

It must be understood first that the simple rule of multiplication does not apply in determining the reliability of a system, except in very rare cases of noninteracting components. Such a rule, applicable to a large system containing 100,000 noninteracting components, would impose stringent requirements on the individual reliabilities. For example, for the system to be operative 99 percent of the time during 24 hours would require, on the basis of equal allocation of reliability, that each component have a reliability of about 99.99999 percent for 24 hours (which corresponds to an *average life* for each component of about 28,000 years)!

The situation is saved by the interaction reliabilities. Effects of the component interactions, measured by what may be termed as "interaction reliabilities," can be either *constructive* (upgrading) or *destructive* (degrading) to the reliability of the overall system. The effect of upgrading or constructive interaction reliabilities may be thought of as the "savings" one receives when he purchases soft drinks by the six pack. One may pay ten cents for a bottle of cola, but only forty-nine cents for the six pack.

Interaction reliabilities cannot exist except among the "pack" of the functioning components of a system, just as one cannot benefit from the "savings" unless he buys the whole six pack of cola. It is the job of the reliability engineer to search in his design for constructive interaction reliabilities, and to suppress destructive ones. The task becomes more important when one realizes that upgrading interaction reliabilities can exist with values *much larger than 100 percent*.

Often, a well-used constructive interaction reliability of a value of over a thousand percent may offset the unreliability of a hundred components! This is not intended as a blow to the important effort of improving components. However, even though we may achieve nearly 100 per cent component reliability, the final *key to system reliability* in computers is found among the more powerful interaction reliabilities.

The upgrading interaction of the automatic volume control in radios is a fine example of a one-million percent constructive interaction reliability. Without AVC, the old radio receivers were operative only over a signal level variation range of perhaps one hundred to one. Modern radios with AVC will play satisfactorily with input signal levels ranging from perhaps ten microvolts up to perhaps ten volts, a million-to-one variation. In addition, AVC provides high tolerance for the degradation of several components.

Destructive interaction reliability can be illustrated by the adverse effect that a screen-grid resistor can have on the life of an electronic vacuum tube. If the screen-grid resistance, initially of a proper value, gradually drops because of aging, the tube may be subjected to unreasonably high screen voltage and current. Screen emission may develop, and the tube will fail. Even new tubes may fail one after another until somebody decides to check and replace the screen resistor or change the design, using a feedback scheme that would desensitize the tube's operation to the value of the screen resistor.

To conclude, we can say that interaction reliabilities are countless in a system. A successful engineering design should take advantage of upgrading interaction reliabilities, either by finding them in existing designs or by inventing new ones. The task is a difficult one, especially since the number of possible interactions in large systems is often unlimited. But the task is also a challenging one, since there exists a chance for "big game" in the electronic wilderness of computer reliability.

4.3
RELIABILITY IN THE NERVOUS SYSTEM

The nervous system of a living organism may be considered as a digital (binary) mechanism, because it communicates information by messages which are physically composed of signals possessing the all-or-none (yes-or-no, 0 or 1) character.[1] Encoding messages which bear information

---

[1] It has been forcefully argued by many in the past that even neurons are not exactly digital organs when one considers their function in considerable detail. The stimulation of a neuron and the development and progress of its impulse, which possesses the all-or-none character, are highly complex electrochemical processes. These processes function in an analog fashion just like a transistor, which is capable of operating as an all-or-none organ *under certain conditions*. This assertion about the neuron is important when we understand the internal functioning of a nerve cell, but is irrelevant when we consider the neuron as a "black box" operating in the all-or-none mode.

about the strength of the stimulus on a neuron seems to be accomplished in a fashion essentially similar to frequency modulation. The neuron responds periodically to a stimulus with a recovery time (dead-time between two successive stimulations) which is a function of the strength of the stimulus, and encoding is thus achieved. This behavior is one of a genuine yes-or-no organ which uses counting in time rather than a radix expansion method (binary or otherwise) for numerical (quantitative) data representation.

Counting the number of pulses per second is a slow and inefficient process, but a highly reliable one. Over one million pulses are required to represent a number of the same size, while only seven decimal or about twenty binary digits would suffice if decimal or binary expansion techniques were used. The counting method appears highly wasteful as compared to our more economical radix-expansion method; nevertheless, the counting method is safer from error. An error of just several pulses in millions of pulses will result in only irrelevant changes on the counted number. If, however, the same number is expressed by decimal or binary expansion, even a single error may deteriorate the entire result.

Accordingly, the high-cost high-reliability demands of our present computing machines may be traced, at least partially, to the radix-expansion system of number representation. Practically error-proof operation of living organisms is, on the other hand, traced to the counting method that they seem to use. In conclusion, a kind of "complementarity principle"[2] in information processing seems to emerge here: "The product of 'safety from error' times the 'efficiency in notation' is constant or remains less than a certain quantity." In other words, one can be improved at the expense of a corresponding loss in the other. Safety from error increases if redundancy is used in notation; this is the case in the counting method.

4.4
REDUNDANCY AND ERROR DETECTION IN CODES

Increased safety from erroneous signal transmission may be achieved by redundancy also in the case that the digital expansion notation is used, repeating every message several times. This, in fact, is done in telegraphy to reduce the contamination of messages by the presence of noise.[3] Theoretically, it may be shown (John von Neumann) that it is possible to reduce the number of errors in message transmissions to as low a level as desired by including sufficient

---

[2] Bohm, D., *Quantum Theory*. Englewood Cliffs, N.J.: Prentice Hall, Inc., 1951.
[3] "I have said it thrice: What I tell you three times is true."—cried the bellman in *The Hunting of the Snark*. See Lewis Carroll's *The Annotated Snark*. New York: Simon & Schuster, 1962.

redundancy. Error control, by redundancy slows down computer processes and for this reason it is used within the computer only in special cases. Several popular error-detection techniques using redundancy in coding will be discussed now.

Error-detection and error-correction power may be invested in special types of codes at the expense of redundancy. A whole new branch of information theory has been formed which aims at the development of new, more powerful and practically implementable error-detection and error-correction codes.

To begin with, a distinction should be made at this point between a "code" like the pure binary code and a special "error-detecting and error-correcting code." The first serves as the official computer language and it may be any one of the codes described in Section 3.11. The "error-detecting and error-correcting code," on the other hand, is a code specially constructed with built-in safeguards in order to achieve increased freedom from errors during the transmission, storing, or processing of information.

Any appropriate symbolic notational system which uses only the essential amount of symbolic representation for relaying information is a "code." As such, the English language, spoken or written, in which combinations from a finite number of available phonetic or written characters are used to relay meaningful information, is a code.

In computer language, binary coded "words" commonly have a fixed length for easy clocking and identification. Thus, binary words of length m (m-bit words) may be seen as m-tuples, that is, m-dimensional vectors formed from the finite field of two elements ("zero" and "one"). Such an example is the 8421 code.

However, when the code word is lengthened by introducing redundant bits for the purpose of increasing the error-detecting power of the code, then such an encoding process leads to an "error-detecting" code, and it may be seen as a mapping of an m-dimensional vector (m-tuple) into an n-dimensional vector (n-tuple) over the same finite field of elements (namely "zero" and "one"), where $n > m$. An example of such a mapping is the conversion of the "8421, code" into the "two-out-of-five code" discussed in the next section. In the transmission of the new code, n minus m bits are redundant and serve only in the recovery of the original message in the event that, due to erroneous transmission some bits of the code are caused to be in error.

Depending on the amount of redundancy introduced, the error-detecting and error-correcting power of the code may vary. For a given amount of redundancy, an appropriate encoding technique which maps m-tuples of zeros and ones into n-tuples of zeros and ones, will lead to a code which has the power of recovering the correct message in the event that a limited number of errors have occurred per coded word. The Hamming code is a classic example of a systematic single-error correcting code, and it is discussed together with other codes later in this chapter.

4.5 ERROR-DETECTING CODES

The two following examples illustrate error-detection principles. The term "combination" refers to the coded sequence of bits for a single character, and the term "code" refers to the set of valid combinations selected to represent a given set of characters.

a. 2-Out-of-5 Code

Of the 40 possible *single errors* in the combinations of the 8421 code (ten valid combinations with four possible single errors per combination) only ten lead to nonvalid, that is, out-of-code, combinations. Consequently, only twenty-five percent of the single errors, assuming that all are equally probable, may be detected by not producing a valid code. If five-bits are used per decimal digit in a code, the fifth bit being not essential but redundant, it is possible to represent the decimal digits by combinations which exclusively use only two 1's. This is possible because the number of five-bit combinations which use two 1's and three 0's are

$$\binom{5}{2} = 5!/(5-2)! \times 2! = 10.$$

A possible arrangement of a 2-out-of-5 code is shown in Table 4.1. A single error may be detected easily in the 2-out-of-5 code, because any single 0 to 1 or 1 to 0 change results in a combination which no longer has two 1's, a fact readily checked with special logic circuits. If two errors occur in the same combination during a transfer, the check may fail. However, the probability of a double error in the same character is very small. This self-checking code has been used in machine designs in spite of being rather inconvenient for doing arithmetic.

b. 2-Out-of-7 Code

Seven bits may be used per coded combination to increase the number of available combinations and to make possible the coding of

Table 4.1. A 2-out-of-5 code (the code is weighted except for 0).

Decimal digit	Code (74210-*weights*)	Decimal digit	Code (74210-*weights*)
0	11000	5	01010
1	00011	6	01100
2	00101	7	10001
3	00110	8	10010
4	01001	9	10100

alphabetic symbols as well. The 2-out-of-5 principle is again applied here. The seven-bit combination is usually divided in two weighted subgroups of two and five bits each, respectively, each of which includes only a single "1" bit. The code in Table 4.2 uses the following weights (5,0), (4,3,2,1,0); it is also referred

Table 4.2. The biquinary code.

Decimal digit	(50)	Code (43210)-*weights*	Decimal digit	(50)	Code (43210)-*weights*
0	01	00001	5	10	00001
1	01	00010	6	10	00010
2	01	00100	7	10	00100
3	01	01000	8	10	01000
4	01	10000	9	10	10000

                      Binary bits   Quinary bits

to as the "biquinary" code. As may be deduced by careful examination of the code, the arithmetic is not too complex.

c. Parity Check

As mentioned in connection with the 8421 code, a parity bit may be affixed to a code, so that the augmented group always includes an odd number of 1's (or always an even number, as may be chosen). See Table 3.5 for example, for the odd parity P8421 code. The added check bit establishes an odd (or even) parity for the binary coded combination, which may be checked readily by logic circuits to diagnose single errors. The probability of double errors per combination in the *same message transfer*[4] *is small* enough to make the parity check quite satisfactory in many cases.

Most of our error-diagnosing techniques today are based on the assumption that only a single error may occur in any one transfer of information. Truly, the probability of a double error (two failures) per transfer is very small, but not zero. To include double-error detection in error diagnosing is a very complex proposition. This emphasizes the point that, in designing computers, we must make code errors and circuit failures as conspicuous as possible. We must also improve our capability to recognize and apprehend the error as soon as it occurs, before further errors have had time to develop.

---

[4] "Message" and "information" are used synonymously.

4.6
SINGLE-ERROR DETECTION

Error-detecting powers may be invested in a code at the price of redundancy. Additional bits added to each combination in an error-detecting code are truly redundant as long as the message is transmitted accurately, because the information contained in the added bits is used only when an error occurs.

Single-error detection by use of redundancy has been demonstrated in the previous two examples and in the use of the parity check. In this section we discuss the principle by which single-error detecting codes are formulated.

If only two given characters, *A* and *B*, are to be coded, the two single bits 0 and 1 should suffice, by assigning, say, 0 to represent *A* and 1 to represent *B*. This code, however, does not possess error-detecting capabilities.

If a redundant bit is used to augment the number of bits per combination from one to two, then the number of possible combinations is raised to four: 00, 01, 10, and 11. We wish to select two out of these four combinations and represent *A* and *B*, so that if a single error occurs in either of them, an out-of-code (invalid) combination results.

Consider the four combinations as corresponding to the coordinates of the vertices of the unit square shown in Fig. 4.1. A single error, that is a single bit-change in any combination, corresponds to a move from one vertex, along the side of the square, to a neighboring one. However, notice that it takes *two* bit-changes to move diagonally. Thus, if we select the diagonal combinations, the ones with the maximum distance between them, to represent *A* and *B*, we have a single-error detecting code. The permissible combinations are either (00, 11), or (01, 10). By paying the price of a redundant bit we have devised a code: (00, 11) or (01, 10), in which each combination differs from the other in two positions. Any single error is detectable, because it moves the valid combination (along the side of the unit square) to an invalid combination.

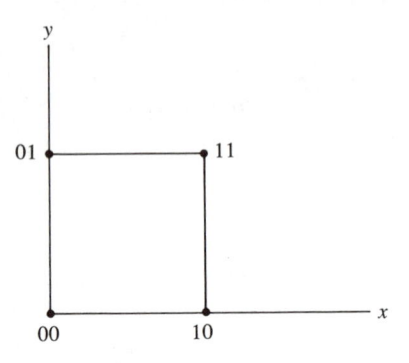

Fig. 4.1 Unit square.

It is of interest to note here that the first code (00, 11) is an "even-parity" code, while the second (01, 10) is an "odd-parity" code. In fact this is the clue to converting an essential code to a single-error detecting code. If an extra bit makes the sum of the bits in each combination even (or odd), then any single bit change leads to a contradiction. It takes at least two such bit changes to maintain the parity. It may be also noted, though, that parity, although sufficient, is not also necessary. As we shall see later, the no-parity code (000, 111) may be used as single-error detecting code for representing the characters *A* and *B*. This code has actually more redundancy than is essential. It also possesses more power; it cannot only detect a single error but also locates it and therefore corrects it.

It may be said then that the basic rule for devising a single-error detecting code is that each combination shall differ from each other in at least two positions. If three characters *A*, *B*, and *C* were to be coded, then apparently two-bit combinations could not possibly provide a single-error detecting code, because no more than two 2-bit combinations may be found, each differing from the others by two positions.

Let us consider next how many valid combinations of a single-error detecting code we may devise using two essential bits ($m = 2$) and one redundant bit ($k = 1$). We refer to the essential information bits as "message bits."

With three available bits we can form eight uniquely distinct combinations, which may be made to correspond to the coordinates of the vertices of a unit cube, as shown in Fig. 4.2. From the eight available combinations, we must select the single-error detecting code. In it, each valid combination must differ from each other in at least two positions. According to our previous discussion, we may select the combinations along the side diagonals, so that no two of them will be on the same edge of the cube. One such choice, marked by crosses in Fig. 4.2, is: (000, 011, 101, 110). It is a simple matter to check and verify that any single-bit change results in an out-of-code combination.

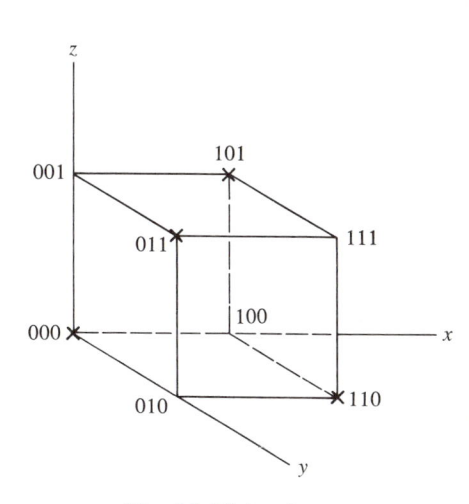

Fig. 4.2 Unit cube.

Again notice that the code is an "even-parity" one. It may be formed by considering the four essential combinations 00, 01, 10, 11, and appending to each a redundant bit so that the sum of the bits in each combination is even. Considering the remaining four combinations in the unit cube we form an odd-parity single-error detecting code. In either case, we have four valid combinations with single-error detecting powers, capable of accommodating four characters.

a. Distance

The devising of a single-error detecting code is best visualized in terms of the concept of "distance" between different combinations of a code. The distance refers to the number of single-bit changes that are needed to change one combination into another. Thus, the two-bit combinations, 00 and 01 are at a distance one from each other since a single-bit change will transform one into the other. In Fig. 4.1 these combinations lie along a side of the square. The distance between 00 and 11, however, is two, as it may be easily verified. These combinations lie along the diagonal in the square of Fig. 4.1. An even easier way of determining distance is by summing up the absolute values of the differences between all the corresponding bits of the two combinations. Thus, if $(a_1 a_2 \ldots a_n)$ and $(b_1 b_2 \ldots b_n)$ represent two combinations in an n-bit code (also the coordinates of two vertex points of the n-cube representation), then the distance between them is

$$d_{ab} = \sum_{i=1}^{n} |a_i - b_i|.$$

For example the distance between the combinations 0010100 and 1110101 is $d = 1 + 1 + 0 + 0 + 0 + 0 + 1 = 3$.

The rule for devising single-error detection codes is stated in terms of distance: "The distance between any two valid combinations must be at least two."

b. Mapping

A convenient means for choosing the valid combinations for a single-error detection code is by using a map. Mapping is discussed further in Chapter 7. It will suffice here to merely demonstrate how the decimal digits from zero to seven inclusive could be coded in a single-error detecting code using the four-variable map shown in Fig. 4.3. The decimal digits are placed in

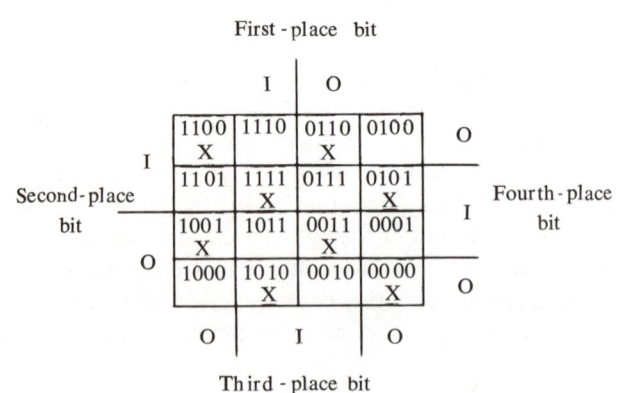

First-place bit

	I		O		
1100 X	1110	0110 X	0100	O
1101	1111 X	0111	0101 X	
1001 X	1011	0011 X	0001	I
1000	1010 X	0010	0000 X	O
	O		I		O	

Second-place bit on left (I above, O below); Fourth-place bit on right (O above, I below).

Third-place bit

Fig. 4.3 A four-variable map for selecting a four-bit error-detecting code. There are sixteen squares in the map, each marked by a 4-bit combination. One may select eight squares, only diagonally adjacent, to form an error-detecting 4-bit code (like the one marked by the X's in the map: 1100, 1111, 0110, 0101, 1001, 0011, 0000), in order to encode the decimal digits from zero to seven inclusive.

non-adjacent squares. Diagonally adjacent squares, which correspond to distance two, are acceptable. The resulting code is also shown in Fig. 4.3. It is an even-parity code.

The student may practice by devising on the map an odd-parity single-error detecting code for the numerals from zero to seven inclusive. Given any two combinations, as for example, 0101 and 0111, one may calculate the distance between them either algebraically or by placing them on the map.

4.7
SINGLE-ERROR CORRECTION

The two combinations along a body-diagonal of the unit cube (Fig. 4.2) differ from each other in three positions (distance 3). There are four such pairs, namely: (000, 111), (010, 101), (001, 110), (011, 100). It can be demonstrated that if one of these pairs is selected to code two given characters A and B, then each pair has two redundant bits per combination, as for example in $0 \rightarrow 010$ and $1 \rightarrow 101$, and constitutes a single-error correcting code. A single error in a combination from such a code is equivalent to a translation along the edge of the cube (Fig. 4.2) to a nonpermitted vertex. However, the new position is still closer to the correct combination than to any other valid combination; the erroneous combination may thus be corrected by selecting the closest valid combination.

We may say that if each combination differs from any other in two positions the code is single-error *detecting*; if each combination differs from any other in three positions the code is single-error *correcting*; it is also double-error detecting as may be easily verified in the above examples.

In general, let us consider a code whose combinations are made up of m-message bits and k-redundancy bits. Each combination has $m + k$ bits. If it differs from any other combination of the code by at least three-bit positions, the code is single-error correcting. A single-error in one of the combinations of the code results in a nonvalid combination, which differs in one position from the correct combination ($d = 1$) but still in at least two positions ($d = 2$) from any other valid combination.

In terms of an $(m + k)$-dimensional cube, a single error is equivalent to translation along the edge of the cube to a neighboring nonpermitted vertex. It takes at least two more edge translations (single-bit changes) to reach a permitted vertex, while only one to return to the original vertex. Thus, when an out-of-code combination is received with a single-error assumed, the correct combination is recovered by choosing the closest valid combination.

It may be noted here that with the above code ($d = 3$) we must assume a single error in order to recover the message correctly. A nonvalid combination received may be the result of a double error, in which case there is only error detection. Only if a single error is assumed may the message also be correctly recovered.

Of course, a triple error will go undetected since it leads to another valid combination.

A few interesting questions are raised at this point: Given an m-bit essential code, how many redundancy bits, k, are needed to turn a single-error correcting code? How is the code formulated? These questions are answered in the next section.

4.8 THE HAMMING CODE

Error-correcting codes which make single-error correction on messages of any length are often referred to as Hamming codes.[5]

Consider first the set of binary combinations, each n-bits in length. All together there will be $N = 2^n$ possible combinations. In principle we may select a subset of $M(<N)$ combinations, which constitute an error-correcting code, if any two combinations in the code differ in at least $2E+1$ positions, where E is the number of bit-errors to be corrected. Thus, the problem of formulating an error-correcting code becomes one of finding the subset of $M(<N)$ combinations in $N = 2^n$ possible ones, each of length n, and each with a minimum distance $d = 2E+1$ from any other combination of the code. Such a subset is often referred to as an "(n,d)-code." In the following we consider only single-error correcting codes, that is, $E = 1$, $d = 3$.

If no redundancy is used, then only m essential message bits are needed per combination in order to code M given characters, where $2^m \geq M > 2^{m-1}$. If $n > m$ bits are used in each combination, then $n - m = k$ bits are redundant.

A single error may occur in any one of the $m + k$ possible positions in a combination, or no error may occur at all. Thus there exists the following $m + k + 1$ single-error conditions:

no single error in the combination
a single error in the 1st bit
a single error in the 2nd bit

.   .   .
.   .   .
.   .   .

a single error in the $(m+k)$th bit.

Hence there are $m + k + 1$ single-error conditions to be identified by using k redundancy bits. It is known that k-binary bits are sufficient to code up to 2^k conditions. It follows then that the number of essential redundancy bits is the smallest number k which satisfies the relation:

$$2^k \geq m + k + 1.$$

---

[5] Hamming, R. W., *Bell System Technical Journal*, vol. 29, 147–160 (1950).

If $m = 1$ (coding only two characters A and B), then $k = 2$. This case corresponds to the single-error correcting codes discussed previously, such as (010, 101). If $m = 2$ or 3 or 4, then $k = 3$. Since $m = 4$ is used in the popular NBCD code for representing the decimal numerals from zero to nine inclusive, in following we exemplify the single-error correcting code formulation with $m = 4$, $k = 3$.

The data messages to be encoded in this case are all four-bit combinations. For instance, a given data message may be 1011. Let d_1, d_2, d_3, d_4 represent the four consecutive data bits, and p_1, p_2, p_3 represent the redundant bits. Each encoded combination must then contain all seven bits, with p_1, p_2, p_3 inserted among the data bits. We shall answer the following questions: "How do we mix the redundant bits with the given data bits?" "How do we assign values to the redundant bits p_1, p_2, p_3?" "How do we locate possible single errors?"

We might place the redundant bits p_1, p_2, p_3 in the first, second and fourth positions of the seven-bit combination, as follows

$$\begin{array}{ccccccc} \uparrow & \uparrow & d_1 & \uparrow & d_2 & d_3 & d_4 \\ p_1 & p_2 & & p_3 & & & \end{array}$$

To assign values to the redundant bits p_1, p_2, p_3, we form three parity relations, each involving four out of all seven bits. The parity relations are formed in this way:
1. the first relation consists of the first, third, fifth, and seventh bit of the combination;
2. the second relation consists of the second, third, sixth, and seventh bit of the combination;
3. the third relation consists of the fourth, fifth, sixth, and seventh bit of the combination.

The values of p_1, p_2, p_3 are determined so that either even or odd parity is observed, as may be chosen. The three parity equations[6] described above are shown below for even parity:

$$\text{1st bit} \oplus \text{3rd bit} \oplus \text{5th bit} \oplus \text{7th bit} = 0$$
$$\text{2nd bit} \oplus \text{3rd bit} \oplus \text{6th bit} \oplus \text{7th bit} = 0$$
$$\text{4th bit} \oplus \text{5th bit} \oplus \text{6th bit} \oplus \text{7th bit} = 0$$

If p_1, p_2, and p_3 are placed at the 1st, 2nd, and 4th bit position respectively, each equation contains only one unknown and the values of the redundant bits are determined easily. In general, the redundant bits and the data bits may be mixed in any order to form a combination.

To locate the error in any one of the seven possible bit positions, the parity is checked at the receiving point and a check-bit "0" is issued if parity is observed; otherwise a check-bit "1" is issued. Three consecutive check-bits are thus read,

---

[6] $\oplus$ means parity summation. If the number of 1's is even, the result is set equal to 0; otherwise it is set equal to 1. It is better known as "sum modulo two."

provided in sequence by the check-bits of the three parity relations above. The three check-bits form a three-bit binary number in the sequence in which they are issued which designates the position of a possible single error.

For the sequence selected previously ($p_1\ p_2\ d_1\ p_3\ d_2\ d_3\ d_4$), the appropriate even-parity equations should be

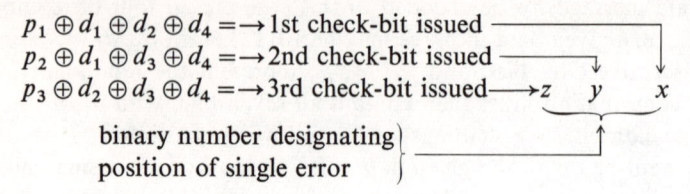

$$p_1 \oplus d_1 \oplus d_2 \oplus d_4 = \rightarrow \text{1st check-bit issued}$$
$$p_2 \oplus d_1 \oplus d_3 \oplus d_4 = \rightarrow \text{2nd check-bit issued}$$
$$p_3 \oplus d_2 \oplus d_3 \oplus d_4 = \rightarrow \text{3rd check-bit issued} \rightarrow z \quad y \quad x$$

binary number designating position of single error

Where x, y, and z denote either 0 or 1, depending on whether or not parity is observed.

If an error occurs, say, in the 5th position, that is, in d_2, the first and third relation above will give a check-bit "1", while the second relation (it does not include d_2) will check "0." The check bits issued in sequence are read as "101," which represents the binary number "five," signifying an error in the 5th position.

The procedure of forming the code and checking for error, is now illustrated in the following example, using the same sequence of data and redundancy bits as before. Let

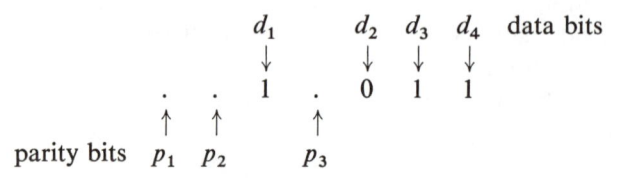

$$\begin{array}{ccccccc} & & d_1 & & d_2 & d_3 & d_4 \quad \text{data bits} \\ & & \downarrow & & \downarrow & \downarrow & \downarrow \\ . & . & 1 & . & 0 & 1 & 1 \\ \uparrow & \uparrow & & \uparrow & & & \\ \text{parity bits} \quad p_1 & p_2 & & p_3 & & & \end{array}$$

The values of p_1, p_2, and p_3 will be selected so that even parity is observed in the three relations shown above.

$$p_1 \oplus 1 \oplus 0 \oplus 1 = 0 \ldots p_1 = 0,$$
$$p_2 \oplus 1 \oplus 1 \oplus 1 = 0 \ldots p_2 = 1,$$
$$p_3 \oplus 0 \oplus 1 \oplus 1 = 0 \ldots p_3 = 0.$$

The coded combination thus becomes:

$$\begin{array}{ccccccc} 0 & 1 & 1 & 0 & 0 & 1 & 1 \\ \uparrow & \uparrow & \uparrow & \uparrow & \uparrow & \uparrow & \uparrow \\ p_1 & p_2 & d_1 & p_3 & d_2 & d_3 & d_4 \end{array}$$

Let us assume that this combination is received as

$$\begin{array}{ccccccc} 0 & 1 & 0 & 0 & 0 & 1 & 1 \\ \uparrow & \uparrow & \uparrow & \uparrow & \uparrow & \uparrow & \uparrow \\ p_1 & p_2 & d_1 & p_3 & d_2 & d_3 & d_4 \end{array}$$

Checking the parity, relations we find

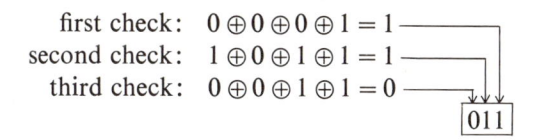

$$\text{first check:} \quad 0 \oplus 0 \oplus 0 \oplus 1 = 1$$
$$\text{second check:} \quad 1 \oplus 0 \oplus 1 \oplus 1 = 1$$
$$\text{third check:} \quad 0 \oplus 0 \oplus 1 \oplus 1 = 0$$

$$\boxed{011}$$

The check bits are read as 011, signifying an error in Position 3.

4.9 ROUNDING-OFF AND TRUNCATION

In the course of information-processing by machine, errors may be produced by faulty operation, faulty transmission, breakdown and accident, or because of *natural limitations* in the physical size of the machine and in the arithmetic procedures being used. Errors caused by size limitations of the machine and by the nature of the arithmetic procedures implemented by the machine can occur in the course of the *normal* operation of all computers. Typical in this category are the errors caused by the "rounding off" and "truncation" of numbers.

"Truncation" refers to cutting-off and ignoring a portion of least significant digits of the number. "Rounding off" consists of placing the radix point to the point of truncation, adding one-half, and then truncating.

In this sense, errors may appear even in the simplest of arithmetic operations performed by a machine, only because of the limited word size in any given machine and of the inherent limitations in the arithmetic procedures used. Because of its finite size, a machine is built to handle numbers of a certain limited length only. Thus, if a machine can handle numbers with up to twenty digits, then it can produce accurately the product of two ten-digit numbers, but will have to round off or truncate the product of two 20-digit numbers.

The rounding-off of errors which may occur in arithmetic operations with large numbers may accumulate over long sequences of such simple, often iterative, computations. The way errors accumulate is not at all clear and not well understood in most cases. If they are assumed to be random, then it follows that in a sequence of N operations, the error will go up not times N but about times $(N-1)/2$. In any way an accumulative effect takes place.

Rounding-off becomes even more important because arithmetic operations performed in the course of a problem may amplify previously introduced errors. This is the main reason for the extremely high accuracy demanded of modern information-processing machines. Accuracy demands of the order of $1:10^{12}$ are commonplace in modern computers in spite of the fact that most problems solved with them involve numbers with precision no better than $1:10^3$.

4.10 COMMENTS

During the operation of a computing system, information is in transit from one place to another practically at all times. Absolutely correct transfer of information is imperative. Because of unavoidable component failures, spurious noise pickups, and similar message-deteriorating factors, reliable operation can become a serious problem. The vulnerability of a computer system to malfunctioning becomes a subject of serious concern for the computer designer. Almost all of our failure-diagnosing techniques are based on the assumption that the machine may contain only one fault at any one time. Iterative subdivisions of the machine into parts permits us to determine which portion contains the fault. These dichotomic methods, however, are powerless as soon as the possibility of several simultaneous faults appears. For this reason, errors are made as conspicuous as possible in order that they can be recognized and apprehended before further errors have had time to develop.

Faulty operation of the computer may also occur because of erroneous transmission of the coded messages from one point of the computer to another. The most common techniques used for error detection and correction during transmission have briefly been analyzed.

Corrective action should be taken by the computer immediately after an error is detected. A parity error may be discovered or forbidden combinations may arise in the course of transferring data from one place to another, and an error check is thus provided under certain circumstances. The usefulness of a correcting scheme is dependent upon its effectiveness to annul errors and upon the amount of equipment necessary to carry out the checking and correcting operations. For example, increase in reliability may be achieved by providing enough duplicate equipment or by repeated transmission of the data, but this proposition is impractical except in special cases.

The all-important faultless transfer of binary data, especially over noisy communication channels in large systems, has been pursued by constructing self-correcting codes like the ones discussed previously. Special error-correcting codes make it possible to instrument the computer with the means to diagnose and correct errors automatically. Some such codes correct for single errors, some for double adjacent errors, some for single bursts of errors during noisy transmission, and others for multiple errors. Self-correcting codes are mainly used in data transmission over noisy communication channels, where some of the code's available room may be sacrificed so that more reliability in error-free transmission may be built-in. However, error-detection and error-correction principles are also applied to codes which are used in arithmetic operations within the computer.

The choice of a code depends on the computer application and is dictated by considerations of economy, compatibility, efficiency in arithmetic, and by reliability requirements.

It is of interest to make a comparison between living organisms and computers about the way in which errors are handled. Malfunctions which occur in living organs are diagnosed and corrected without any significant outside intervention. But the approach followed by nature is different. The organism goes through a readjustment to *first minimize* the effects of a malfunction; it then permanently corrects or isolates the faulty component. In contrast with nature, computers are designed in such a way as to make errors as conspicuous as possible and to allow apprehension and correction of errors as soon as they occur. More techniques of analysis rather than of synthesis are needed for information handling in digital computers, and the Theory of Information seems to provide the only analytical tool available today.

BIBLIOGRAPHY

Myers, R. H., *Reliability Engineering for Electronic Systems*. New York: John Wiley and Sons, Inc., 1964.
Peterson, W. W., *Error-Correcting Codes*. Cambridge, Mass.: MIT Press, 1961.
Piernschka, E., *Principles of Reliability*, Englewood Cliffs, N.J.: Prentice-Hall, Inc., 1963.
Roberts, N. H., *Mathematical Methods for Reliability Engineering*. New York: McGraw-Hill, Inc., 1964.

Problems

4.1 Given the code (0000, 0010, 0101, 0111, 1000, 1010, 1101, 1111) for 0 to 7 inclusive. Demonstrate that it is not a single-error detecting code by
(a) calculating distances,
(b) using the four variable map.

4.2 Given an n-cube, consider any one of its vertices. How many vertices are found at $d=1$, $d=2$, ...$d=k$, ...$d=n$, from it?

4.3 Seven-bit words $d_1 d_2 d_3 d_4 d_5 d_6 d_7$ may be used to code over one hundred symbols. In order to invest such seven-bit combinations ($m=7$) with single-error correcting power, we need to lengthen the word size by four redundant bits ($k=4$). Given the seven-bit message word 1001101, place the redundant bits p_1, p_2, p_3, p_4, in the first, second, fourth and eighth positions respectively, of the encoded eleven-bit combination. Use even parity to form the needed parity relations, and determine the values of the redundant bits p_1, p_2, p_3, p_4. By deliberately inserting a single error in the encoded eleven-bit combination, perform a complete check in order to demonstrate the single-error correcting power of the coded message.

4.4 Given the eleven-bit combination $1001101p_1p_2p_3p_4$, which is encoded with four redundant bits for single-error correction, use odd parity to determine the values of the redundant bits. Demonstrate the single-error correcting power of the resulting eleven-bit combination by deliberately inserting a single error in the sixth position and performing a complete check.

4.5 If p_x and p_y are the even parity bits for the binary numbers $X = x_n x_{n-1} \ldots x_1 x_0$ and $Y = y_n y_{n-1} \ldots y_1 y_0$ respectively then we may write:
$$p_x = x_n \oplus x_{n-1} \oplus \ldots \oplus x_2 \oplus x_1 \oplus x_0$$
$$p_y = y_n \oplus y_{n-1} \oplus \ldots \oplus y_2 \oplus y_1 \oplus y_0.$$
Given p_x and p_y for two numbers X and Y, prove that the parity bit p_s for their sum $S = X + Y$ is
$$p_s = p_x \oplus p_y \oplus c_n \oplus c_{n-1} \oplus \ldots \oplus c_1$$
where $c_n, c_{n-1}, \ldots c_1$ are the carries, that is $s_i = x_i \oplus y_i \oplus c_i$.

4.6 Consider the augmented single-error correcting Hamming code

$$p_1 p_2 x_1 p_3 x_2 x_3 x_4 p_0$$

where p_1, p_2, p_3, are the Hamming parity check bits and p_0 is an even parity bit added after the Hamming code is completed. Form the error-checking algorithm for the above code to demonstrate that the addition of p_0 makes it possible to recognize error conditions with two or more errors, which the Hamming check $(p_3 p_2 p_1)$ alone cannot locate.

4.7 Encode the decimal digits from zero to nine using 4-bit binary groups with the following weights:

(a) 5,3,2,–1
(b) 7,–4,–2,1
(c) 8,–4,2,–1

Which of these weight-sets provides a code that yields the complement of a number when one interchanges the zeros and the ones?

4.8 You are at the receiving end of a binary channel, and you receive the ten-bit word: 1001011000, which contains error-locating bits of the Hamming Code. Assuming that at most only a single error has occurred, determine whether it is possible to tell if the word has been received correctly, and if it is possible, show how and determine the correct word.

5
Logic Design

The overall objective of machine logic design is the analysis and synthesis of interconnected small hardware blocks, each of which does some specific function of information processing, into larger functioning units. Information processing specifications established for the machine and its major functioning units by the systems engineer are met by the logic designer by making use of the small hardware blocks (switching circuits and devices) that the circuit and the device engineers make available to him.

The binary system of numeration was singled out in Chapter 3 as being particularly suited for the hardware implementation of arithmetic operations because of reasons of economy and reliability. Switching devices, circuits or signals, which exhibit two discrete conditions of operation —henceforth referred to as "binary devices, circuits or signals"—become of principal importance in designing information-processing hardware. A variety of small hardware "blocks" are built using binary devices and elementary switching circuits and then organized by the logic designer at various levels of complexity to carry out diversified tasks of information processing.

Logic engineers have found it convenient to adopt a language and procedures for the description of the interrelationships between functioning blocks, which is formally identical to the language and procedures of *symbolic logic* applied to some forms of logical reasoning and to describing relations among "classes" of objects or verbal "propositions." Certain data transformations are also labelled as "logical" because they relate to "logical connectives" found in symbolic logic. For this reason, an introduction to the formal symbolic logic of propositions is essential for the comprehension of the methods involved in the logic design of information-processing machines. Symbolic logic is discussed in the following chapter.

The deductive system of a binary algebra, called Boolean algebra, is developed from the premises of symbolic logic and its ramifications, such as the calculus of classes and the algebra of propositions. It is extensively used in the logic organization of information processing machines. The engineering version of a Boolean algebra, better known as *switching algebra*, is the subject of Chapter 7.

The operation of networks of interconnected switches of various types may be expressed in terms of Boolean functions and equations. Furthermore, switching algebra provides the means for analyzing the logic of an arithmetic task, for optimizing the procedural sequence in information-processing, and for synthesizing the switching circuits and systems to perform the task. Problems of analysis, optimization, and synthesis of switching circuits, are found in the design of such systems as telephone-dialing exchanges, control installations, and information-processing machines.

Logic design is constrained by the prevailing systems specifications and by motivation for economy, reliability, speed, capacity, and functional versatility. The design proceeds in an evolving manner from the design of primary "building blocks"[1] such as adders and shift registers that perform elementary arithmetic operations as well as editing and transporting of information to higher-level logic design of "functional units" that perform large fractions of a machine instruction. This is followed by a still higher level logic design of full machine units which carry out full instructions.

5.1
THE DESIGN PROCEDURE

The logic engineer is faced with the problem of converting a set of verbal specifications about a piece of digital equipment into the best possible circuit diagram. Regardless of the complexity of the equipment he must proceed in a methodical way through a series of procedural steps.

The first step is analysis of the task. This involves studying the specifications, making sure that no ambiguities or contradictions exist. Then he proceeds by dissecting the task into a number of subtasks, much simpler but sufficiently compact to be tractable in terms of the available methods. He then formulates

---

[1] The design of gates, which perform elementary logical transformations such as "AND," "OR," and so on, are considered to be the work of circuit designers.

logical equations which express the necessary operations involved in the subtasks. A critical process of minimization and optimization follows, which aims in reducing the set of logical equations to the minimum possible complexity that meets the design requirements. Finally a wiring diagram is prepared and the circuits are implemented and checked. During all steps of logic design the engineer performs necessary checks to verify that the logical equations meet the original verbal specifications.

The procedure of logic design then involves:
1. analysis of the task and critical examination of the specifications,
2. dissection of the task into tractable parts,
3. formation of logical equations for the parts,
4. minimization and optimization of the logical equations,
5. block diagramming, and
6. implementation with hardware and testing.

In this chapter we examine only briefly the essentials of the above-outlined logic design procedure. A more analytical view of the methods and means employed in the design of digital equipment is undertaken in the chapters which follow. It should be noted again that the methods of logic design are applicable to a wide variety of machines, including the general categories of control equipment, telephone-switching systems, and large digital information-processing machines.

5.2
CALCULUS OF PROPOSITIONS AND NETWORKS OF SWITCHES

There exists an analogy between the condition of a network of interconnected switches and the inference of a set of binary propositions. This very valuable analogy is stressed and exemplified in this section.

The term "switch" is used in this instance to mean devices which can exist in either of two discrete, well-defined conditions. For example, a simple electric switch found in the house, may be ON or OFF, CLOSED or OPEN, UP or DOWN. Similarly, a ring made of magnetic material (a "flip-flop") may be magnetized in the CLOCKWISE ("set") or in the COUNTERCLOCKWISE ("reset") sense. Electronic switching components are discussed further in Chapter 9.

The calculus of propositions (Chapter 6) provides the rules and methods for manipulating complexes of binary propositions for the purpose of determining their inference. A binary proposition is a declarative linguistic statement, free of ambiguity, which may be either TRUE or FALSE.

By using the methods of the calculus of propositions it is possible to reduce a given set of compound statements to one declarative inference. The truth or falsity of a compound statement, composed of several propositions, depends directly on the truth or falsity of the component propositions and on the manner of their interconnections. The specific meanings of the statements and propositions involved are immaterial, and only the truth or falsity of each one is of importance. For example, the statement: "Birds fly AND two times two equals four," is TRUE because each of the component propositions is TRUE and they are connected by the logical connective "AND." Since we are only concerned about the truth or falsity of the propositions and compound statements, we may use symbols to simplify notation. Thus, the above compound statement may be written as: $f = a \cdot b$, where a signifies the proposition "Birds fly," b signifies the proposition "two times two equals four" and "$\cdot$" signifies the logical connective "AND." If the truth of a statement is represented by 1, and the falsity by 0, then the above statement corresponds to the symbolic operation $1 \cdot 1$ and the inference is true, that is, $f = 1$.

The statement "Birds fly AND two times two equals five" has a truth value 0 (it is FALSE) because one of the components has a truth value 0 (it is FALSE). However, the statement "Birds fly OR two times two is five" is TRUE, that is, it has a truth value 1. In the case of the logical connective OR the compound statement is TRUE if any or all of the component propositions are TRUE. In terms of the previous notation the second statement is written as: $f = a + c$, where c signifies the new proposition "two times two equals five," and "$+$" signifies the logical connective OR. In this case $f = 1$ if either $a = 1$, or $c = 1$, or both.

Propositions can be made about the states of switching components in circuits. The truth or falsity of propositions such as: "the circuit is closed" or "the switch is ON," or "the flip-flop is SET," or "the voltage is at the HIGH level," is examined in order that the truth value of a compound statement be determined. For example, the statement: "The circuit between points A and B is closed" (Fig 5.1) is equivalent to the compound statement: "the switch X is closed AND the switch Y is closed." In accordance with previous definitions about the meaning of the logical connective AND, the circuit between points A and B is closed

Fig. 5.1 Two switches in series.

when, and only when, switch X is closed *and* switch Y is closed. If the truth values of the propositions: "the switch X is closed," and "the switch Y is closed" are x and y, respectively, then the truth value f of the compound statement may be tabulated as shown in Table 5.1. This is referred to as a "truth table."

For each combination of the values x and y of the component propositions, the table gives the value of the compound statement, in accordance with the definition of the logical connective AND. The truth value 0 designates that the statement is FALSE, and the truth value 1 designates that the statement is TRUE. Notice again that only the truth values of the statements and the manner of their interconnection (logical connectives) is of importance; the meaning of each proposition does not concern us here.

Table 5.1. Truth table for the logical connective AND.

x	y	$f = x \cdot y$
0	0	0
0	1	0
1	0	0
1	1	1

With reference to the circuit shown in Fig. 5.2, the statement: "the circuit between points A and B is closed," is equivalent to the compound statement: "the switch X is closed OR the switch Y is closed, OR both are closed." The meaning of the logical connective OR is apparent from the circuit, since the circuit between A and B is closed if either X or Y or both are closed. The truth value of the compound statement: $f = x + y$, is given in Table 5.2.

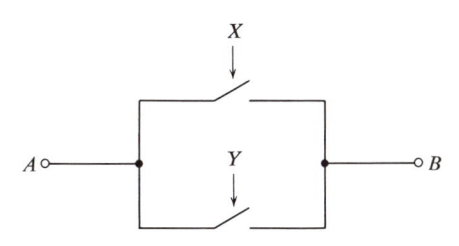

Fig. 5.2 Two switches in parallel.

To complete our set of essential logical operations we should include for the moment the operation of "complementation" or "negation," often referred to also as the "NOT" operation. Thus, if X is a true proposition, the negation of X, denoted by $\bar{X}$, is a false proposition. And if x is the truth value of X, then $\bar{x}$ is the truth value of $\bar{X}$. Therefore, if $x = 1$, then $\bar{x} = 0$, and vice versa.

Table 5.2. Truth table for the logical connective OR.

x	y	$f = x + y$
0	0	0
0	1	1
1	0	1
1	1	1

5.3 GATES

Gates are small blocks of hardware, which can perform elementary logical operations, such as the operations denoted by the logical connectives "AND," "OR," and "NOT." A great variety of electronic or other

108 LOGIC DESIGN

devices are used in the construction of logical gates. Several such components are discussed in Chapter 9.

The most elementary gates are shown with their symbols and truth tables in Fig. 5.3. It is important to notice here that the condition at the output of each gate is determined solely by the conditions at the inputs at any particular instance. This is characteristic of a class of circuits known as "combinational circuits," discussed further in Chapter 8.

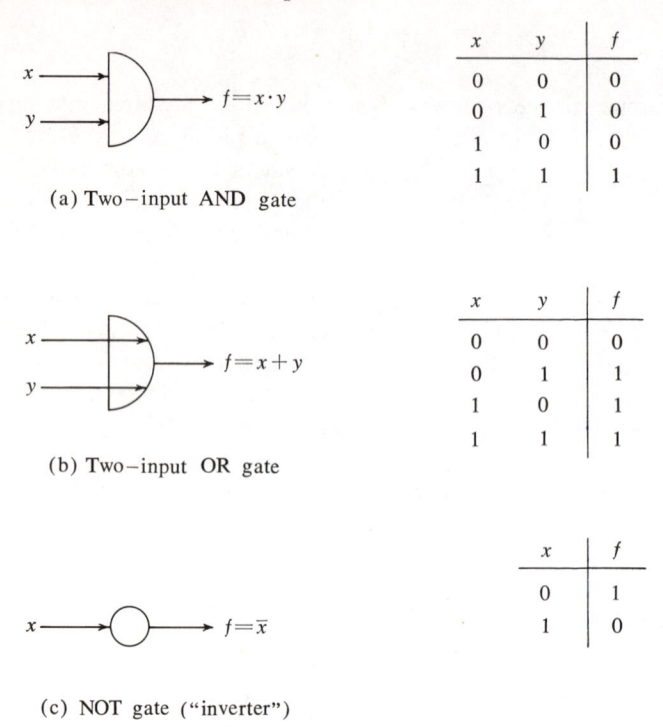

$f = x \cdot y$

x	y	f
0	0	0
0	1	0
1	0	0
1	1	1

(a) Two-input AND gate

$f = x + y$

x	y	f
0	0	0
0	1	1
1	0	1
1	1	1

(b) Two-input OR gate

$f = \bar{x}$

x	f
0	1
1	0

(c) NOT gate ("inverter")

Fig. 5.3 Elementary gates.

Electronic gates operate with electrical signals, such as voltages or currents, which may be found in either one of two recognizable conditions. If they are voltage-operated gates, for example, any terminal found in "HIGH" voltage condition is considered as having a value "1". If the voltage is "LOW," the value is 0. The voltage of each terminal is measured with reference to some common "ground." Thus, the output of an AND-gate is at HIGH voltage, (value 1) if, and only if, both inputs at the time are at HIGH voltage (each at value 1).

The nature of digital equipment which make use of "all-or-none" types of signals and of gates like the AND, OR, NOT blocks discussed before invite the use of a special type of algebra. This algebra, often referred to as "switching algebra," is one of a class of algebras, known as Boolean algebras, discussed in Chapter 7. The mathematical means used in switching algebra have evolved

5.4 LOGICAL EQUATIONS FROM CIRCUIT DIAGRAMS

To see how the operation of a circuit of various gates may be expressed and analyzed in terms of logical equations, consider the circuit

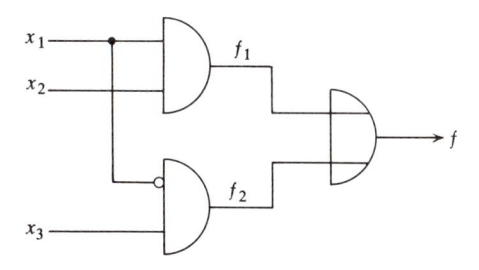

Fig. 5.4 Circuit diagram for $f = x_1 \cdot x_2 + \bar{x}_1 \cdot x_2$.

diagram shown in Fig. 5.4. Observe that the output of the OR-gate may be written in terms of the two inputs as:

$$f = f_1 + f_2$$

Also we have

$$f_1 = x_1 \cdot x_2$$

and

$$f_2 = \bar{x}_1 \cdot x_3.$$

Therefore

$$f = x_1 \cdot x_2 + \bar{x}_1 \cdot x_3.$$

Notice that the NOT operation on x_1 is shown by a small circle attached to the AND symbol.

As another simple example, consider the diagram shown in Fig. 5.5(a). The logical equation for this circuit is

$$f = f_1 + f_2 = x_1 \cdot x_2 + \bar{x}_1 \cdot x_2.$$

110   LOGIC DESIGN

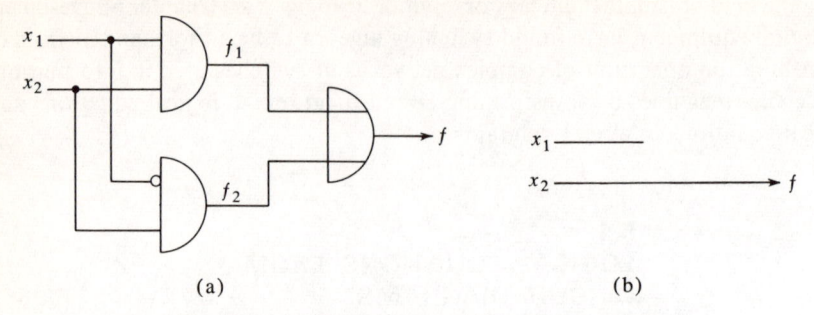

Fig. 5.5 (a) Circuit diagram; (b) minimization.

As we shall see in Chapter 7, the distributive rule of ordinary algebra is applicable in switching algebra, so that

$$f = (x_1 + \bar{x}_1)x_2.$$

However $x_1 + \bar{x}_1$ is identically equal to 1 (tautology), because for either value of x_1 (0 or 1), one of the two terms is equal to 1. Thus,

$$f = x_2.$$

This process of "minimization" of the logical equation has led to a simplification of the circuit, which saves the designer four gates, and reveals that the output is not affected by the condition of one of the inputs.

Three more circuits are shown in Fig. 5.6. The first two, in parts (a) and (b) of the figure, are equivalent. The derivation of the logical equations is left as an exercise for the student.

Notice that the AND gate of the diagram in Fig. 5.6(c) is a three input gate. Actually, the definitions given previously for two input AND and OR gates may be extended easily to n-inputs. Thus, "an n-input AND gate is one which will have an output value equal to 1 (HIGH voltage for example) if, and only if, all n-inputs have a value equal to 1 (HIGH voltage). An n-input OR gate is one which will have an output value equal to 1 (HIGH voltage) if any one or more of the inputs has a value equal to 1 (HIGH voltage)."

The logic diagrams presented in the illustrations of this section are relatively easy and offer little challenge. The subject of logical equations and minimization is discussed in more detail in later chapters and more complex problems are presented there.

5.5 CIRCUIT DIAGRAMS FROM LOGICAL EQUATIONS

Logical equations may be derived by analyzing the specifications imposed on a digital piece of equipment. The operation of the equipment is first broken down into smaller parts. This process of analyzing the overall

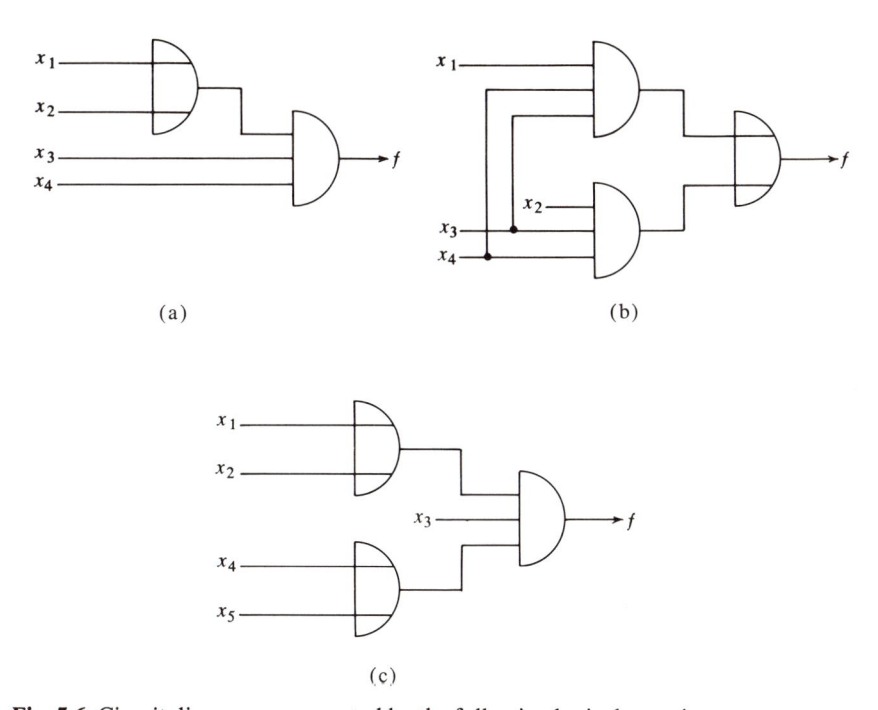

Fig. 5.6 Circuit diagrams represented by the following logical equations:
(a) $f = (x_1 + x_2) x_3 \cdot x_4$, (b) $f = x_1 \cdot x_3 \cdot x_4 + x_2 \cdot x_3 \cdot x_4$, (c) $f = (x_1 + x_2) \cdot x_3 \cdot (x_4 + x_5)$.

specifications and dissecting each information-processing task down to its essential parts is continued until each partial subtask is tractable in terms of a logical equation. These logical equations are then synthesized and minimized before physical implementation takes place.

When a logical equation is as simple as, for example, $f = x_1 \cdot x_2 + x_3 \cdot x_4 \cdot x_5$, drawing out the correct diagram presents very little problem. The above equation should be recognized immediately as an OR gate with two inputs, each of which is the output of an AND gate.

If the logical equation involves only AND, OR, and NOT operations, drawing the circuit diagram in general is facilitated by following the step-by-step procedure illustrated in the following example.

Consider the problem of drawing the logical circuit diagram for the equation

$$f = \overline{(x_1 \cdot x_2 + x_3)} \cdot (\bar{x}_1 + x_2 \cdot x_4) + x_3 + \overline{x_2 \cdot \bar{x}_3}.$$

We may write the equation as follows:

$$f = \bar{y} \cdot z + x_3 + \bar{w},$$

where

$$y = x_1 \cdot x_2 + x_3; \quad z = \bar{x}_1 + x_2 \cdot x_4; \quad w = x_2 \cdot \bar{x}_3.$$

Notice that if each one of the three new variables is realized separately, the

112   LOGIC DESIGN

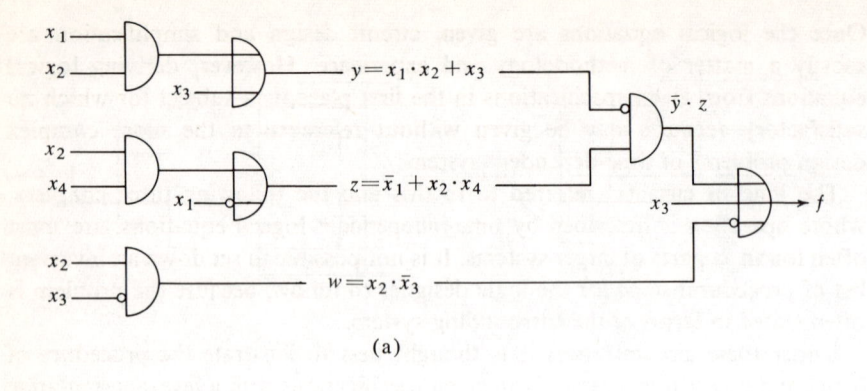

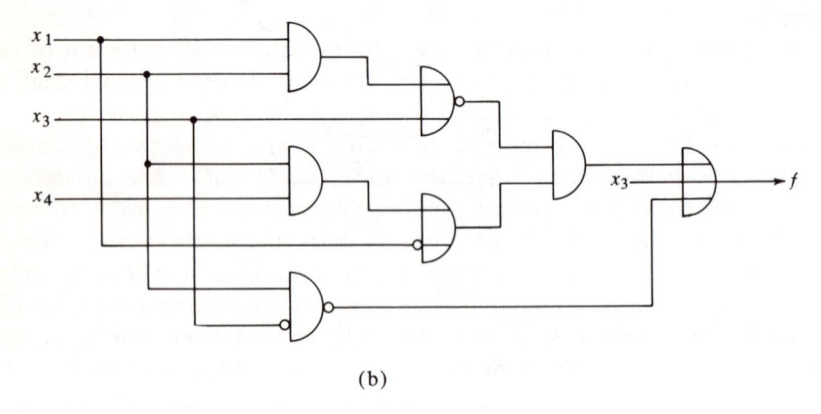

Fig. 5.7 Logical circuit diagram for the logical equation:
$f = \overline{(x_1 \cdot x_2 + x_3)} \cdot \overline{(\bar{x}_1 + x_2 \cdot x_4)} + x_3 + \overline{x_2 \cdot \bar{x}_3}$; (a) step-by-step drawing; (b) complete diagrams.

interconnection that will give f is now more apparent. The procedure of drawing the logical circuits is shown in Fig. 5.7. In the case of an inversion of a variable the inversion symbol (small circle) may be placed anywhere along the line carrying the variable. Thus, in the cases of y and w the inversion circles were transferred from the end of the lines [Fig. 5.7(a)] to the beginnings [Fig. 5.7(b)].

The transformation of a logical equation into a circuit diagram and vice versa is a most useful tool for the design and simplification of digital, information processing, or other equipment. Only the essential features were illustrated here. Further discussion is reserved for the chapters which follow.

5.6 FORMATION OF LOGICAL EQUATIONS

Transformations between logical equations and circuit diagrams and simplification were illustrated briefly in the last two sections.

FORMATION OF LOGICAL EQUATIONS

Once the logical equations are given, circuit design and simplification are mostly a matter of methodology and experience. However, deriving logical equations from given specifications in the first place, is a subject for which no satisfactory account may be given without reference to the more complex design problems of time-dependent systems.

The kind of circuits[2] referred to in this and the following three chapters, whose operation is described by time-independent logical equations, are most often found as parts of larger systems. It is not possible to set down an invariant list of procedural steps for the logic designer to follow, because the problem is often stated in terms of the surrounding system.

Under these circumstances it is thought best to illustrate the procedure of forming the logical equations from given specifications with a few representative examples.

Often the logic designer is given a sometimes vague verbal statement of the specifications of the problem. This gives the logic designer a great deal of freedom in choosing a specific design. Consider the following problem:

"Design a digital alarm equipment, which will flash a red warning light inside the automobile whenever the driver speeds beyond 60 mph while his door is unlocked and the turning lights are 'on,' or whenever he speeds above 60 mph, with the turning lights 'on,' the door locked, unless the steering wheel is within 30° from center position, or whenever the steering wheel is within 30° from center position and the turning lights are 'on.' Also the warning red light will flash whenever the driver speeds above 60 mph, with the steering wheel turned beyond 30° from center position and while the turning lights are 'off'."

Analyzing the statement of the problem we find that there are four elementary binary propositions involved, namely:

$x = $ the automobile speed is greater than 60 mph,
$y = $ the steering wheel is turned more than 30° from center position,
$z = $ the turning lights are 'off,'
$w = $ the door is locked.

The logical equation which describes the circumstances in which the warning light will flash, is

$$f_1 = x \cdot \bar{z} \cdot \bar{w} + x \cdot y \cdot \bar{z} \cdot w + \bar{y} \cdot \bar{z} + x \cdot y \cdot z.$$

Before we implement this logical equation, an attempt is made to simplify it. Several techniques of simplification are discussed in Chapter 7. The simplified form of this equation is[3]

$$f_2 = x \cdot y + \bar{y} \cdot \bar{z}.$$

---

[2] "Combinational" circuits: their output values are uniquely and solely determined by the input values. Circuits whose output values at any time depend on present input values, but also on past circuit states, such as the preceding sequence of input values (sequential circuits) are discussed in Chapter 10.

[3] The proof is illustrated in Chapter 7, Section 7.8.

114 LOGIC DESIGN

Truth tables, which may be used to facilitate simplification, are also used for verification, as demonstrated in Table 5.3. The four variables (elementary

Table 5.3 Construction of truth tables for checking that $f_1 = f_2$†.

Input combinations				Identical logical functions: $f_1 = f_2$							
x	y	z	w	\multicolumn{5}{c|}{$x\cdot\bar{z}\cdot\bar{w}+x\cdot y\cdot\bar{z}\cdot w+\bar{y}\cdot\bar{z}+x\cdot y\cdot z = f_1$}	\multicolumn{3}{c|}{$x\cdot y+\bar{y}\cdot\bar{z}=f_2$}
0	0	0	0	0	0	1	0	1	0	1	1
0	0	0	1	0	0	1	0	1	0	1	1
0	0	1	0	0	0	0	0	0	0	0	0
0	0	1	1	0	0	0	0	0	0	0	0
0	1	0	0	0	0	0	0	0	0	0	0
0	1	0	1	0	0	0	0	0	0	0	0
0	1	1	0	0	0	0	0	0	0	0	0
0	1	1	1	0	0	0	0	0	0	0	0
1	0	0	0	1	0	1	0	1	0	1	1
1	0	0	1	0	0	1	0	1	0	1	1
1	0	1	0	0	0	0	0	0	0	0	0
1	0	1	1	0	0	0	0	0	0	0	0
1	1	0	0	1	0	0	0	1	1	0	1
1	1	0	1	0	1	0	0	1	1	0	1
1	1	1	0	0	0	0	1	1	1	0	1
1	1	1	1	0	0	0	1	1	1	0	1

† f_1 and f_2 are identical logical functions.

binary propositions) form sixteen (2^4) possible combinations. For each combination (input combination) the above functions f_1 and f_2 are determined and compared. Notice that the corresponding two columns are identical.

The two circuits implementing the logical functions f_1 and f_2 are shown in Fig. 5.8. Notice the savings that are effected by simplification.

Before the input–output relationships are expressed in terms of logical equations, it is often desirable or even necessary to construct a truth table. The truth table provides a detailed pictorial representation of input–output combinations. Consider for example the task of designing a "translator" circuit which will convert from NBCD to two-out-of-five code. The translator circuit in this case is a four-input, five-output circuit, since four-bit combinations (8421) of the NBCD code constitute the input and five-bit combinations (74210) of the two-out-of-five code constitute the output. These codes may be found in Tables 3.4 and 4.1, respectively.

The input-output relations are shown in Table 5.4. Only ten of the sixteen possible input combinations are meaningful. If the logic designer can assume that

FORMATION OF LOGICAL EQUATIONS 115

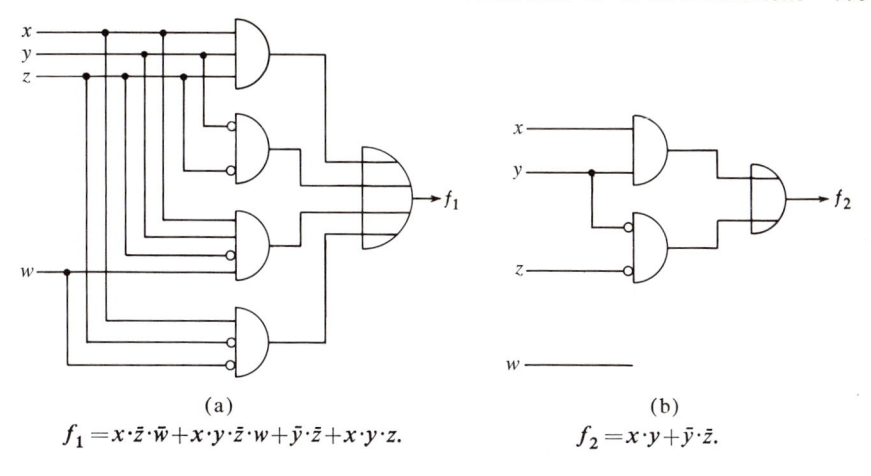

(a)
$f_1 = x \cdot \bar{z} \cdot \bar{w} + x \cdot y \cdot \bar{z} \cdot w + \bar{y} \cdot \bar{z} + x \cdot y \cdot z.$

(b)
$f_2 = x \cdot y + \bar{y} \cdot \bar{z}.$

Fig. 5.8 Circuits for the automobile alarm. The simplified circuit is shown in (b).

Table 5.4. Truth table for code translator.

Decimal number	NBCD-Code 8 4 2 1 w z y x	Two-out-of-five code 7 4 2 1 0 f_5 f_4 f_3 f_2 f_1	
0	0 0 0 0	1 1 0 0 0	
1	0 0 0 1	0 0 0 1 1	
2	0 0 1 0	0 0 1 0 1	
3	0 0 1 1	0 0 1 1 0	
4	0 1 0 0	0 1 0 0 1	
5	0 1 0 1	0 1 0 1 0	
6	0 1 1 0	0 1 1 0 0	
7	0 1 1 1	1 0 0 0 1	
8	1 0 0 0	1 0 0 1 0	
9	1 0 0 1	1 0 1 0 0	
Input combinations not used	1 0 1 0 1 0 1 1 1 1 0 0 1 1 0 1 1 1 1 0 1 1 1 1	d d d d d d d d d d d d d d d d d d d d d d d d d d d d d d	"Don't care" entries

none of the remaining six combinations will ever appear at the input of the translator, he can assign any output bits corresponding to those six input combinations without affecting the circuit's performance. These "don't care" entries should be filled in such a way as to make the resulting circuit as simple as possible. The derivation of the logical functions for the outputs f_1, f_2, f_3, f_4, f_5

is not discussed here. Instead, the code translator is discussed in Chapter 8, Section 8.6. The point demonstrated with this example is that truth tables often offer a most convenient tool for deriving logical equations.

Occasionally the specifications may be implicit in the name of the equipment. For example, to design one stage of a binary adder, one may assume three inputs (addend, augend, and carry) and two outputs (sum and carry). The necessary logical relationships are formed by proper reference to the rules of binary addition. A truth table may be constructed so that: (1) the sum-output is 1 if, and only if, an odd number of 1's appears at the three inputs, and (2) the carry-output is 1 if, and only if, there are two or three 1's at the three inputs. The complete circuit of a binary adder involves time delays (memory operations) and therefore it cannot be discussed fully here. A specific design of a full-binary adder is illustrated in Chapter 10.

5.7
LOGICAL DIAGRAMS FOR RESOLUTION OF ALTERNATIVES

In the course of normal operation within a processor a number of decisions may have to be made by the machine, as for example in cases of conditional branching. The classical problem of symbolic logic (Chapter 6) is to infer the truth or falsity of a proposition from a number of statements, each known to be true.

Let us, for example, assume that in the process of computing an algorithm, the machine is programmed to reach a certain point, to ask a question, and depending on the answer to take a different course of action. Let us further assume that the question asked is: "What is the value of N?". If N is always a positive integer not greater than eight, there are eight alternative answers to the question, (that is, $N = 1$, $N = 2$, ..., $N = 8$) and thus there are eight different courses of action.

To resolve a question with many alternatives requires great logic resolution and therefore is time-consuming. The above question: "What is the value of N?" may be replaced by a set of N binary-decision questions, such as: "Is $N = 1$?," "Is $N = 2$?," ... "Is $N = 8$?" Each binary question requires minimum resolution and may be diagrammed using a diamond-shaped decision box. Eight such boxes are required in the diagram shown in Fig. 5.9 in order to resolve the previous question about the value of N. All decision boxes are entered from the same point and they may be processed in parallel.

The least number of binary decisions required to resolve the question, "What is the value of N?" in the logic system shown in Fig. 5.9 is equal to the number of alternatives N. The number of essential binary decisions to be taken in order to resolve the question may be reduced if a *time sequence* is employed as shown in Fig. 5.10, where each decision follows in time a previous one. In the logic

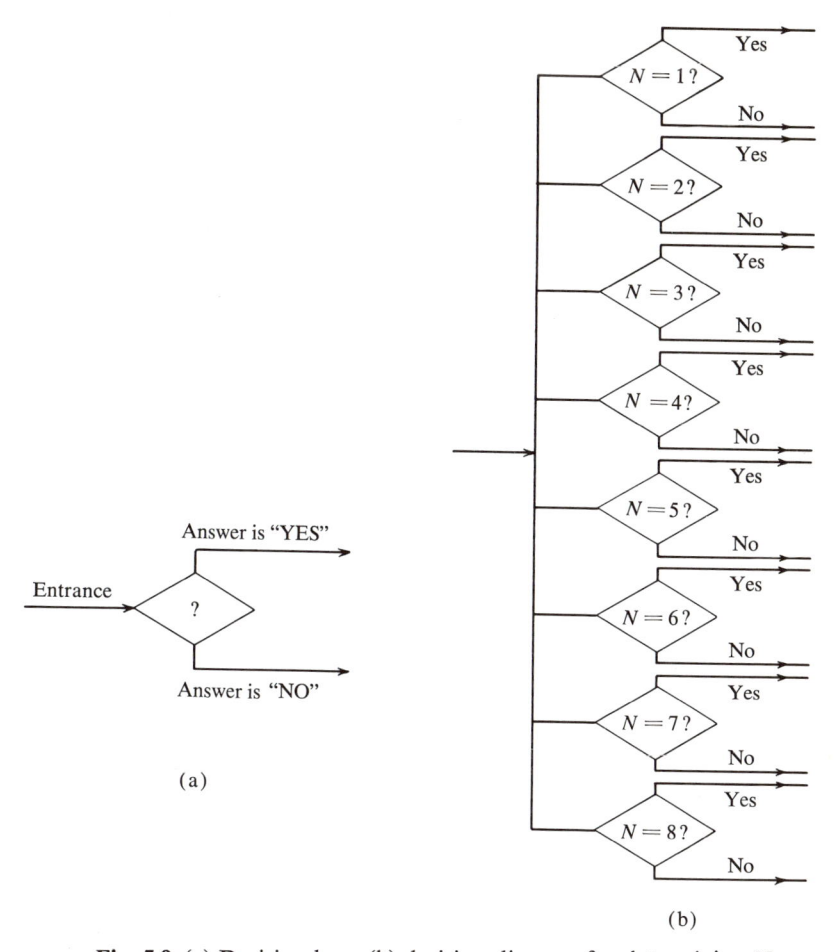

Fig. 5.9 (a) Decision box; (b) decision diagram for determining N.

system of Fig. 5.10 the question is resolved in three time-sequenced binary decisions.

According to the method of "halving," the number of alternatives is subdivided by two, down to the last decision box. The essential number of binary decisions required to resolve a question with N alternatives becomes a *minimum* and it is equal to n, where $2^{n-1} < N \leq 2^n$.

If N is not equal to an exact power of two the first binary decision may not necessarily involve half of the N alternatives. This point and the method of "halving" is illustrated by the following two examples:

EXAMPLE ONE: "Offer a deck of playing cards to a person, asking him to memorize one card. Determine the memorized card with

118   LOGIC DESIGN

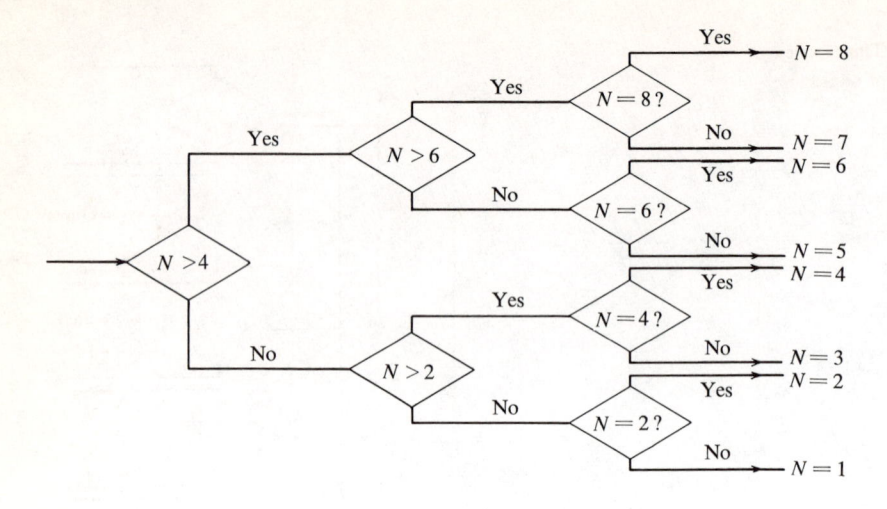

Fig. 5.10 Determination of N by the method of "halving".

the *minimum* number of binary-decision questions."

Since the number of alternatives is 52, the minimum number of decisions needed is $n = 6$ ($2^5 < 52 < 2^6$). One possible logic diagram is shown in Fig. 5.11. Observe that the player could be changing the memorized card during the game and you will not require more than 6 binary questions, as long as his answers remain consistent.

Only the decision branch beginning at "hearts" is completed because the questions are simply repeated identically for any other card suit. After the card suit has been established, the next question is: "Above 8?" This is so because there are only 13 choices left and with the remaining four questions we can determine any one of 16 choices. Therefore we use the power of 2 just below 13, that is, we use 8, as the new breaking point and continue subdivision from that point on. If the playing cards were numbered from 1 to 52 we could have started the questioning by asking first: "Is it greater than 32?" since 32 is the highest power of 2 just below 52. The minimum number of required questions would have been the same. The student should clarify these points for himself with a few exercises of his own. Here is another example.

EXAMPLE TWO: "The telephone directory of Palo Alto and vicinity has 319 pages. Each page includes up to 416 names, (over 120,000 names in all). Let the player open the directory at random

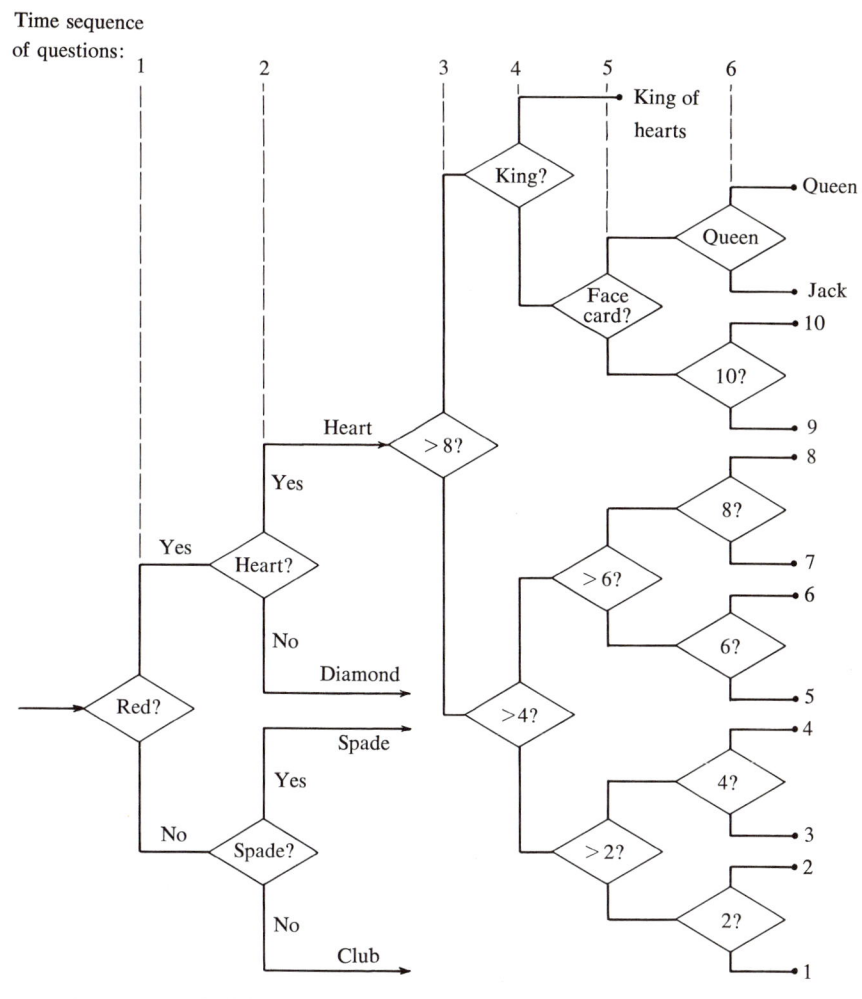

LOGICAL DIAGRAMS FOR RESOLUTION OF ALTERNATIVES 119

Fig. 5.11 Logical diagram (decision tree) for the playing-card example.

and select a name. Assume that the names in each page are numbered consecutively. You are asked to determine the exact location of the selected name with the *minimum* number of binary-decision questions."

Since $2^8 < 319 < 2^9$ and $2^8 < 416 < 2^9$, we need no more than 18 questions in all; 9 questions in order to determine the correct page, and 9 questions in order to determine the correct location on the page. Assume that the name selected, "Gray, John T." is on page 121 and the 261st name on that page. The logic diagramming for the solution of this problem is shown in Fig. 5.12.

120   LOGIC DESIGN

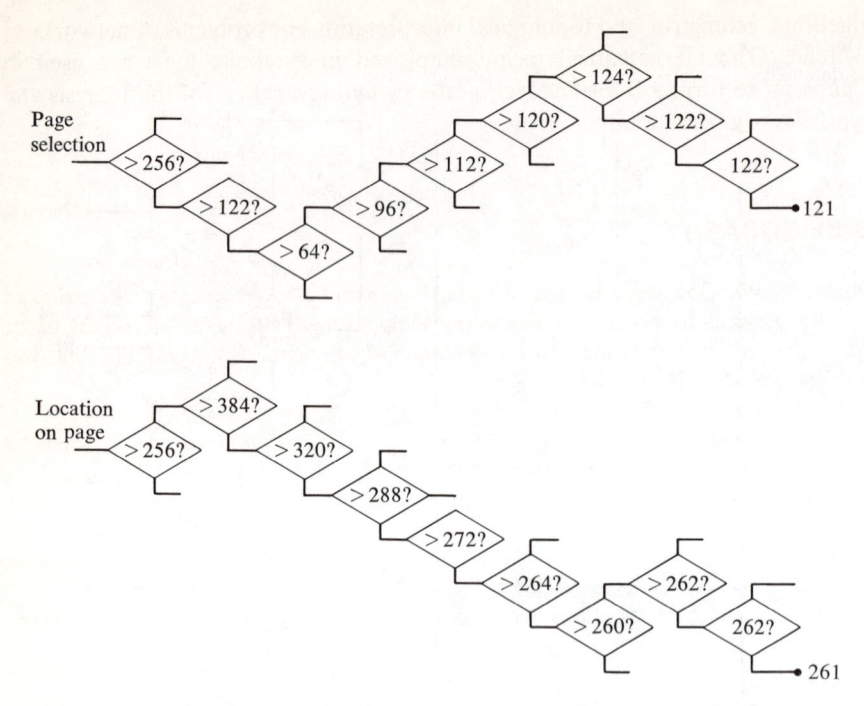

Fig. 5.12 Logical diagrams (decision trees) for the telephone directory example.

5.8
COMBINATIONAL AND SEQUENTIAL LOGIC

Circuits used for information processing may be of the "combinational" or the "sequential" type and often of a mixed kind with combinational and sequential portions interconnected in various fashions. Combinational circuits perform logic operations without the benefit of memory. Their output is solely and uniquely determined by their input without regard to past history of either the input signals or the states of the circuit. Sequential circuits employ the advantage of memory, thus permitting a more efficient decision process to be used, as was exemplified in the previous section. The output of a sequential circuit is a function of the input, the state of the circuit, plus the immediate history of events at the input and states of the circuit. Sequential circuits are discussed in Chapter 10.

The following three chapters include discussions leading to the investigation of combinational circuits. The subject of Chapter 6 (Symbolic Logic) is to discuss the means by which the truth value of a statement is inferred from the truth values of a number of propositions. This is directly analogous to the operation of a combinational circuit, in which the value of its output is determined by the existing values at its inputs. The calculus of propositions (symbolic logic) leads to algebraic, geometric, and topological interpretations of problems of networks of switches. The mathematical means employed in symbolic logic are used in Chapter 7 to form a Boolean algebra, the switching algebra, for the analysis and synthesis of combinational circuits.

BIBLIOGRAPHY

Burks, A. W., Goldstein, H. H., and Von Neumann, J., "Preliminary Discussion of the Logical Design of an Electronic Computing Instrument," (classical paper reprinted in abbreviated form by *Datamation*, Sept., Oct. 1962; reprint was annotated by P. Armer).

Flores, I., *Computer Design*. Englewood Cliffs, N.J.: Prentice-Hall, Inc., 1967.

Flores, I., *Computer Logic—The Functional Design of Digital Computers*. Englewood Cliffs, N.J.: Prentice-Hall, Inc., 1960.

Phister, Montgomery, Jr., *Logical Design of Digital Computers*. New York: John Wiley & Sons, Inc., 1963.

Shannon, C. E. and Weaver, W., *The Mathematical Theory of Communication*. Urbana, Ill.: Univ. of Illinois Press, 1949.

Shannon, C. E., "A Symbolic Analysis of Relay and Switching Circuits," *Trans. of the Am. Institute of Electrical Engineering*, vol. 57, 1938.

Problems

5.1 The circuit shown in Fig. 5.1.1 has five inputs x_1, x_2, x_3, x_4, x_5. If $x_1 = 1$, $x_2 = 0$, $x_3 = 0$, $x_4 = 1$, $x_5 = 0$, will the signal at the output represent a 0 or a 1? What happens to the output f if x_2 changes periodically from 0 to 1 and back, while the other input variables maintain their previous values? Repeat now by considering that x_4 changes periodically from 0 to 1 and back.

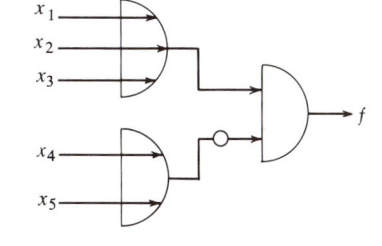

Fig. 5.1.1

5.2 Write a logical equation for a logical circuit which will have a 1 output when its inputs are $x_1 = 0$, $x_2 = 1$, $x_3 = 0$, or when $x_1 = 1$, $x_2 = 1$, $x_3 = 0$. The output of the circuit is 0 for all other input combinations. Draw a block diagram for the circuit.

5.3 Prepare the truth table for the following equation:

$$f = \overline{x_1 \cdot x_2} \cdot (x_1 \cdot \bar{x}_2 \cdot x_3 + \bar{x}_1 \cdot \bar{x}_2 \cdot x_3 + x_1 \cdot \bar{x}_3).$$

Construct a logical block diagram for the above equation.

5.4 Write the following expressions using only symbols for the logical connectives:
(a) x_1 OR (x_2 AND x_3)
(b) (x_1 OR x_2) AND x_3
(c) NOT (NOT x_1 OR x_2) AND NOT (x_3 OR NOT x_2)
(d) {(NOT (x_1 AND NOT x_2) OR x_3) AND (x_4 OR NOT x_3)} OR x_5.

5.5 In a certain game with four other persons you pose "true" or "false" questions, and then you ask the four participants to respond simultaneously, each indicating his answer by flipping a switch to either the "true" or the "false" position. You want to design three logical circuits, connected as illustrated in the block diagram of Fig. 5.5.1, so that they will turn "on" in-

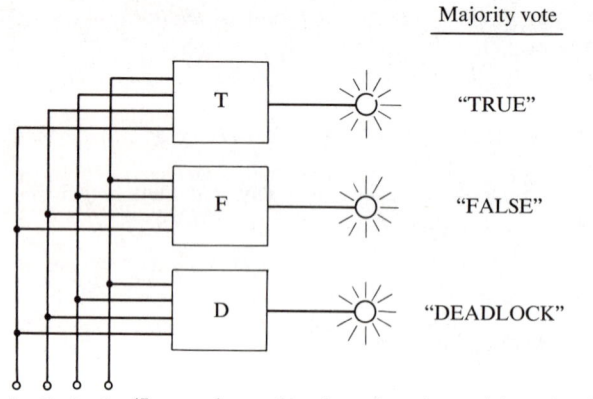

Fig. 5.5.1

dicator lights showing the majority vote. If the majority of the participants have voted "true", the T-logical circuit must activate its output and turn "on" the "TRUE" indicator light. If the majority vote is for "false", then the F-logical circuit must turn "on" the "FALSE" indicator light. If there is a deadlock in the voting (two "true" and two "false" answers), then the D-logical circuit must turn "on" the "DEADLOCK" indicator light.

(a) Construct the truth tables for the logical circuits by representing "true" with "1" and "false" with "0".
(b) Write the logical equations for the outputs of the logical circuits.
(c) Draw the logical block diagrams using AND, OR and NOT gates.

5.6 You are playing the "true-false" game described above in problem 5.5. but with only three other persons. Each time you

                T
                                SW   Output
                                        indicator
                F   light

x_1 x_2 x_3 (Inputs whose values depend on the participants' voting)

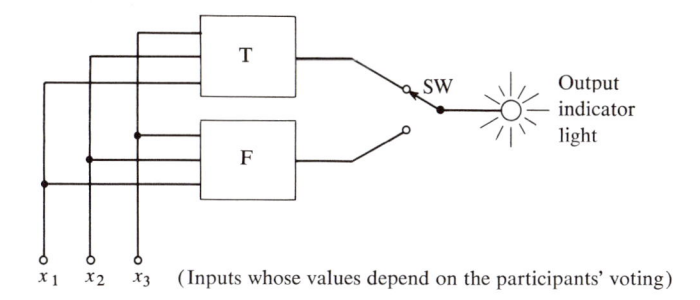

Fig. 5.6.1

pose a question and just before you ask the participants to respond, you set your switch SW, shown in Fig. 5.6.1, into the appropriate position; that is you set SW into the T-position if the correct answer to the question posed is "true", or you set SW into the F-position if the correct answer is "false". You want the output indicator light to turn "on" each time the majority of the participants have voted for the correct answer. Design the T- and F-logical circuits by:

(a) Constructing the truth tables (again represent "true" with "1" and "false" with "0").

(b) Writing the logical equations for the outputs of the logical circuits.

(c) Drawing the logical block diagrams using AND, OR and NOT gates.

5.7 Repeat problem 5.5 assuming that two of the participants (say x_1 and x_2) always vote identically.

5.8 Repeat problem 5.6 assuming that two of the participants (say x_1 and x_2) always vote identically.

5.9 Construct the truth table, write the logical equation for the output, and draw the logical block diagram of a five-input logical circuit whose output is 1 only if any four of the inputs are 1.

5.10 A certain chemical process is checked at three locations by thermometers which measure the temperatures of the three locations at all times. The process will self-correct itself if anyone of the measured temperatures *alone* deviates from the set normal operating ranges. However, if any two, or if all three of the measured temperatures deviate from the norm simultaneously, then the process must be stopped immediately for the necessary corrective action. An alarm system is needed for such occurrences. You are asked to design an alarm logical electronic circuit. You may assume that an abnormal temperature deviation sets a "high" voltage level (representing a "1") at the corresponding input of the logical circuit.

(a) Construct the truth table.

(b) Write the logical equation for the output of the logical circuit.

(c) Draw the logical block diagram using AND, OR and NOT gates.

5.11   Write the minimum decision tree for determining the color of a car if there are four choices: red, blue, green, and black.

5.12   In the example of Section 5.7 (Example 2), the three digit page number N is to be determined by finding each of the three digits separately. What is the minimum number of decisions which are required in this process?

6
Symbolic Logic

The average individual expects logical exactness in mathematics. However, the affirmation that logic is a mathematical science is a relatively new idea, existing only a little over a hundred years. Logic as a ramification of the mathematical science has come to be known as mathematical or symbolic logic. Symbolic logic is of fundamental importance in digital computation because the digital computer is essentially a "logic machine." The digital computer is able to perform quickly a vast number of logical operations upon complexes of propositions and then determine what proposition is implied by these complexes.

Logic analyzes and systematizes the process of reasoning, argumentation, and inference. As a separate discipline, it was introduced by Aristotle, and to this day it has outworn and outlasted all the other philosophical systems. Aristotle used logic to demolish the dialectic tricks of the Sophists. His system was later elaborated by the medieval thinkers and has remained perfectly valid, for the most part, until the present day. It is remarkable, however, that some of the errors of Aristotle remained undetected until recent developments.

An outstanding contribution (although not the first) in the development of logic was made by George Boole[1] about

---

[1] Boole, George "*An Investigation of the Laws of Thought on Which are Founded the Mathematical Theories of Logic and Probabilities.*" London: Walton & Maberly, 1854.

1850, when he introduced extensive mathematical methods and symbolism into logic. The growth of logic, which was practically at a standstill for two thousand years (from Aristotle to Boole) has progressed since with amazing vitality. The reason for this phenomenon may be found in the tremendous augmentation of the power of reasoning that resulted from the introduction of symbolic methods. This made it possible to lead from elementary premises of extreme simplicity to conclusions far beyond the reach of unaided reason.

6.1
THE USE OF SYMBOLS

The use of symbols has characterized logic from the beginning, but only in representing the familiar phonetic terms used in syllogisms. Leibnitz also used appropriate "ideograms" in place of the phonograms of ordinary language to facilitate the processes of logical analysis. In mathematics, improvements in notation have always had a profound effect on its progress. The far-reaching consequences of Maxwell's equations may have never been reached were it not for the suggestiveness of symbols. Had it not been for the adoption of new and more versatile ideographic symbols, many branches of mathematics could never have been developed, because the human mind is unable to grasp the essence of their operation in terms of ordinary language phonograms.

The generality which is reached by the extensive use of symbols is one of the essential characteristics of mathematical logic. Its laws and inferences do not belong exclusively to rational procedures dealing with numbers and quantities. The use of ideographic symbols has sharpened meanings and provided powerful new methods of analysis. New and important implications of accepted logical principles have come about through studies which are immensely deeper and wider than those possible in traditional logic. Subtle ambiguities and errors which have previously passed unnoticed are now detected and removed. Ideograms also enable us to concentrate on what is essential in a given context without being distracted by irrelevancies. The simplicity and conciseness of purely symbolic expressions have undisputable psychological importance.

A fact to be stressed is that the subject matter of symbolic logic is still logic. Both mathematical (symbolic) and traditional logic are concerned with the analysis and the general principles of thinking and reasoning. Their difference is essentially one of presentation and means, and not one of substance. Thus, the word "mathematical" in logic refers to the formal nature of presentation and procedure rather than to substance. In symbolic logic the emphasis is not necessarily on numerical relations. However, propositions may be made about the values of numbers, and logic operations may substitute for ordinary operations of arithmetic. Arithmetic results are then found in the propositions implied. These aspects of symbolic logic, ramified through the logic of classes and the logic of propositions are examined next. The whole subject is not covered, just the fundamentals necessary for understanding the discussion of special computer algebra.

6.2
THE CALCULUS OF CLASSES—DEFINITIONS

Elements of a "class" or "set" are identified either by some common delineating property or by enumeration. The delineating property may be a physical characteristic of the elements in the class, as in the case of "all men." Or, it may be some common mathematical property, as in the case of geometric loci which are sets of geometric points that have some definite common geometric property. For the discussion which follows, let us represent classes by X, Y, and Z. The symbolic relation $X = Y$, then, represents an "identity" equation which implies that the two classes X and Y have the *same* elements; that is, every element of X belongs to Y, and vice versa. The symbolic relation $X \leq Y$ means that X is a subset of Y; that is, every element of X belongs to Y. If Y has elements which do not belong to X, then the set X is a "proper" subset of Y. This is the property of "inclusion."

A *universal class* is one which consists of every element under discussion, all related by some "universal" property or by enumeration. For example, the class X of all positive integers below 51 may belong to the universal class of all positive integers. The number 49 belongs to X, the number 186 does not. The number $\sqrt{2}$ or the color of my car are elements on which no judgment may be taken, because they do not belong to the universal class defined by all positive integers. The universal class is denoted by 1, and every class is a subclass of the universal class.

The *complement of X* is defined as the class which consists of all the elements of the universal class which are not elements of X. The complement of X is symbolically represented by $\bar{X}$. Thus, as before, if X is the class of all positive integers below 51, then the number 186 belongs to $\bar{X}$. The "complement of X" cannot be defined unless the universal class is known. Thus, the complement of the class of all positive integers might be negative integers, fractional positive numbers, or perhaps the residents of San Francisco. It all depends upon the particular universe chosen.

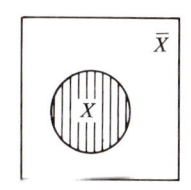

Fig. 6.1 Venn diagram.

Venn diagrams may be used to visualize the relationship between classes. The elements of a universal class are represented by the geometrical points within a square, and a class which belongs to this universe is represented by the points within a circle lying within the square. Figure 6.1 is a Venn diagram which shows the classes X and $\bar{X}$ as corresponding to the points within and around the indicated circle, respectively.

A *null class* is one that contains no elements at all. It is an empty class and is denoted by the symbol 0 and is a subclass of any other class. The 1 and 0 are not used here with the conventional connotation of quantity, but as symbols of two special classes.

A *unit class* consists of a single element. In operations with classes, we deal exclusively with classes and not with elements. Thus, the concept of "unit class" is necessary to replace the concept of "element" in the equations. A unit class is often represented by U.

6.3 OPERATIONS

A class consisting of a cat and a boy in "union" with a class consisting of an apple alone equals another class consisting of a cat, a boy, and an apple as shown in Fig. 6.2(a). A number is a property of a class. Thus, in the

(a)

No common elements   With common elements

(b)

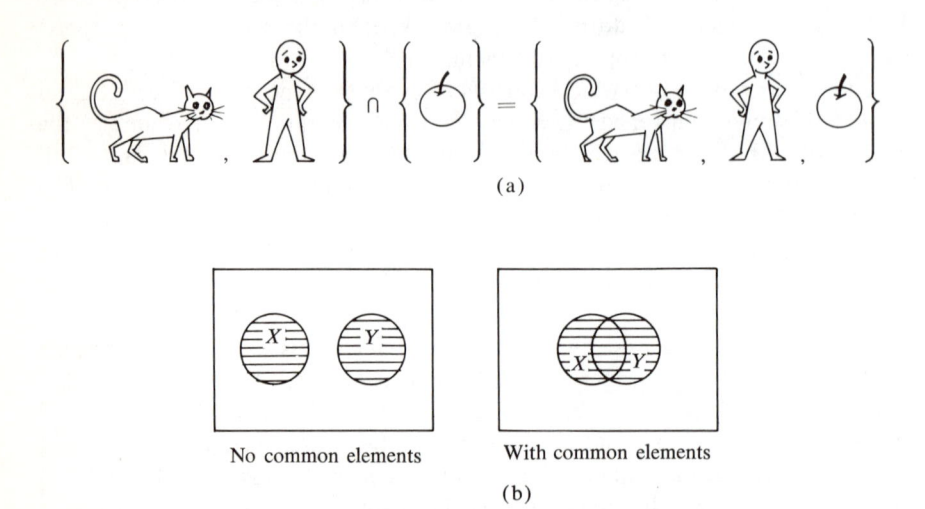

Fig. 6.2 Union of classes.

above example, the number "two" is a property of the first class, the number "one" is a property of the second class and the number "three" is a property of the implied class. The class inferred by the operation of union, symbolically represented by $X \cup Y$, consists of elements which are members of *X or Y or* both. The operation is illustrated by Venn diagrams in Fig. 6.2(b).

"Intersection" between two classes, *X* and *Y*, is an operation which leads to a class that contains all the common elements. These elements are members of

(a)

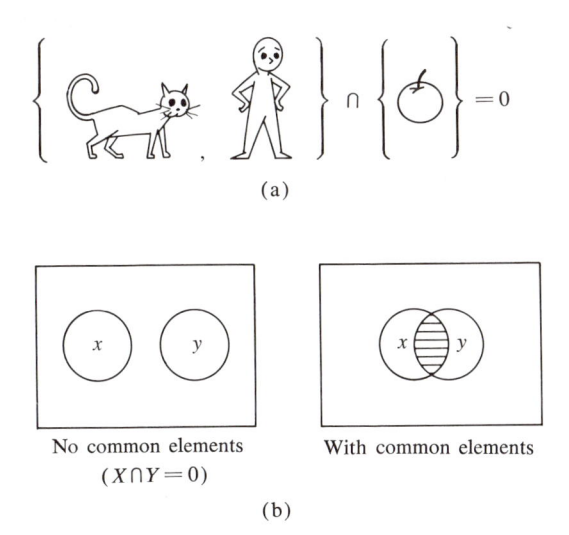

(b) No common elements ($X \cap Y = 0$) / With common elements

Fig. 6.3 Intersection of classes.

both X and Y. The operation is presented symbolically by $X \cap Y$ and is illustrated by Venn diagrams and an example in Fig. 6.3. Note that the union of X and $\bar{X}$ makes up the universal class, that is, $X \cup \bar{X} = 1$. The intersection of X and $\bar{X}$ (common parts), is null, that is, $X \cap \bar{X} = 0$. Also note that the following operations with the 0 and 1 classes hold true:

$$\bar{1} = 0, \quad 1 \cap X = X, \quad 0 \cap X = 0, \quad 1 \cup X = 1, \quad 0 \cup X = X.$$

The inferences of complexes of classes rest upon the words used to connect the statements about the classes. These connecting words, such as "and," and "or," are called "logic connectives" and will be represented throughout symbolic logic by their respective symbols.[2]

The properties of "membership," "inclusion," "complementation," and the basic operations of union and intersection are defined again in Table 6.1.

The symbolism displayed in Table 6.1 helps in expressing functions of classes; for example the function

$$X \cap \bar{Y} = \{x \mid x \in X \quad \text{and} \quad x \notin Y\}.$$

which corresponds to arithmetic "subtraction" of two classes, may be easily verified with a Venn diagram. A function may also be expressed conditionally, as for example by saying that $Z(x) = 0$ if $x \notin X$ and $Z(x) = 1$ if $x \in X$.

---

[2] Different sets of symbols have been proposed. We use the ones most commonly used by computer designers.

Table 6.1. Basic operations, properties, and symbols in the Calculus of Classes.

$x \in X$	*Membership.* Here, x is a member of X, that is, x possesses the identifying characteristics of the members of class X.
$x \notin X$	*Nonmembership.* Here, x is not a member of X, that is, x does not possess the identifying characteristics of the members of class X.
$X \leq Y$	*Inclusion.* Class X is included in class Y, that is, every member of X is also a member of Y.
$\bar{X}$	*Complement.* This class is formed by the elements of the universal class that are not members of X. (Also called "negation of X.") $\bar{X} = \{x \mid x \notin X\}$.
$\{x \mid y\}$	*Class.* Each member x possesses the property y. For example, $X = \{x \mid x \in X\}$.
$X \cup Y$	*Union of X and Y.* This class is formed by all elements which are members of X or Y or of both, that is, $\{x \mid x \in X \text{ or } x \in Y\}$.
$X \cap Y$	*Intersection of X and Y.* This class is formed by all elements that are members of *both* X and Y, that is, $\{x \mid x \in X \text{ and } x \in Y\}$.

6.4 EQUATIONS

As mentioned before, the symbolic relation $X = Y$ represents our "identity" equation, in that the two classes X and Y have the same elements and no more. Two different complexes of classes related with the equality sign also constitute two identical classes, as in the case of equation $\bar{X} \cap \bar{Y} = \overline{X \cup Y}$ which may be verified easily by a Venn diagram.

A useful property of the calculus of classes is the "duality" property, by which the dual of an identity equation is also a true identity equation. The "dual" of any equation may be obtained if each union ($\cup$) is replaced by an intersection ($\cap$) and vice versa, while each 0 is replaced by 1 and vice versa. Thus, for example the dual of $\bar{X} \cap \bar{Y} = \overline{X \cup Y}$, is $\bar{X} \cup \bar{Y} = \overline{X \cap Y}$, also a true identity equation.

6.5 FUNDAMENTAL LAWS

Several fundamental laws of the calculus of classes are stated next. The reader should justify them by constructing the appropriate Venn diagrams. Proofs are offered in specialized texts on symbolic logic.

The operations of union and intersection are commutative, distributive, and associative. The laws are stated in the following dual pairs of equations:

$X \cup Y = Y \cup X,$ $\qquad\qquad X \cap Y = Y \cap X \qquad$ *Commutative law*
$X \cup (Y \cup Z) = (X \cup Y) \cup Z,$ $\qquad X \cap (Y \cap Z) = (X \cap Y) \cap Z$ *Associative law*
$(X \cup Y) \cap (X \cup Z) = X \cup (Y \cap Z), (X \cap Y) \cup (X \cap Z) = X \cap (Y \cup Z)$
$$\qquad\qquad\qquad\qquad\qquad\qquad\qquad\qquad\qquad\qquad \textit{Distributive law}$$

A class in union or in intersection with itself equals itself:

$X \cup X = X, X \cap X = X.$ $\qquad\qquad\qquad\qquad$ *Law of Tautology.*

Using the law of Tautology and the distributive law, we may easily prove:

$X \cap (X \cup Y) = X,$ $\qquad X \cup (X \cap Y) = X$ $\qquad\qquad$ *Absorption law.*

The Venn diagrams in Fig. 6.4 demonstrate the very important DeMorgan's theorems:

$\overline{X \cup Y} = \overline{X} \cap \overline{Y}, \qquad \overline{X \cap Y} = \overline{X} \cup \overline{Y} \qquad\qquad$ *DeMorgan's theorems.*

Note that some of the above laws show no similarity with the laws of real numbers. There is no real reason why they should.

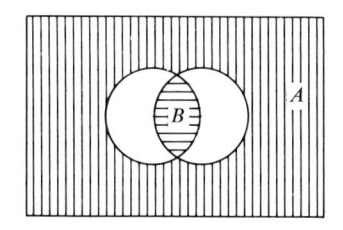

Fig. 6.4 Demonstration of the verity of De Morgan's theorems. The classes $\overline{X \cup Y}$ and $\overline{X} \cap \overline{Y}$ are identical to A (vertically shaded area) while the classes $\overline{X \cap Y}$ and $\overline{X} \cup \overline{Y}$ are identical to B (the complement of the horizontally shaded area).

6.6
CALCULUS OF INFERENCE

Different classes $X, Y, \ldots$, belonging to the same universal class, may be defined by stating their diagnostic properties. The fundamental question in the logic of classes is: "Is element x (of the universal class), a member of the class X?" A statement then, such as "x is a member of $X \cup Y$" may be examined for its truth or falsity, that is, for its "truth value." The truth value is 1 if x is a member of the "inferred class" in question, and 0 if it is not.

Investigations on the membership of an element x to the class inferred by a complex of marked operations are greatly facilitated by the construction of appropriate Venn diagrams. For more details on the use of the diagrams, the students may refer to the available textbooks on symbolic logic.[3]

6.7 EXAMPLES OF INFERENCE

An example from the logic of classes may best demonstrate and review the concepts introduced up to this point.[4]

In a very hotly fought battle, at least 70 percent of the combatants lost an eye, 75 percent an ear, 80 percent an arm, and 85 percent a leg. How many lost all four members?

The four classes in the problem are represented by W, X, Y, and Z. The class of those who lost all four members is $W \cap X \cap Y \cap Z$. The maximum percentage of the members of the universal class, (those who fought in the battle), belonging to the class $W \cap X \cap Y \cap Z$, is obviously 70 percent. To determine the minimum percentage, we base our thinking on the fact that the number of members in the intersection of two classes equals the sum of the members in each class, minus the number of members in their union. (A simple Venn diagram will verify this fact.) The maximum percentage of members belonging to any union of the above classes is, of course, 100 percent. Thus,

minimum percent of members in $W \cap X$ =
$70 + 75 - 100 = 45$;
minimum percent of members in $(W \cap X) \cap Y$ =
$45 + 80 - 100 = 25$;
minimum percent of members in $[(W \cap X) \cap Y] \cap Z =$
$W \cap X \cap Y \cap Z = 25 + 85 - 100 = 10$.

Thus, the percentage of the combatants who lost all four members of their body in the battle may be as much as 70 percent but no less than 10 percent.

The solution of this problem is vividly demonstrated in Fig. 6.5. Let the universal class (all of the combatants) be represented by a unit circle, that is by a circle with radius equal to one. The class W (those who lost an eye) constitutes at least 70 percent of the universal class and is represented by a circle with a

---

[3] Hilbert, D., and Ackerman, W., "*Principles of Mathematical Logic*" (1928), reprinted in English translation by Chelsea Publishing Company, New York, 1950. Stabler, E. R., "*An Introduction to Mathematical Thought*," Cambridge, Mass.: Addison-Wesley Publ. Co., 1953. Rosenbloom, P. C., "*The Elements of Mathematical Logic*," New York: Dover Publications, 1950.

[4] Birkhoff and MacLane attributed this problem to Lewis Carroll.

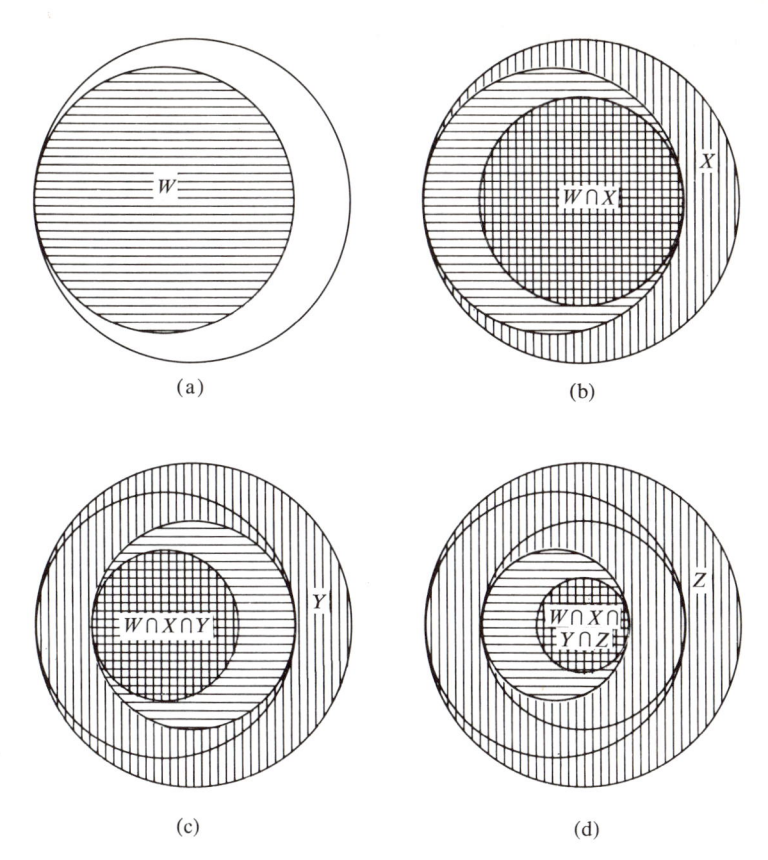

Fig. 6.5 Solution demonstration by Venn diagrams.

radius equal to $\sqrt{0.70}$, placed anywhere within the universal circle [Fig. 6.5(a)]. For minimum intersection $(W \cap X)$, the area of the class X (those who lost an ear) is chosen in such a way as to use all 30 percent of the universal circle outside W. Thus, the intersection $W \cap X$ is represented by a circle with an area equal to 45 percent of the universal circle (with a radius equal to $\sqrt{0.45}$) lying within the circle W [Fig. 6.5(b)]. The classes Y and Z are then placed similarly to form minimum intersections $W \cap X \cap Y$ and $W \cap X \cap Y \cap Z$, respectively, as shown in Fig. 6.5(c) and (d).

Another illustration of applying symbolic calculus to facilitate the determination of logic inference from a set of logic statements is the following exercise, also taken from a textbook on symbolic logic by Lewis Carroll.[5] The logic inference is sought from the following set of five statements:

---

[5] Dodgson, C. L. (Lewis Carroll), *Symbolic Logic and the Game of Logic, Mathematical Recreations of Lewis Carroll*, vol. 2, p. 119. New York: Dover Publications, 1958.

No kitten that loves fish is unteachable.
No kitten without a tail will play with a gorilla.
Kittens with whiskers always love fish.
No teachable kitten has green eyes.
No kittens have tails unless they have whiskers.

The universe of discourse is "kittens." The following six classes are then defined by "immediate reference."

A. Kittens that love fish,
B. Kittens that are teachable,
C. Kittens that have tails,
D. Kittens that are willing to play with a gorilla,
E. Kittens that have whiskers,
F. Kittens that have green eyes.

The five statements may then be rearranged and written symbolically as follows. The arrow symbolizes material inference.

(1) $A \to B$: If kittens love fish, then they are teachable.
(2) $\bar{C} \to \bar{D}$: If kittens have no tails, then they will not play with a gorilla.
(3) $E \to A$: If kittens have whiskers, then they love fish.
(4) $B \to \bar{F}$: If kittens are teachable, then they do not have green eyes.
(5) $C \to E$: If kittens have tails, then they have whiskers.

We now use two theorems, intuitively justified, whose proof may be found in a text on symbolic logic.

THEOREM 1   If $A \to B$ and $B \to C$, then $A \to C$.

THEOREM 2   If $A \to B$, then $\bar{B} \to \bar{A}$.

From the inferences of numbers (1) and (4) above, we obtain $A \to \bar{F}$. From (5) and (3) we obtain $C \to A$. By Theorem 2 above, $\bar{A} \to \bar{C}$. Combine this with (2) and we receive $\bar{A} \to \bar{D}$, or $D \to A$.

Now from $A \to \bar{F}$ and $D \to A$ the inference is that $D \to \bar{F}$ or $F \to \bar{D}$; verbally it is expressed as "Kittens with green eyes will not play with a gorilla."

These particular brain-teasers may seem artificial or even trivial. However, they are typical of problems that may arise in digital computer design, as well as in business, government, scientific, and engineering activities. The problems may be so complex that their solution by the processes of verbal logic becomes a hopeless proposition. Problems in the logic organization of large digital computers have become too intricate for the human brain to analyze with words alone. Symbolic logic offers an extremely versatile tool for doing the job. Given certain premises each time, the operations of symbolic logic can reduce the complexity of the problem to manageable proportions and provide the answer to the validity of certain conclusions.

6.8 ALGEBRA OF PROPOSITIONS

The formal treatment of logic evolves around the concept of "proposition." A proposition is a declarative statement which is free of ambiguity,[6] and is either "true" or "false." Examples of propositions are "sea water is heavier than distilled water," "three times two is six," "151 is a smaller number than 51," "iron is softer than velvet." The last two are certainly false propositions, but still valid as propositions. "Gold is faster than green + 5" is not a proposition, but a meaningless assortment of words and symbols.

The question of "membership" of the element x in a given class or in the inferred class by a complex of classes is substituted in the logic of propositions by the question of "truth" or "falsity" of simple propositions or of complexes of propositions. Symbolic logic, in the form of "propositional calculus" interesting to us, deals with the interactions in complexes of two-valued propositions; that is, it studies the implications from a complex of logical statements. Each statement can only invite one-out-of-two possible logical answers, one affirmative and one negative ("yes" and "no," or "true" and "false"), designated by the symbols "1" and "0," respectively (truth values). The complexes are formed by propositions combined or manipulated by logical operations, expressed verbally by the connective words AND, OR, NOT, and so forth. For example, "5 is a prime number" and "the sun is cold" are two propositions which may be combined with the connective AND to form the compound proposition "5 is a prime number AND the sun is cold." The truth values of the propositions in the complexes or of the implied propositions are of primary interest.

Propositions constitute elements of information[7] and may be represented with two-valued variables denoted by the symbols $a, b, c, \ldots$. Each of the variables may take only one-out-of-two possible values; that is, it is a "binary" variable. A compound proposition may be constructed from elementary propositions $a, b, c, \ldots$, by interactions similar to the ones found in the logic of classes (intersection, union, negation). The "algebra of propositions" treats operations performed on propositions and the resulting complexes of propositions. Table 6.2

---

[6] Complete freedom from ambiguity is a very difficult requirement to justify for any given statement. We can, however, be as tolerant in saying that a statement is a proposition, as we are in saying that a pencil dot on paper is a geometric point without dimensions. The discussion here aims to help intuitively in selecting suitable statements for use as propositions in applications, rather than to give an explicit definition of "proposition."

[7] To be called "bits" of information, borrowing the name from the field of information theory.

136 SYMBOLIC LOGIC

Table 6.2. Basic symbols and operations in the algebra of propositions.

a, b	Binary variables (simple propositions) representing elements of information (bits of information.)	
$\bar{a}$	"not-a": complementation, negation	The statement $\bar{a}$ is true if a is false.
$a + b$	"a or b": disjunction, logical sum	The statement $a + b$ is true if either a or b or if both a and b are true.
$a \cdot b$	"a and b": conjunction, logical product	The statement $a \cdot b$ is true if, and only if, both a and b are true.
$f_1(a, b, \ldots) = f_2(c, d, \ldots)$†	"f_1 is equivalent to f_2: equivalence, logical equality	The statement $f_1 = f_2$ denotes that the complexes f_1 and f_2 have the same truth value, that is, they are logically equivalent.

† $f_1(a, b, \ldots)$ denotes a complex of propositions, constructed by combining several propositions.

Comments:

(1) The name "logical sum" is justified by a correspondence between the OR-operation and the plus-operation found in ordinary arithmetic. Similarly, the name "logical product" is justified by a correspondence between the AND-operation and the operation of multiplication of ordinary arithmetic. Thus, there is some similarity between the calculus of propositions and ordinary arithmetic.

(2) Alternative symbols are also used for the above operations. The reader may refer to textbooks on the subject for more detail.

(3) Several operations and symbols, although basic, have been omitted in the Table 6.2, because these operations may be represented by complexes of the above fundamental operations. Two such basic operations are:

a "exclusively or" b, is equivalent to $(\bar{a} \cdot b) + (a \cdot \bar{b})$;

"a implies b," (material implication, symbolically represented as $a \rightarrow b$), is equivalent to $\bar{a} + b$, and should be read as "either not a, or b" to avoid confusion resulting from the use of the word "implies" which here does not mean logical deduction.

---

shows some of the most basic operations and symbols in the calculus of propositions. According to this table, the most primitive concepts in the calculus of propositions are those of "negation," "conjunction" and "disjunction." They are equivalent to the "complementation," "intersection," and "union" of the calculus of classes, respectively. The terms AND and OR will be used in reading the symbols "·" and "+," respectively.

Let a compound proposition, represented by $f(a, b, \ldots)$, imply the proposition q. Then

$$q = f(a, b, \ldots).$$

The relation between the variables $a, b, \ldots$ and the implied proposition q may be displayed by constructing a table of values called a "truth table." All possible combinations of truth values for the variables (2^n combinations, where n is the number of variables) are recorded together with the truth values for the implied proposition. Examples are provided in Table 6.3 for several basic operations.

Table 6.3. Truth tables for several basic operations

a b	$q=a+b$	a b	$q=a \cdot b$	a b	$q=\bar{a}+b$	a b	$q=(\bar{a} \cdot b)+(a \cdot \bar{b})$	a	$q=\bar{a}$
0 0	0	0 0	0	0 0	1	0 0	0	0	1
0 1	1	0 1	0	0 1	1	0 1	1	1	0
1 0	1	1 0	0	1 0	0	1 0	1		
1 1	1	1 1	1	1 1	1	1 1	0		

Construction of a truth table for complicated complexes of propositions is best carried out by breaking the complex into simpler combinations, constructing the respective truth tables, and then combining the results into one truth table. If, in such a table, the implied proposition q has the value of 1 (is true) independently of the values of the variables, the complex is a tautology. Examples of tautologies are the complexes $q = a + \bar{a}$ and $q = a \cdot b + \bar{a} \cdot b + a \cdot \bar{b} + \overline{a \cdot b}$, because $a + \bar{a} = 1$ and $a \cdot b + \bar{a} \cdot b + a \cdot \bar{b} + \overline{a \cdot b} = 1$.

Propositions may be made about situations, events, or the values of numbers, and regular arithmetic operations may be performed by combining operations in the calculus of propositions. The need of a special algebra to do this, one which will be based on the binary concepts of the logic of propositions, is satisfied by a Boolean algebra, to be discussed next.

BIBLIOGRAPHY

Aristotle, *The Works of Aristotle*. Translated under Ed. W. D. Ross, 11 vols. Oxford: The Clarendon Press, 1931.

Arnold, B., *Logic and Boolean Algebra*. Englewood Cliffs, N.J.: Prentice-Hall, Inc., 1962.

Asser, G., *Einführung in die Mathematische Logik*. Leipzig: Teubner, 1959.

Berkeley, E. C., *A Summary of Symbolic Logic and its Practical Applications*. Newtonville, Mass.: Berkeley Enterprises, 1957.

Boole, G., *Investigation of the Laws of Thought*. London, England: Walton, 1854. (Reprinted many times—See for example reprint New York: Dover Publ.).

Boole, G., *The Mathematical Analysis of Logic*. Cambridge, 1847, reprinted by Oxford Press, New York, 1948, and by Philosophical Library, New York, 1951.

Curry, H. B., *Foundations of Mathematical Logic*. New York: McGraw-Hill, Inc., 1963.

De Morgan, A., "On the Foundations of Algebra," *Cambridge Philosophical Trans.* VII (1841, 1842); VIII (1844, 1847).

Dodgson, C. L. (Lewis Carrell), *Symbolic Logic*. Part I, Elementary, 4th ed. London, England: The MacMillan Co., 1897.

Hilbert, D. and Ackermann, W., *Principles of Mathematical Logic*. New York: Chelsea Pub. Co., 1950.

Meigne, Maurice, *Recherches sur une logique de la pensée créatrice en Mathématiques*. Paris: Blanchard, 1964.

Mendelson, E., *Introduction to Mathematical Logic*. Princeton: Van Nostrand, 1964.

Meschkowski, H., *Einfuhrung in die Moderne Mathematik*. Mannheim: Bibliographisches Institut, 1964.

Novikov, P. S., *Elements in Mathematical Logic*. Edinburgh: Oliver and Boyd, 1964.

Phister, Montgomery, *Logical Design of Digital Computers*. New York: John Wiley & Sons, Inc., 1958.

Reichenbach, H., *Elements of Symbolic Logic*. New York: The MacMillan Company, 1948.

Russell, R. and Whitehead, A. N., *Principia Mathematica*. 3 vols. Cambridge, England: The University Press, 1910–13.

Schmidt, H. A., *Mathematische Gesetze der Logik*. Berlin: Springer-Verlag, 1960.

Venn, J., *Symbolic Logic*. London, England: The Macmillan Company, 1894.

Whitehead, A. N., *An Introduction to Mathematics*. New York: Oxford Univ. Press 1948.

Problems

6.1 Show with the use of a Venn diagram that
$$X + Y \cdot Z = (X + Y) \cdot (X + Z).$$

6.2 Use Venn diagrams to demonstrate the validity of the following operations:

(a) $X \cup (Y \cup Z) = (X \cap Y) \cup (X \cap Z)$

(b) $(X \cap Y) \cup (X \cap \bar{Z}) \cup (\bar{Y} \cap \bar{Z}) = (X \cup \bar{Y}) \cap (Y \cup \bar{Z})$

(c) $\{X \cup (\bar{Y} \cap Z)\} \cap \{(\bar{X} \cap Y) \cup Z\} = (X \cap Z) \cup (\bar{Y} \cap Z)$

6.3 Express F_1 and F_2 areas in the Venn diagram (Fig. 6.3.1) in terms of union ($\cup$), intersection ($\cap$), and complement operations among X, Y, and Z.

6.4 In a certain high-school, at least 55 percent of the students play soccer, 65 percent play tennis, 85 percent play football, and 90 percent like swimming. How many do all four sports?

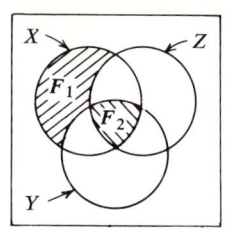

Fig. 6.3.1

6.5 Determine the logic inference from the following set of four logic statements:
(1) People that watch a lot of television are not very busy.
(2) People that like spinach do not argue a lot.
(3) People that are not very busy argue a lot.
(4) People that do not watch a lot of television have strong muscles.

(Many games of logic inference like the above, with solutions, are available in: C. L. Dodgson (Lewis Carroll) *Symbolic Logic and the Game of Logic, Mathematical Recreations of Lewis Carroll*, New York: Dover Publications, 1958.)

6.6 Form and solve a problem of logic inference similar to the one presented in the text (Section 6.7) by making use of six classes and forming five inherently and mutually consistent logic statements. (Note that inherent and mutual consistency in the logic statements will guarantee that there will be no conflicting logic inferences derived by the manipulation of the logic statements).

7

Switching Algebra

From the premises of symbolic logic, an algebraic discipline was developed known as Boolean algebra in commemoration of its prime originator.[1] Boolean algebra is the generic name for a class of algebras, all fulfilling a given set of fundamental and independent postulates and conditions. It is characterized as an "algebra" because of its structural resemblance to ordinary algebra. It follows rules different from those of conventional algebra, but is easily understood because of the simplicity of its fundamental concepts and postulates. For that reason it may be introduced in basic algebra courses. A formal presentation of Boolean algebra should be abstract. In the following discussion, however, reference to the algebra of classes and propositions is maintained constantly, because they are perhaps the most intuitive of all particular cases, and at the same time complete enough to reveal the essential nature of a Boolean algebra.

A Boolean algebra deals with binary variables in the same manner that the logic of propositions deals with the binary notion of affirmative or negative propositions. For a generalized mathematical treatment, a binary variable is used, with its values defined as "1" and "0". It should be pointed here that Boolean algebra should not be confused operationally

---

[1] Boole, G., *The Mathematical Analysis of Logic*, Cambridge, 1847; *An Investigation of the Laws of Thought on Which Are Founded the Mathematical Theories of Logic and Probabilities*, London, 1854.

with binary arithmetic. While in binary arithmetic we have $1 + 1 = 10$, in Boolean algebra, $1 + 1 = 1$, and also $a + a = a$.

Modern switching algebra was originally constructed by C. Shannon[2] specifically for the problems encountered in the analysis and synthesis of circuits with binary switches. It is only one of a class of similar algebras known today (Boolean algebras), and it is often referred to by engineers as *the* Boolean algebra. It is particularly suited to the problems encountered in the mathematical analysis and design of digital equipment. No doubt other algebras will be invented, bringing greater advantages to the design of logic circuits. Insofar as such applications are concerned, however, none is known today in more general use than the switching (Boolean) algebra which is discussed in this chapter.

Shannon's formulation of the switching algebra is based on the postulates of the algebra of propositions. Some writers in the field refer the switching algebra to the calculus of classes rather than to the calculus of propositions. Since both of them derive from the original work of Boole and from the common grounds of symbolic logic, either one can be considered as the mathematical origin of switching algebra.

Great proficiency must be developed by the logic designer in manipulating Boolean algebra equations and deriving equivalent logical circuits, and in minimizing and generally simplifying logical equations and circuits, so that the most economical implementation may be achieved. This chapter should provide considerable help to the beginner in this field.

7.1
DEFINITIONS

The development of Boolean algebra will proceed here with what is perhaps a minimum of definitions and notations. Only the essentials are used to show explicitly the fundamental importance of Boolean algebra in computer design. Lower-case alphabetic symbols—a, b, c, ... —are used to represent *binary variables*. The notion of a variable needs some clarification. The symbols a, b, c, ... do not bear the notion of quantity (number) like the variables of conventional algebra, but they correspond directly to the "bits" of information theory. Each variable is binary in nature and its two possible *truth values* bear information about the condition of the variable. The symbols 0 and 1 are used here to signify the two possible conditions of a variable. The importance of these symbols is not one of *magnitude* but one of *condition*; the symbols T (true) and F (false) or any other pair of dichotomous symbols could be used equally well.

---

[2] Shannon, Claude E., "A Symbolic Analysis of Relay and Switching Circuits," *Tran. AIEE* vol. 57, pp. 713–723, 1938; "The Synthesis of Two-Terminal Switching Circuits," *Bell Syst. Tech. J.*, vol. 28, pp. 59–98, 1949.

Binary variables may refer to the truth or falsity of dichotomous propositions, to mention one particular case, or they may bear information about the membership of a given element in a considered class. Whatever system of interpretation is used for the symbols which are employed to represent the binary variables, the system is admissible as a Boolean algebra as long as the truth of some accepted postulates is withheld. Theorems and operational rules for combining symbols are derived from the original accepted postulates.

7.2 AXIOMATIC DEFINITION OF A BOOLEAN ALGEBRA

A Boolean algebra, like any other deductive mathematical discipline, begins from a set of unprovable axioms or postulates and a set of undefined operations. Within the framework of these postulates and operations one proceeds to derive consistent rules and theorems.

For our definition of a Boolean algebra, we will follow the postulates stated by E. V. Huntington[3] in 1904. A set of six postulates is used, each postulate being independent in the sense that it cannot be derived from the others. Any one of several others, slightly different sets of postulates could have been used equally well.

Consider a class B of elements $a, b, \ldots$, together with two undefined operations ("connectives") represented by the symbols " $+$ " and " $\cdot$ ". (For the particular case of the algebra of propositions, the elements of the class B are dichotomous propositions and the two operations are the operations of "disjunction" and "conjunction," respectively.) The class B together with the two operations constitute a Boolean algebra if, and only if, the following postulates are held valid:

P1 If a is in B and b is in B, then also $a + b$ and $a \cdot b$ are in B. ("The class B is closed under the " $+$ " and " $\cdot$ " operations".)

P2 The class B contains distinct *identity elements* relative to the operations " $+$ " and " $\cdot$," respectively. Let the identity elements be represented by the symbols 0 and 1. Then for any element in B, $a + 0 = a$ and $a \cdot 1 = a$.

P3 The operations are *commutative*; that is, for the elements a and b in B, it is true that $a + b = b + a$ and $a \cdot b = b \cdot a$.

P4 Each operation is *distributive* over the other; that is, for the elements a, b, and c in B, it is true that $a \cdot (b + c) = (a \cdot b) + (a \cdot c)$ and $a + (b \cdot c) = (a + b) \cdot (a + c)$.

P5 If the identity elements 0 and 1 are unique, then for every element a in the class B there exists an element $\bar{a}$ also belonging to B, such that $a + \bar{a} = 1$ and $a \cdot \bar{a} = 0$. Also when $a = 0$, then $\bar{a} = 1$ and vice versa.

P6 There exist at least two elements a and b in B such that $a \neq b$.

[3] Huntington, E. V., "Sets of Independent Postulates for the Algebra of Logic," *Trans. Amer. Math. Soc.*, vol. 5, pp. 288–309, 1904.

Any mathematical system conforming to the above set of postulates is a Boolean algebra. Of the possible Boolean algebras defined by the above postulates, the one which makes use of just two elements, 0 and 1, and of the binary operations of disjunctions or logical sum, "$+$," and conjunction or logical product, "$\cdot$," is of particular interest to us because of its direct application to the arithmetic and logic design of information-processing machines. This is the switching algebra, also often referred to by engineers as the Boolean algebra.

In switching algebra each variable can only take the values 0 or 1. The fundamental operations of "negation," "logical sum," and "logical product," directly derivable from the above postulates, are stated in Table 7.1.

Table 7.1. Basic operations.

Negation		Logical Sum			Logical Product		
a	$\bar{a}$	a	b	$c=a+b$	a	b	$c=a\cdot b$
0	1	0	0	0	0	0	0
1	0	0	1	1	0	1	0
		1	0	1	1	0	0
		1	1	1	1	1	1

It is of interest to point out here that if (0 and 1) and ($+$ and $\cdot$) are interchanged in any one of the above set of postulates, another postulate of the same set results. This is an example of the general principle of *duality*, which is true for all switching algebra postulates and theorems. The duality is due to the fact that the values 1 and 0 are assigned to physical conditions arbitrarily and the assignments may be interchanged without hurting the validity of the equations of switching algebra.

7.3 THEOREMS AND OPERATION RULES

A few basic theorems in switching algebra, which may be helpful in subsequent discussions and exercises, are stated in the following. These theorems may be easily verified on appropriate truth tables or they may be proven using the previously defined postulates. In truth tables both sides of each equation are checked to yield identical results for all possible combinations of the variables involved.

First, the following theorems for a single element may be proven easily:

$$a + a = a, \quad a \cdot a = a,$$
$$1 + a = 1, \quad 0 \cdot a = 0.$$

Then, the *absorption* rule

$$a + (a \cdot b) = a, \quad a \cdot (a + b) = a$$

and the *associative* rule

$$(a + b) + c = a + (b + c) = b + (a + c) = a + b + c,$$
$$(a \cdot b) \cdot c = a \cdot (b \cdot c) = b \cdot (a \cdot c) = a \cdot b \cdot c$$

may also be proven by simple algebraic manipulations.

Two especially useful theorems which can be applied only to Boolean algebra are the *DeMorgan's theorems*:

$$\overline{a + b} = \bar{a} \cdot \bar{b},$$
$$\overline{a \cdot b} = \bar{a} + \bar{b}.$$

The statement of these theorems may be extended to any number of variables. (See Section 7.6.)

DeMorgan's theorems may be proven by making reference to the founding postulates. We shall prove here only the first of them and leave the second to the student for exercise. To prove that $\overline{a + b}$ and $\bar{a} \cdot \bar{b}$ are identical, it is enough to show that $a + b$ and $\bar{a} \cdot \bar{b}$ are the complements of one another. (It is easy to show that the complement of a variable is uniquely determined.)

To prove that $a + b$ and $\bar{a} \cdot \bar{b}$ are the complements of one another, we must show that

$$(a + b) + (\bar{a} \cdot \bar{b}) = 1$$

and that

$$(a + b) \cdot (\bar{a} \cdot \bar{b}) = 0.$$

PROOFS

$$\begin{aligned}
(a + b) + (\bar{a} \cdot \bar{b}) &= a \cdot (b + \bar{b}) + b + (\bar{a} \cdot \bar{b}) \\
&= (a \cdot b) + (a \cdot \bar{b}) + b + (\bar{a} \cdot \bar{b}) \\
&= b \cdot (a + 1) + (a \cdot \bar{b}) + (\bar{a} \cdot \bar{b}) \\
&= b + \bar{b} \cdot (a + \bar{a}) \\
&= b + \bar{b} \\
&= 1
\end{aligned}$$

and

$$\begin{aligned}
(a + b) \cdot (\bar{a} \cdot \bar{b}) &= (a \cdot \bar{a}) \cdot \bar{b} + \bar{a} \cdot (b \cdot \bar{b}) \\
&= 0 + 0 \\
&= 0.
\end{aligned}$$

Similarly we may prove the following theorems:

THEOREM 1
$$a + (\bar{a} \cdot b) = a + b$$
$$= b + (\bar{b} \cdot a).$$

PROOF
$$a + (\bar{a} \cdot b) = a \cdot (b + \bar{b}) + (\bar{a} \cdot b)$$
$$= (a \cdot b) + (a \cdot \bar{b}) + (\bar{a} \cdot b)$$
$$= b \cdot (a + \bar{a}) + (a \cdot \bar{b})$$
$$= b + (\bar{b} \cdot a);$$

also

$$a + b = a + b \cdot (a + \bar{a})$$
$$= a + (a \cdot b) + (\bar{a} \cdot b)$$
$$= a + (\bar{a} \cdot b)$$

(the last operation involved the absorption law).

THEOREM 2 $(a \cdot b) + (\bar{a} \cdot c) + (b \cdot c) = (a \cdot b) + (\bar{a} \cdot c).$

PROOF
$$(a \cdot b) + (\bar{a} \cdot c) + (b \cdot c) = (a \cdot b) + (\bar{a} \cdot c) + b \cdot c \cdot (a + \bar{a})$$
$$= (a \cdot b) + (\bar{a} \cdot c) + (a \cdot b \cdot c) + (\bar{a} \cdot b \cdot c)$$
$$= a \cdot b \cdot (1 + c) + \bar{a} \cdot c \cdot (1 + b)$$
$$= (a \cdot b) + (\bar{a} \cdot c).$$

Up to this point, for reasons of formality, we have maintained the symbol "·" in designating logical products. It is suggested now that for the rest of the book we substitute "·" by juxtaposition of the symbols of the variables.

The above-mentioned theorems are quite useful for deriving a multitude of other theorems and for simplifying algebraic expressions. The subject of simplification of algebraic expressions is an important one and is discussed in some detail later on in this chapter.

7.4 COMBINATIONS AND FUNCTIONS OF BINARY VARIABLES

To facilitate understanding the notions of "combination" and "function" of binary variables, let each binary variable be represented by the state of a corresponding switch. The variable has the value 1 if the switch is ON (closed), and 0 if it is OFF (open). Then, two variables, like two switches, can result in only four combinations: ON-ON (1-1), ON-OFF (1-0), OFF-ON (0-1), and OFF-OFF (0-0). In general n binary variables (like n switches) will result in 2^n possible combinations.

To represent the concept of a function of binary variables, which is somewhat more difficult to grasp, again let three variables a, b, c be represented by the states of three switches. There are eight possible combinations of OFF and ON configurations for the three switches (binary variables). Using the symbols 0 and 1, these combinations are listed in Table 7.2.

Table 7.2. Combinations and Functions of Three Variables.

Combination	a b c	Function f_1	Function f_2
0	0 0 0	0	0
1	0 0 1	1	1
2	0 1 0	0	1
3	0 1 1	1	0
4	1 0 0	0	1
5	1 0 1	0	0
6	1 1 0	1	0
7	1 1 1	0	1

Now let some arbitrary action be predicated on the occurrence of some particular combinations from those listed. For example, the action could be initiated if either combination 1 or 3 or 6 occurs (function f_1), or if an odd number of 1's is present in the combination, which is true for combinations 1, 2, 4 and 7 (function f_2).

Such "combinations of combinations" represent functions; that is, we may simply say that a function is some particular grouping of combinations of binary variables. Since the binary variable can have only two possible states, a finite number of binary variables must yield a finite number of possible combinations and of possible functions. In general, n binary variables yield 2^n possible different combinations and 2^{2^n} possible different functions. Thus, in particular, three variables yield eight combinations and 256 functions.

Boolean functions may be defined and represented by constructing their respective "truth tables," like the ones shown for the functions f_1 and f_2 in Table 7.2.

A Boolean function may be read from a given truth table by forming a logical product for each combination of the variables which produces a 1 in the function, and then by taking the logical sum of all those products. For example, the function f_1 in Table 7.2 is determined by reading the combinations 1, 3, and 6 as $\bar{a}\bar{b}c$, $\bar{a}bc$, and $ab\bar{c}$, respectively. Since each one of these combinations results in $f_1 = 1$, we should have

$$f_1 = \bar{a}\bar{b}c + \bar{a}bc + ab\bar{c}.$$

Similarly, it may be easily verified that

$$f_2 = \bar{a}\bar{b}c + \bar{a}b\bar{c} + a\bar{b}\bar{c} + abc.$$

COMBINATIONS AND FUNCTIONS OF BINARY VARIABLES   147

In general, equivalent functions may be derived by algebraic manipulations, all such functions having identical truth tables. An important task for the logic designer is to place the function in a form which leads to the most economical hardware implementation. Algebraic, topological, and tabular techniques may be employed in this pursuit. However, the designer depends heavily also upon his experience and his intuition.

All sixteen possible functions of two variables are shown in Table 7.3. Any

Table 7.3. Truth tables for all sixteen possible functions of two variables.

a b	0	$\overline{ab}$	$\overline{ab}$	$\bar{a}$	$a\bar{b}$	$\bar{b}$	$a\bar{b}+\bar{a}b$	$\bar{a}+\bar{b}$	ab	$ab+\overline{ab}$	b	$\bar{a}+b$	a	$a+\bar{b}$	$a+b$	1
0 0	0	1	0	1	0	1	0	1	0	1	0	1	0	1	0	1
0 1	0	0	1	1	0	0	1	1	0	0	1	1	0	0	1	1
1 0	0	0	0	0	1	1	1	1	0	0	0	0	1	1	1	1
1 1	0	0	0	0	0	0	0	0	1	1	1	1	1	1	1	1
Function Number	0	1	2	3	4	5	6	7	8	9	10	11	12	13	14	15

truth table with two variables should be matched with one of the sixteen tables shown. The functions AND and OR defined previously are by no means the only functions of two binary variables. The following functions are also much in use, and therefore they have been given characteristic names.

$f_8 = ab$ AND
$f_{14} = a + b$ OR
$f_7 = \overline{ab} = \bar{a} + \bar{b}$ NAND, also STROKE FUNCTION
$f_1 = \overline{a + b} = \bar{a}\bar{b}$ NOR, also DAGGER FUNCTION
$f_6 = a\bar{b} + \bar{a}b = a \oplus b$ EXCLUSIVE-OR, also NON-EQUIVALENCE
$f_9 = ab + \bar{a}\bar{b} = (a \equiv b)$ EQUIVALENCE[4]
$f_3 = \bar{a}$ NEGATION
$f_5 = \bar{b}$ NEGATION

Less frequently used functions are

$f_0 = 0$ ZERO, also NULL FUNCTION
$f_2 = \bar{a}b$ INHIBITION
$f_4 = a\bar{b}$ INHIBITION
$f_{10} = b$ ASSERTION
$f_{11} = \bar{a} + b = a \rightarrow b$ IMPLICATION
$f_{12} = a$ ASSERTION
$f_{13} = a + \bar{b} = b \rightarrow a$ IMPLICATION
$f_{15} = 1$ ONE, also IDENTITY FUNCTION

[4] Notice that $f_6 = \overline{f_9}$.

The NOR and NAND functions are of particular interest because any other function of two variables can be synthesized using only nor and nand connectives, and because simple transistor circuits exist for implementing these functions (Chapter 8). In the earlier days of logical circuit design, however, the AND and OR gates supplemented by the INVERTER which realizes the operation of "negation," were used almost exclusively. Their popularity was mainly due to the relay and diode circuits available at that time. Logical circuit design still uses AND, OR, and NOT circuits. The sixteen functions of two variables (Table 7.3), implemented with AND, OR gates, and inverters are illustrated in Fig. 7.1. An "inverter" is a one input-one output device which provides at its output the complement of the binary signal received at its input.

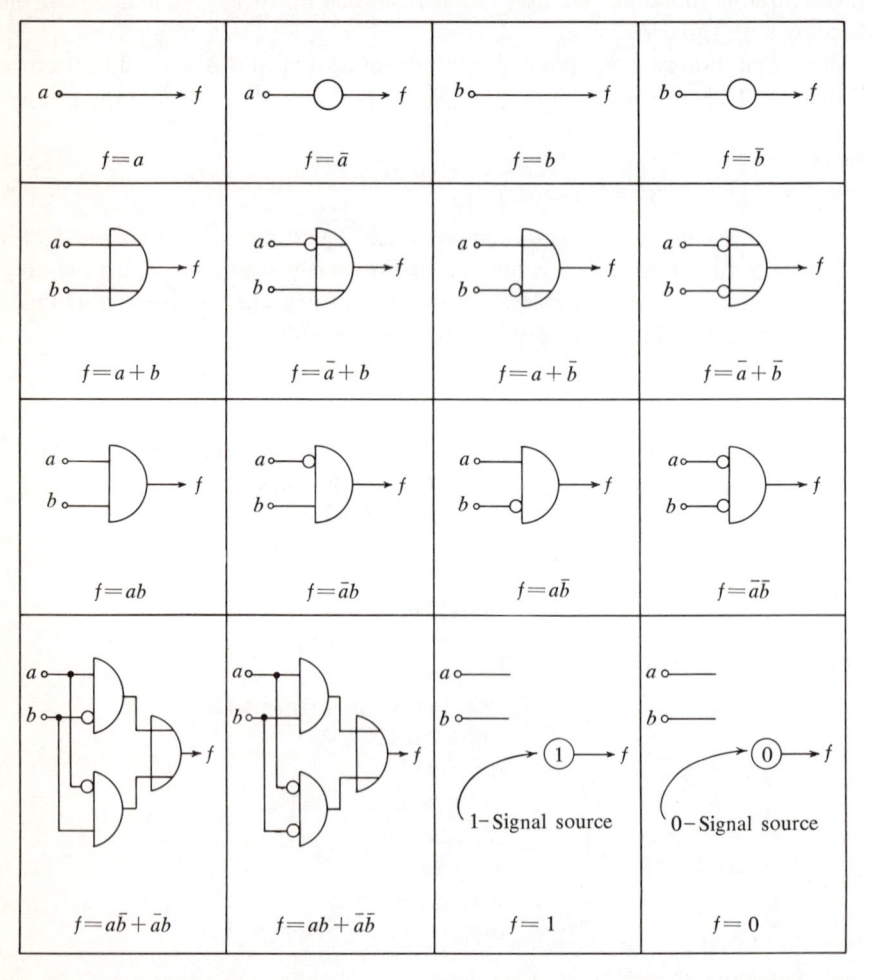

$f = a$	$f = \bar{a}$	$f = b$	$f = \bar{b}$
$f = a + b$	$f = \bar{a} + b$	$f = a + \bar{b}$	$f = \bar{a} + \bar{b}$
$f = ab$	$f = \bar{a}b$	$f = a\bar{b}$	$f = \bar{a}\bar{b}$
$f = a\bar{b} + \bar{a}b$	$f = ab + \bar{a}\bar{b}$	$f = 1$	$f = 0$

Fig. 7.1 Hardware implementation of the sixteen Boolean functions of two variables.

7.5 EQUALITY OF BOOLEAN FUNCTIONS— BOOLEAN EQUATIONS

Two functions f_1 and f_2, each a function of the n binary variables $a_1, a_2, a_3, \ldots, a_n$, are said to be equal if, and only if, they have the same "truth value" for all possible combinations of values of the n binary variables. The equality of two given Boolean functions may be checked, then, by constructing and comparing their respective "truth tables."

To secure that every possible assignment of 0 and 1 has been accounted for, one may count from zero up to $(2^n - 1)$ inclusive in the pure binary system using n digits for each number. For example, in order to construct the truth table of a three-variable function, one may count from 000 up to 111, as is done for the functions of Table 7.2.

Verifying Boolean equations by constructing truth tables is illustrated in Table 7.4. DeMorgan's theorem for two variables, $\overline{a+b} = \bar{a}\bar{b}$, and the absorp-

Table 7.4. Verification of boolean equations by constructing truth tables

(a) $\overline{a+b} = \bar{a}\bar{b}$ DeMorgan's theorem

a	b	$\overline{a+b}$	$\bar{a}\bar{b}$
0	0	1	1
0	1	0	0
1	0	0	0
1	1	0	0

(b) $(a+b)(a+c) = a+bc$ Absorption law

a	b	c	$(a+b)(a+c)$	$a+bc$
0	0	0	0	0
0	0	1	0	0
0	1	0	0	0
0	1	1	1	1
1	0	0	1	1
1	0	1	1	1
1	1	0	1	1
1	1	1	1	1

tion rule for three variables, $(a+b)(a+c) = a+bc$, are checked. If the truth table technique is used to verify equations involving complicated functions, the construction of the truth table is best carried out progressively in steps.

150 SWITCHING ALGEBRA

A Boolean equation may also be checked by constructing the appropriate Venn diagrams. A number of circles, equal to the number of variables which appear in the equation, appear in the diagram, each circle intersecting all others. The validity of the equation is checked by verifying that the area on the diagram which is suggested by each side of the equation is the same one. This corresponds to finding the class which is suggested by either side of the equation, as discussed in Chapter 6. Each variable represents a class.

An illustrative example is provided in Fig. 7.2. The three functions which appear in the equations

$$x + \bar{x}yz = x + yz = (x + y)(x + z)$$

all suggest the same area on the Venn diagram, which is indicated in the diagram as a shaded area.

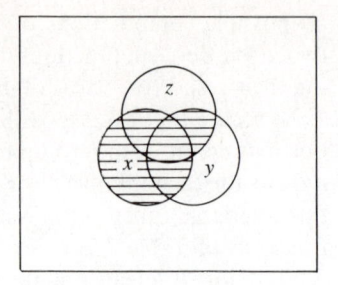

Fig. 7.2 Venn diagram illustrating the equality of the Boolean functions: $x + \bar{x}yz$, $x + yz$, $(x + y)(x + z)$. Each function is represented by the shaded area.

Another way used to verify a Boolean equation is to begin with the function on one side of the equation, and derive the other side by algebraic manipulations, making use of accepted postulates or proven theorems. For example one of the equations shown in Fig. 7.2 may be verified as follows:

$$(x + y)(x + z) = x + xy + xz + yz$$
$$= x(1 + y + z) + yz$$
$$= x + yz.$$

In the above algebraic manipulations the identities

$$xx = x$$
$$1 + y + z = 1$$

and

$$1x = x$$

were used.

7.6 THE OPERATION OF COMPLEMENTATION

The concept of complementation is of great importance in the manipulation of switching[5] functions, and therefore also in the design of

---

[5] The terms "switching," "logical," and "Boolean" are used as synonyms. Therefore expressions like "switching functions" or "logical functions," and so forth, are used interchangeably.

THE OPERATION OF COMPLEMENTATION 151

switching circuits. To take the complement of a given function, we apply the following theorem suggested by Shannon:

$$\overline{f(x_1, x_2, \ldots, x_n, +, \cdot)} = f(\bar{x}_1, \bar{x}_2, \ldots, \bar{x}_n, \cdot, +).$$

This is a generalized form of DeMorgan's theorem, which suggests that the complement of a given function f of n binary variables is obtained by replacing each variable by its complement and, at the same time, by interchanging the operations of logical addition and logical multiplication.

As an illustrative example consider taking the complement of the switching function

$$\overline{\bar{x}(y + x\bar{y}z) + (\bar{z}y)} = [x + \bar{y}(\bar{x} + y + \bar{z})](z + \bar{y}).$$

It is best to begin taking the complement from the "center" of the function and move outwards. In the above example this corresponds to taking first the complement of the product $x\bar{y}z$. Care should be exercised to preserve the groupings of the variables during the operation.

The feature of complementation is inherent to two-valued systems and binary algebraic expressions. It confirms the expectation that for every two-valued system or function there exists another one which has exactly complementary characteristics. This basic idea is often very useful in simplifying a switching function or circuit. At times it becomes simpler to design a switching circuit which does the opposite of what we want, and then achieve our objective by inverting the output. This feature is now illustrated in the following example.

Input Combination	x	y	z	f	$\bar{f}$
0	0	0	0	1	0
1	0	0	1	0	1
2	0	1	0	1	0
3	0	1	1	1	0
4	1	0	0	1	0
5	1	0	1	0	1
6	1	1	0	1	0
7	1	1	1	1	0

Fig. 7.3 Functions f and $\bar{f}$ specified by truth tables.

Consider that as a result of the specifications of a logic design problem a switching function f is specified by its truth table, shown in Fig. 7.3. The function, f, may have resulted by specifying the desired input-output combinations, as it was required, for example, in the design of a code translator in Section 5.6 (Table 5.4).

Different ways of realizing a switching function, f, from its truth table are discussed with illustrative examples in the following sections of this chapter. A simple way of realizing this particular switching function is by first realizing its complement, $\bar{f}$, and then complementing $\bar{f}$. The complement $\bar{f}$ is also shown in the truth table of Fig. 7.3. Notice that only two input combinations lead to $\bar{f} = 1$. Therefore

$$\bar{f} = \bar{x}\bar{y}z + x\bar{y}z$$
$$= \bar{y}z(\bar{x} + x)$$
$$= \bar{y}z$$

and

$$f = \overline{(\bar{f})}$$
$$= \overline{(\bar{y}z)}$$
$$= y + \bar{z} \quad \text{(by DeMorgan's theorem.)}$$

The switching circuits realizing f are shown in Fig. 7.4. In this case, realizing first the complement and then the function itself has resulted directly in a simplified algebraic expression for the function f. The subject of simplification of

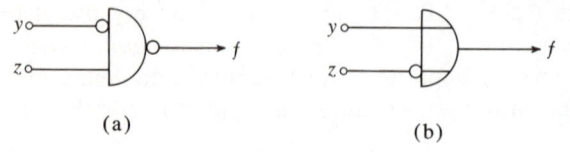

Fig. 7.4 Switching circuit implementations of f; (a) by implementing $\bar{f} = \bar{y}z$ and (b) by implementating $f = y + \bar{z}$.

switching functions will be discussed further in a later section of this chapter. It may be added here that, during simplification of a switching function, the use of complementation may ease the algebraic manipulations greatly.

7.7 CANONICAL FORMS

An important theorem of any two-valued Boolean algebra states that "Any Boolean function may be expressed as a disjunctive combination of conjunctions, or alternatively, as a conjunctive combination of disjunctions." In terms of switching circuits the theorem means that any switching function may be realized either by a group of AND gates whose outputs are fed into an OR gate, or by a group of OR gates whose outputs are fed into an AND gate.

CANONICAL FORMS   153

The above theorem may be stated in another way, in terms more akin to the subject of switching algebra. It may be said that any switching function may be expressed algebraically either in the form of a "sum-of-products" or in the form of a "product-of-sums." Both these forms are known as *canonical* forms of switching functions. To explain better, let us consider again the three-variable switching function, f, given by the truth table in Fig. 7.3. The function f can always be realized in either one of the two canonical forms.

a. Sum-of-products

We may notice that each of the input combinations 0, 2, 3, 4, 6, and 7, gives a truth value 1 to the function f. Each combination leading to $f = 1$ may be read as a product which includes every one of the variables. Each variable appears either in the uncomplemented or in the complemented form, but only once in each combination. If a variable appears in the combination as 0, then the complemented form of the variable must be used in the algebraic product. For example, the combination 2, (010), which results in $f = 1$, should be read as $\bar{x}y\bar{z}$. Using all the different input combinations which provide a truth value 1 to the function f, that is, by satisfying the 1 values of the function f, we have:

$$f = \bar{x}\bar{y}\bar{z} + \bar{x}y\bar{z} + \bar{x}yz + x\bar{y}\bar{z} + xy\bar{z} + xyz.$$

The function f has been placed in the sum-of-products *canonical* form. The truth value of f is 1 if, and only if, any one of the products takes a truth value of 1. Notice that no more than one of the products may become 1 for a given input combination of values. For example if $x = 0$, $y = 0$, $z = 0$, then only the combination $\bar{x}\bar{y}\bar{z}$ becomes 1. Since the combination $\bar{x}\bar{y}\bar{z}$ appears in f, the values $x = 0$, $y = 0$, $z = 0$ also make $f = 1$. However, the values $x = 0$, $y = 0$, $z = 1$, which make $\bar{x}\bar{y}z = 1$, also make $f = 0$, since the combination $\bar{x}\bar{y}z$ does not appear in f.

Each product in the sum-of-products canonical form is called a *minterm* or a *fundamental product*. By definition, "a minterm of n variables is a logical product of these n variables, with each variable appearing only once in either the complemented or in the uncomplemented form." For example, the four possible minterms of two variables are: $\bar{x}\bar{y}$, $\bar{x}y$, $x\bar{y}$, xy. The eight possible minterms of three variables are shown in Table 7.5.

b. Product-of-Sums

The combinations 1 and 5 in the truth table of function f (Fig. 7.3) lead to $f = 0$, and therefore to $\bar{f} = 1$. In other words we may write $\bar{f}$ in the form of a sum-of-products, as follows:

$$\bar{f} = \bar{x}\bar{y}z + x\bar{y}z.$$

By the generalized DeMorgan's theorem of complementarity, discussed in Section 7.6 we may write

$$\bar{f} = \bar{x}\bar{y}z + x\bar{y}z = \overline{(x+y+\bar{z})(\bar{x}+y+\bar{z})}.$$

Therefore

$$f = (x+y+\bar{z})(\bar{x}+y+\bar{z}).$$

The function f has now been written in the other *canonical* form, that of a product-of-sums. This expression may be read directly from the truth table of f, *by satisfying the 0 values of the function f.*

Each sum which appears in the product-of-sums canonical form is called a *maxterm*. By definition, "a maxterm of n variables is a logical sum of these n variables, with each variable appearing only once in either the complemented or in the uncomplemented form." For example, the four possible maxterms of two variables are: $\bar{x}+\bar{y}$, $\bar{x}+y$, $x+\bar{y}$, $x+y$. The eight possible maxterms of three variables are shown in Table 7.5.

Table 7.5. Minterms and maxterms for three variables.

x	y	z	Minterms	Maxterms
0	0	0	$\bar{x}\bar{y}\bar{z}$	$\bar{x}+\bar{y}+\bar{z}$
0	0	1	$\bar{x}\bar{y}z$	$\bar{x}+\bar{y}+z$
0	1	0	$\bar{x}y\bar{z}$	$\bar{x}+y+\bar{z}$
0	1	1	$\bar{x}yz$	$\bar{x}+y+z$
1	0	0	$x\bar{y}\bar{z}$	$x+\bar{y}+\bar{z}$
1	0	1	$x\bar{y}z$	$x+\bar{y}+z$
1	1	0	$xy\bar{z}$	$x+y+\bar{z}$
1	1	1	xyz	$x+y+z$

To illustrate the expansion of a given function into its canonic forms, consider the function

$$f = \overline{x + \bar{y}\bar{z}}$$

The truth table of this function may be easily constructed in a step-by-step fashion as shown in Fig. 7.5. The sum-of-products form of the function is

$$f = \bar{x}\bar{y}z + \bar{x}y\bar{z} + \bar{x}yz.$$

The product-of-sums form is

$$f = (x+y+z)(\bar{x}+y+z)(\bar{x}+y+\bar{z})(\bar{x}+\bar{y}+z)(\bar{x}+\bar{y}+\bar{z}).$$

It should be observed that each one of the canonical forms is implemented

CANONICAL FORMS 155

Input Combination	x y z	$\bar{y}\bar{z}$	$x+\bar{y}\bar{z}$	$f=\overline{x+\bar{y}\bar{z}}$
0	0 0 0	1	1	0
1	0 0 1	0	0	1
2	0 1 0	0	0	1
3	0 1 1	0	0	1
4	1 0 0	1	1	0
5	1 0 1	0	1	0
6	1 1 0	0	1	0
7	1 1 1	0	1	0

Fig. 7.5 Truth table construction for the function $f = \overline{x + \bar{y}\bar{z}}$.

as a *two-level* logical circuit. This means that the input signal needs to go through two gates before it reaches the output. Thus, the sum-of-products form is implemented by a group of AND gates whose outputs feed an OR gate, as demonstrated in Fig. 7.6(a) for the above example. The other canonical form is implemented by a dual circuit configuration.

A two-level circuit is desirable if the signal propagation "time delay" is to be minimized. The propagation delay is the time it takes for a change in the input of a circuit to propagate to the output. The fewer the number of gates the signal has to go through before it reaches the output, the shorter the time delay is, and therefore the faster the response of the circuit. In computer systems where

$f = \bar{x}\bar{y}z + \bar{x}y\bar{z} + \bar{x}yz$

(a)

$f = \overline{x + \bar{y}\bar{z}}$

(b)

$f = \bar{x}(y+z)$

(c)

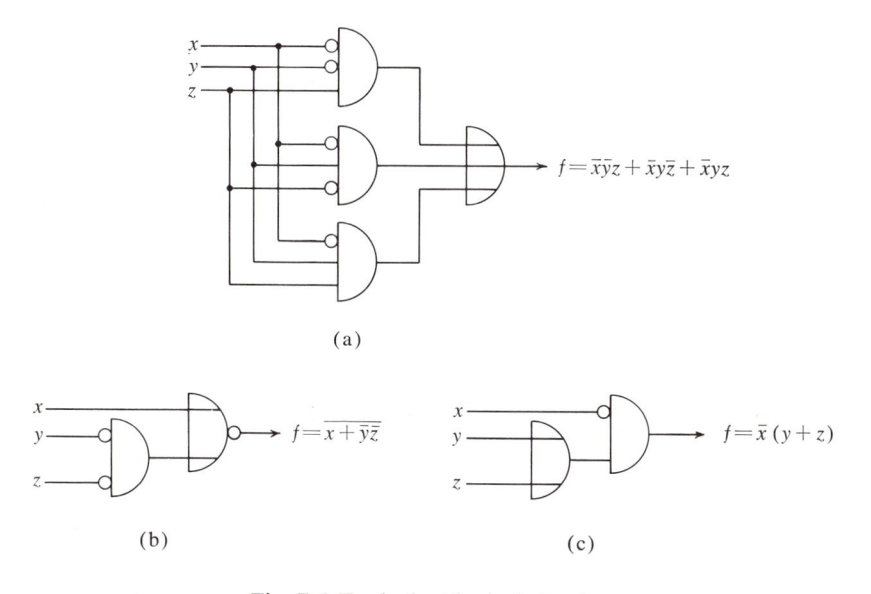

Fig. 7.6 Equivalent logical circuits.

many repetitive logical transformations may be required for doing simple arithmetic operations, the minimization of time delay may be of great importance.

The canonical circuit configurations are not the only two-level realizations possible for a given switching function, and often they are not the ones using the least number of gates. If the economy of constructing a circuit is dictated by the number of gates needed, the design must simplify the circuit configuration within the constraints imposed by other demands as well. For example, the function

$$f = \overline{x + \bar{y}\bar{z}}$$

whose canonical two-level realization is shown in Fig. 7.6(a), may also be realized by the two-level circuit configurations shown in Figs. 7.6(b) and (c), using fewer number of gates. The configuration shown in Fig. 7.6(c) is derived from the algebraic transformation

$$f = \overline{x + \bar{y}\bar{z}} = \bar{x}(y + z)$$

based on the generalized DeMorgan's theorem (Section 7.6).

7.8
SIMPLIFICATION OF SWITCHING CIRCUITS

A switching function is first determined from the specifications of the design problem by methods discussed before in Chapter 5. Several economic and reliability considerations enter the design then, which constrain the logic designer in the choice of an "optimum" logical circuit configuration.

The various demands about the cost of circuit construction, about the safeguarding against certain failures during operation, or about the speed of response to input signals, may be conflicting. No general method exists then for reaching the "optimum" circuit design, and a compromise solution involving various trade-offs must be sought in each particular case. Methods for reaching an optimum design taking into account only a limited range of demands, such as minimization of time-delay and perhaps of some cost figure, are known.

The cost, reliability or other aspects in the logical design of a specific circuit must be judged with an overview of the host system as a whole. A particular single-output switching circuit is often part of a larger system, and the optimization of the circuit configuration should be carried out considering the circuit in its relations to the host system. Such is the case, for example, in the implementation of the switching functions $f_1, f_2, \ldots, f_5$, which were defined in Table 5.4 (Chapter 5) in connection with the design of a code-translator. This problem is discussed further in Chapter 8. Overall system optimization must be taken into account when several switching circuits are judiciously combined to form a larger switching system.

If no other constraints are known, the logic designer would aim for the simplest possible circuit configuration, that is, the one with the minimum number of gates and interconnections. It must be pointed out here, however, that present integrated-circuit technology could not care less about minimizing the number of gates on each silicon wafer, but it does indeed care about the number of interconnections between blocks belonging to different wafers.

With all these considerations in mind let us assume now that a logical circuit is to be constructed from individual electronic circuit components, and therefore that the cost is determined by the cost per input of each gate used. The circuit is to be constructed exclusively from AND, OR, and NOT gates. Let us also, for further reference, assume that the OR gates cost $2.00 per input, the AND gates cost $3.00 per input and the NOT circuits (inverters) cost $4.00 each. Circuit simplification is then sought, which results in minimum cost for circuit construction. In general, reduction of the complexity of the Boolean functions would lead to fewer blocks and inputs in the logical circuit configuration, and thus to more economical designs. Often this problem of designing the "best" possible switching circuit has two or more alternative solutions, all of which may be of about equal merit.

The field of "optimization" in the design of switching circuits is an important field being continuously researched. Algebraic procedures, graphical aids, and algorithms, all based on the conceptual framework of switching algebra, have been devised to facilitate the reduction of switching functions to optimum forms. Design problems of greater complexity are being handled today using digital computers to perform the tedious chores of repetitive calculations and tabulations.

In the following discussions we are primarily interested in demonstrating the use of algebraic manipulations and of mapping techniques in the simplification or optimization of Boolean functions. The choice between algebraic or mapping methods is up to the individual, but also depends on the complexity of the function at hand. The mapping methods have the advantage of being formalized to a greater extent, while the algebraic method depends greatly on the individual's ingenuity and experience.

Simplification achieved by algebraic manipulations make use of the theorems and the operational rules of switching algebra. The theorems and postulates we have discussed earlier in this chapter, especially DeMorgan's theorems, are of great use in these algebraic manipulations. It is thought best to illustrate simplification by algebraic manipulations with a few examples.

EXAMPLE ONE $\quad f = \overline{y\bar{z} + xy}$
$\quad\quad\quad\quad\quad\quad = \overline{y\bar{z}(\bar{x} + \bar{y})}$
$\quad\quad\quad\quad\quad\quad = \overline{\bar{x}y\bar{z} + y\bar{y}\bar{z}}$
$\quad\quad\quad\quad\quad\quad = \overline{\bar{x}y\bar{z}}$
$\quad\quad\quad\quad\quad\quad = x + \bar{y} + z.$

158 SWITCHING ALGEBRA

The student should verify that DeMorgan's generalized theorem (Section 7.6) and the identity $y\bar{y} = 0$, were used in the above algebraic simplification. The truth tables verifying the equivalence of these functions, and the circuit configurations are shown in Fig. 7.7.

The cost for implementing each of the three circuit configurations shown in Fig. 7.7 is, respectively,

Fig. 7.7(b): Cost = (2) (2) ($3) + (1) (2) ($2) + 3($4) = $28,
Fig. 7.7(c): Cost = (1) (3) ($3) + 2($4) = $17,
Fig. 7.7(d): Cost = (1) (3) ($2) + 2($4) = $14 (minimum cost).

x y z	xy	$y\bar{z}$	$\overline{y\bar{z}}$	$\overline{y\bar{z} + xy}$	f $y\bar{z} + xy$
0 0 0					
0 0 1					
0 1 0		1	0	0	1 → $f = \bar{x}y\bar{z}$
0 1 1					$f = x + \bar{y} + z$
1 0 0					(Canonical forms)
1 0 1					
1 1 0	1	1	0		
1 1 1	1				

(a) Truth table for $f = \overline{y\bar{z}} + xy$.

The complements of the truth values shown in each column of the truth tables are omitted for simplification of presentation.

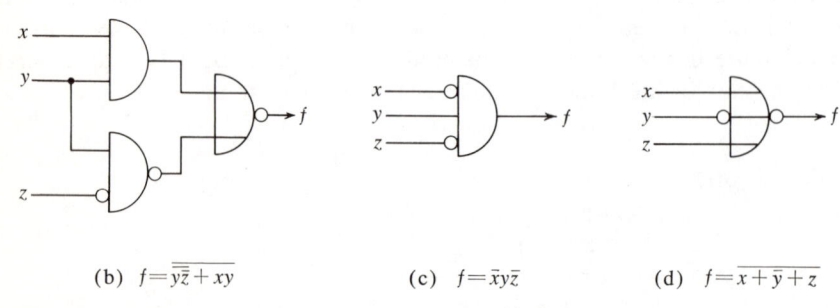

(b) $f = \overline{y\bar{z}} + xy$ (c) $f = \bar{x}y\bar{z}$ (d) $f = x + \bar{y} + z$

Fig. 7.7 Truth table and circuit configurations for Example one.

SIMPLIFICATION OF SWITCHING CIRCUITS 159

EXAMPLE TWO: In Section 5.6 the design of a logical circuit for use with the alarm system of an automobile was discussed. The algebraic simplification of the switching function derived in connection with that problem will now be demonstrated.

$$f = x\bar{z}\bar{w} + xy\bar{z}w + \bar{y}\bar{z} + xyz$$
$$= x\bar{z}\bar{w}(y + \bar{y}) + xy\bar{z}w + \bar{y}\bar{z} + xyz$$
$$= xy(\bar{z}\bar{w} + \bar{z}w + z) + x\bar{y}\bar{z}\bar{w} + \bar{y}\bar{z}$$
$$= xy[\bar{z}(w + \bar{w}) + z] + \bar{y}\bar{z}(1 + x\bar{w})$$
$$= xy + \bar{y}\bar{z}$$

(reduced from a four- to a three-variable function).

Notice that each one of the parentheses which resulted from factoring has a truth value 1. The circuit configurations are shown in Fig. 7.8. The costs are as follows:

Fig. 7.8(a): Cost = (2) (3) ($3) + (1) (2) ($3) + (1) (4) ($3)
$$+ (1) (4) (\$2) + (5) (\$4) = \$64,$$

Fig. 7.8(b): Cost = (2) (2) ($3) + (1) (2) ($2) + (2) ($4)
$$= \$24 \text{ (minimum cost).}$$

EXAMPLE THREE: $f = \overline{\bar{w} + z(\bar{x} + y)}$
$$= w(\bar{z} + x\bar{y}) \qquad (DeMorgan's \text{ theorem was used here})$$
$$= x\bar{y}w + \bar{z}w$$

The circuit configurations are shown in Fig. 7.9. The student should verify the equality of these functions by

$f = xyz + \bar{y}\bar{z} + xy\bar{z}w + x\bar{z}\bar{w}$

(a)

$f = xy + \bar{y}\bar{z}$

(b)

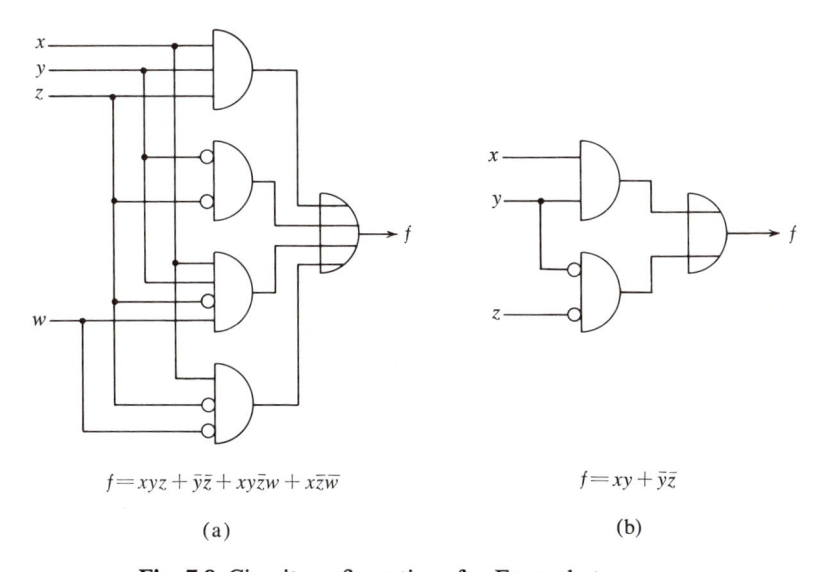

Fig. 7.8 Circuit configurations for Example two.

160 SWITCHING ALGEBRA

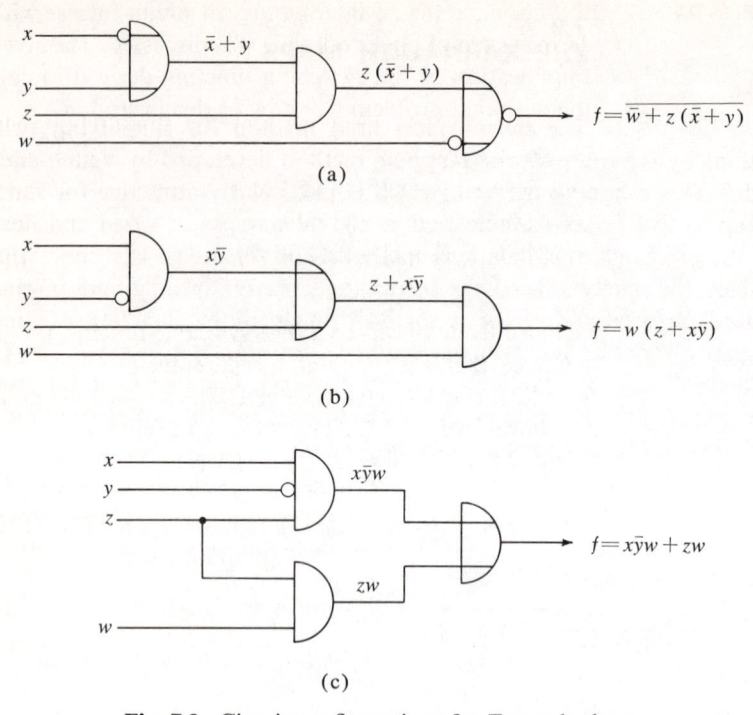

Fig. 7.9 Circuit configurations for Example three.

constructing the respective truth tables. The cost of the circuits are as follows:

Fig. 7.9(a): Cost = (1) (2) ($3) + (2) (2) ($2) + (3) ($4) = $26,

Fig. 7.9(b): Cost = (2) (2) ($3) + (1) (2) ($2) + (1) ($4) = $20,

Fig. 7.9(c): Cost = (1) (3) ($3) + (1) (2) ($3) + (1) (2) ($2) + (1) ($4) = $23.

Notice that the configuration shown in Fig. 7.9(b) corresponds to minimum cost. However, if minimization of time delay is important one may decide to accept a small additional cost and select the two-level circuit configuration shown in Fig. 7.9(c), over the three-level circuit of Fig. 7.9(b). Simplification of algebraic functions may also be achieved by graphical techniques. Venn diagrams may be used for example to facilitate or supplement algebraic simplification. However, by far the most effective graphical technique employed for simplifying logical functions is a mapping method making use of the Karnaugh maps, described in the next section.

7.9
THE KARNAUGH MAPS

The most widely used method for simplifying switching functions by far is a pictorial mapping method developed by Veitch and Karnaugh.[6] This mapping method, which is particularly attractive for functions with up to five or six variables, offers the advantages of speed and flexibility over the algebraic manipulations method. For functions with more than six variables the method becomes increasingly more difficult and impractical. A mixed pictorial and algebraic method which allows handling of functions with many variables has been devised by Quine and McCluskey.[7] The Quine-McCluskey method, although simple in principle, is rather long. It is based on first expanding the function in a canonical form, and then combining the terms judiciously so that a simplified form may result.

In this text we will be concerned primarily with functions of five or fewer variables which comprise the majority of those found in practical logic design problems. Therefore we will concentrate our attention on the discussion of the Karnaugh mapping method alone.

In a way, the Karnaugh maps are a form of truth tables. Each map has a number of small squares, called "cells." The number of cells is equal to 2^n, where n is the number of variables in the switching function. The cells are laid out in a rectangular configuration of 2×2, 2×4, and 4×4 cells, for two, three and four variables, respectively.

Divided in different ways to take care of all the variables, half of the map (with 2^{n-1} cells) is assigned to one of the variables, while the other half is assigned to its complement, as shown in Fig. 7.10. For example, the two-variable Karnaugh map may be divided as shown in Fig. 7.10(a) in two equal vertical slices, each containing two cells, to take care of the X-variable and its complement. The map is then divided into two equal horizontal slices, with two cells per slice, to take care of the Y-variable and its complement. The division of the map for three and four variables is shown in Figs. 7.10(b) and (c), respectively. Notice that in the case of the three- and four-variable maps the variables z and z and w, respectively, are accommodated by having the uncomplemented variable occupy a center half-slice of the map, while the complemented variable occupies the two quarter-slices on both sides of the center half-slice. This is done so as to avoid the ambiguity which would result if any two variables are assigned identically the same half portions of the map. The student may easily

---

[6] Veitch, E. W., "A Chart Method for Simplifying Truth Functions," *Proc. Pittsburgh Assoc. Com. Mach.*, 1952, pp. 127–133; Karnaugh, M., "The Map Method for Synthesis of Combinational Logic Circuits," *Trans. AIEE, Part I, Commun. & Elect.*, vol. 72, Nov., 1953, pp. 593–598.

[7] McCluskey, E. J., Jr., "Minimization of Boolean Functions," *Bell Syst. Tech. J.*, vol. 35, Quine, W. V., "A Way to Simplify Truth Functions," *Am. Math. Monthly*, vol. 62, Nov., 1955,pp. 627–631; Nov., 1956, pp. 1–28.

162 SWITCHING ALGEBRA

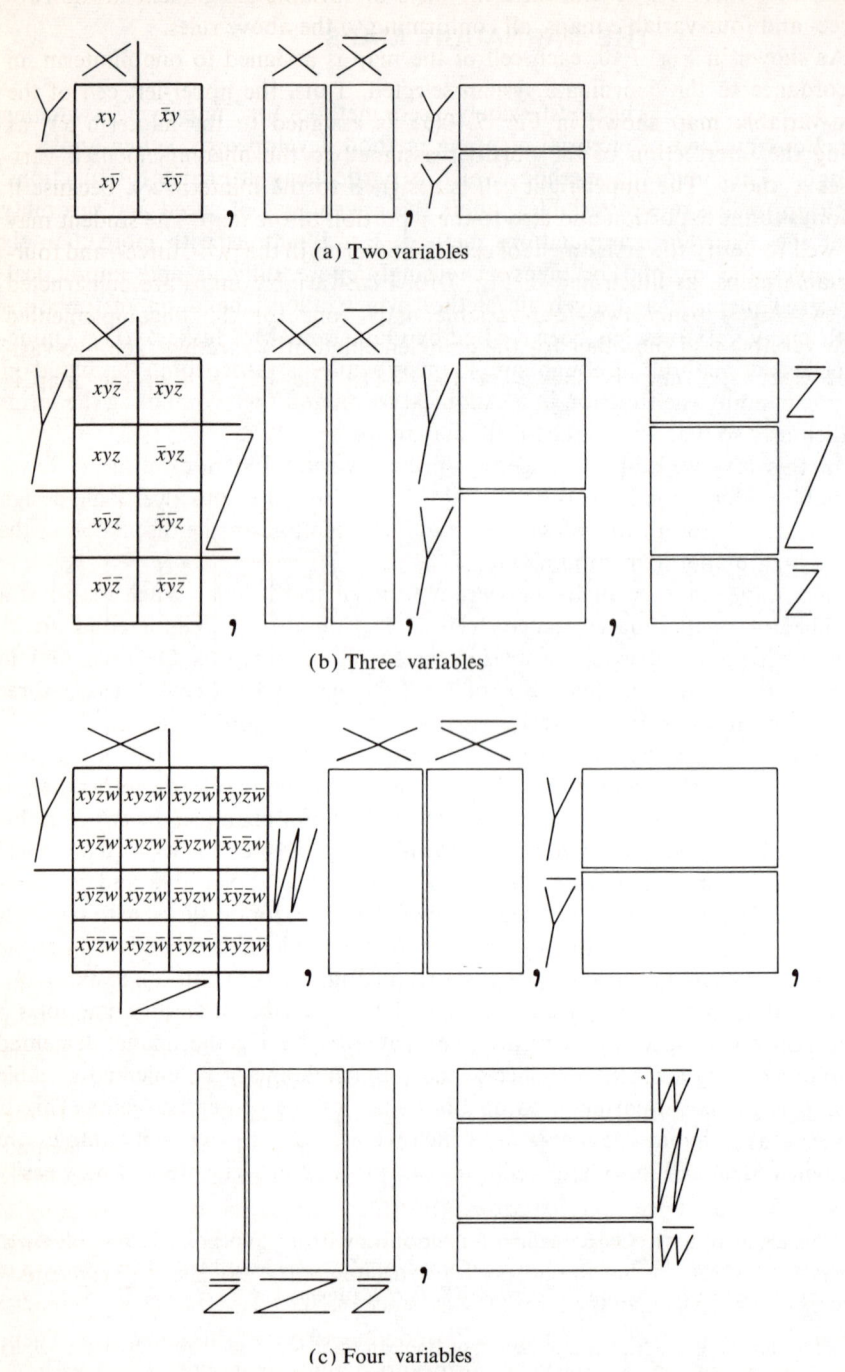

(a) Two variables

(b) Three variables

(c) Four variables

Fig. 7.10

verify that there are several different ways of variable assignment in the two- three- and four-variable maps, all conforming to the above rules.

As shown in Fig. 7.10, each cell of the map is assigned to one minterm, in accordance to the coordinate system selected. Thus, the upper-left cell of the two-variable map shown in Fig. 7.10(a), is assigned to the minterm xy, as being the intersection of the portions assigned to the uncomplemented variables x and y. The upper-right cell is assigned to the minterm $\bar{x}y$, because it belongs to the $\bar{x}$-portion and also to the y-portion of the map. The student may do well to verify the assignment of cells to minterms in the two-, three-, and four-variable maps, as illustrated in Fig. 7.10. Five-variable maps are constructed in two levels using two four-variable maps, one for the uncomplemented fifth variable and the other for the complemented fifth variable. The five-variable Karnaugh map is shown in Fig. 7.11. The left half of the map is

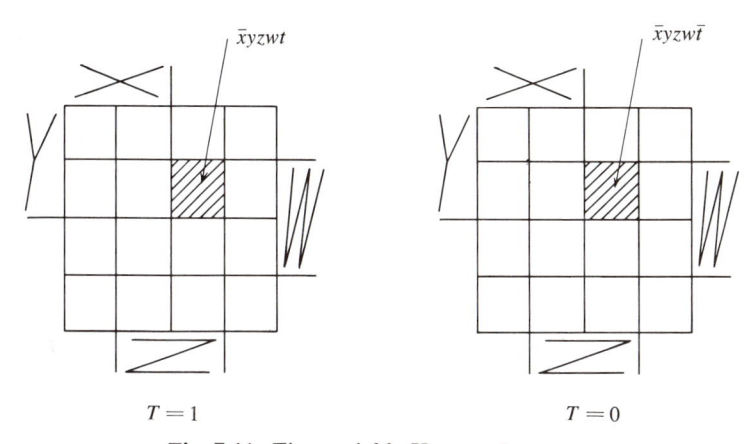

Fig. 7.11 Five-variable Karnaugh map.

assigned to the uncomplemented fifth variable, that is, to $t = 1$, while the right half of the map is assigned to $t = 0$. Thus, the minterm $\bar{x}yzwt$ is assigned to a cell in the left half as indicated in the figure. The minterm $\bar{x}yzw\bar{t}$ is assigned to the corresponding cell in the right half of the map.

A six-variable map is made up of two five-variable maps, one assigned to the uncomplemented sixth variable ($q = 1$) and the other to the complemented sixth variable ($q = 0$). A six-variable map is shown in Fig. 7.12, together with examples of minterm representations. In order to find the cell representing a given minterm, it is best to proceed beginning with the sixth variable and then the fifth variable, so that the proper four-variable portion of the map is located. For example, let us map the term $\bar{x}\bar{y}z\bar{w}\bar{t}q$. Since the sixth variable, q, is uncomplemented, the upper half of the map, corresponding to $q = 1$, should be used. Then, we observe that the fifth variable, t, appears complemented. The four-variable

164 SWITCHING ALGEBRA

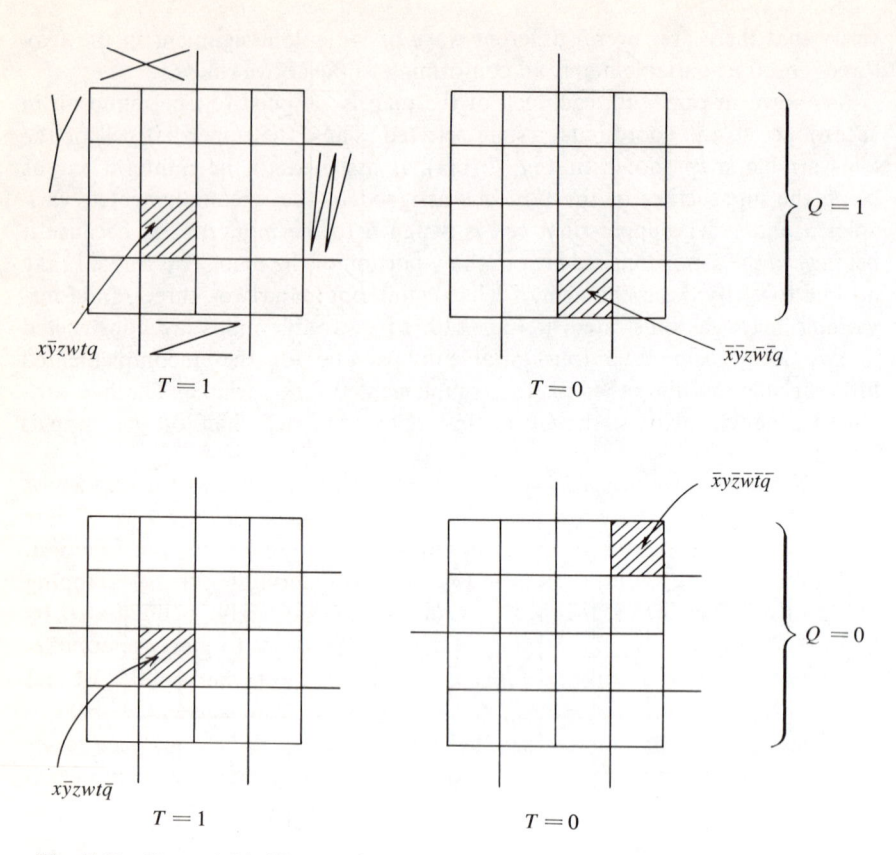

Fig. 7.12 Six-variable Karnaugh map with examples of minterm representations.

portion of the upper half, which corresponds to $t = 0$, is then chosen. Finally, in this four-variable portion of the map we locate the cell which corresponds to the remaining combination $\bar{x}\bar{y}z\bar{w}$, as illustrated in Fig. 7.12. Other examples are also illustrated.

7.10
THE MAPPING METHOD

The mapping of a Boolean function into a Karnaugh map is based on the following three simple rules:

1. "*Each single variable in its uncomplemented or complemented form* (for example, x or $\bar{x}$), *is assigned a half-portion of the area of the map.*" This rule, which has been mentioned before in Section 7.9, is illustrated in Fig. 7.13 by mapping the single variable z and $\bar{z}$ in a three-variable map. Each n-variable map is subdivided in n different ways in order to accommodate all the uncomplemented and complemented variables.

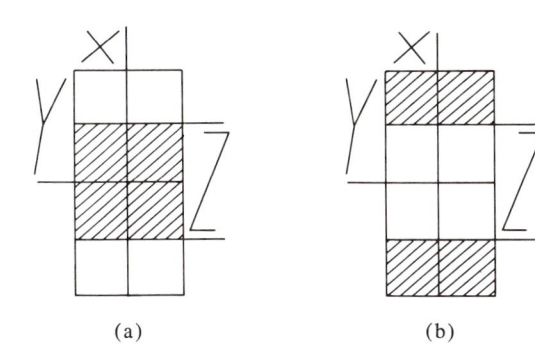

Fig. 7.13 Mapping of single variables.

2. "*In order to map a logical product, we determine the area of the map which is common to all the areas assigned to the components of the product.*" For example, if the product $x\bar{z}$ is part of a four-variable switching function, it may be mapped in a four-variable map by determining the overlapping portion of the areas assigned to x and to $\bar{z}$, respectively, as illustrated by the shaded area in Fig. 7.14(a). Notice that in order to map the complement of the above product, that is, $\overline{x\bar{z}}$, we map first the product $x\bar{z}$ and then take the complement, as shown also in Fig. 7.14(a) (shaded area).

3. "*In order to map a logical sum, we determine the total area assigned to the components of the sum.*" For example, in a four-variable map the logical sum $x + \bar{z}$ is mapped by determining the total area occupied by x or $\bar{z}$, as shown in Fig. 7.14(b). The complement of the above sum, that is, $\overline{x + \bar{z}}$, is again mapped by determining the complement of the area which corresponds to the sum $x + \bar{z}$, as shown in Fig. 7.14(b).

By applying these rules as needed, we may map a switching function of any complexity. In order to illustrate the mapping of functions let us consider the function

$$f = x\bar{y}(\overline{\bar{x} + z}).$$

This is a three-variable function and a three-variable map should be used. In sequence we may map $\bar{x} + z$, $\overline{\bar{x} + z}$, $x\bar{y}$, and finally the product $x\bar{y}(\overline{\bar{x} + z})$, as shown in Fig. 7.15(a).

It is of interest to notice here that the given function and the simple logical product $x\bar{y}\bar{z}$ map identically into the same area of the map. Thus, they are equivalent and $x\bar{y}\bar{z}$ is the simplified form of the given function f. This equivalence may also be proven by algebraic manipulations as follows:

$$\begin{aligned} f &= x\bar{y}(\overline{\bar{x} + z}) \\ &= x\bar{y}(x\bar{z}) \quad (DeMorgan's\ theorem\ was\ used\ here) \\ &= x\bar{y}\bar{z}. \end{aligned}$$

166 SWITCHING ALGEBRA

(a)

(b)

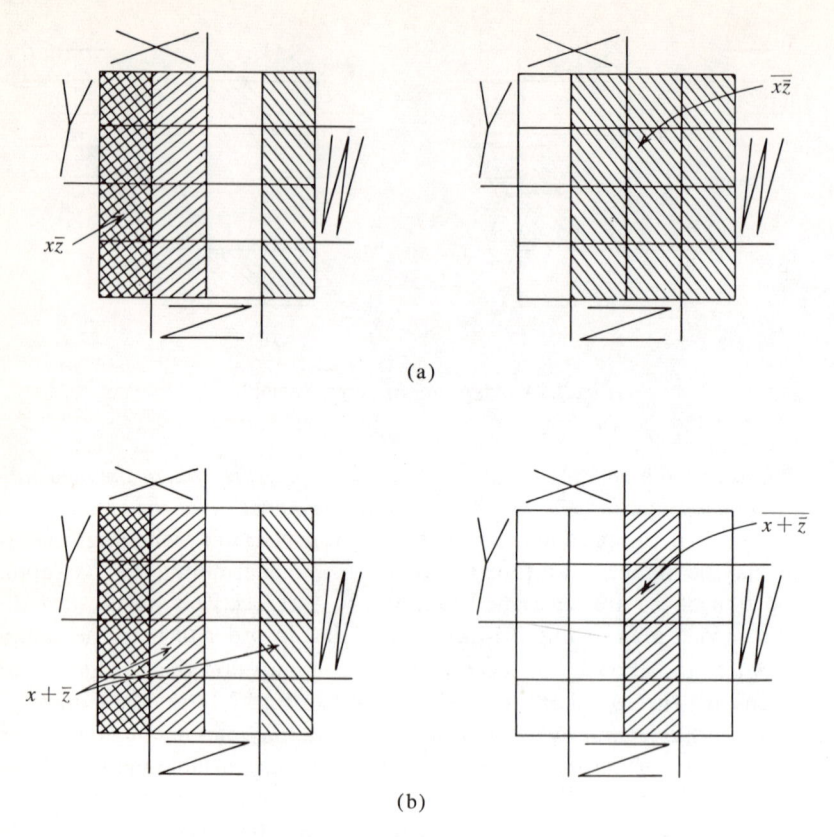

Fig. 7.14 Mapping of products and sums.

The fact that the function f has been mapped into the single cell [Fig. 7.15(a)], which is identified by the product $x\bar{y}\bar{z}$, is used to simplify f. This outstanding property of the Karnaugh maps will be discussed further in this chapter.

As another example, consider the function

$$f = x\bar{z} + \bar{x}y\bar{w} + xyzw.$$

This is a four-variable function, whose mapping is illustrated in Fig. 7.15(b).

The mapping of a switching function is accomplished by entering information in the cells of the map. With a little practice this may be done in one step. In a way, the map is used like a truth table. The cells which are marked (shaded) during the mapping process correspond to standard combinations of the variables (their coordinates), which in turn give to the function a truth value of 1. For example, in the case of the function mapped in Fig. 7.15(b), we may read the standard combination (minterm) of the variables for each shaded cell, beginning at the top of the map and moving from left to right. We have then:

$$xy\bar{z}\bar{w}, \; \bar{x}yz\bar{w}, \; \bar{x}y\bar{z}\bar{w}, \; xy\bar{z}w, \; xyzw, \; x\bar{y}\bar{z}w, \; x\bar{y}\bar{z}\bar{w}.$$

(a) $\bar{x}+z$ $\bar{x}+z$ $x\bar{y}$ $x\bar{y}(\bar{x}+z)$

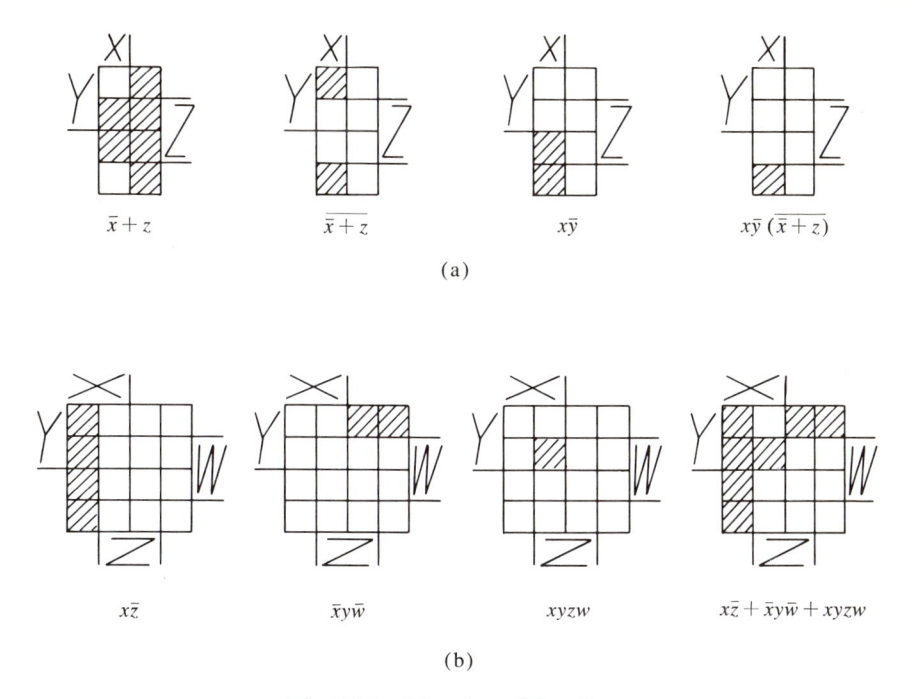

(b) $x\bar{z}$ $\bar{x}y\bar{w}$ $xyzw$ $x\bar{z}+\bar{x}y\bar{w}+xyzw$

Fig. 7.15 Mapping of functions.

Since each of the above combinations provides the function with a truth value of 1, the function may be put in its canonical form by forming the logical sum of all the above minterms. The complement of this function may also be written in its canonical form by forming the sum of the minterms corresponding to the blank cells of the map. Obviously then $f+\bar{f}=1$.

In the following, we shall discuss the demapping of a function. By using some very simple rules, a mapped function may be read off the map in its most simplified form.

7.11 ADJACENT CELLS

The special assignment of variables to map regions (Fig. 7.10) makes possible special groupings of "adjacent" cells, which in turn are used in the simplification of a mapped switching function. The definition of adjacent cells is based on the following rule: "Only one variable can change value between two adjacent cells." Thus, any two cells side by side are adjacent, as may be verified by the cell assignments shown in Fig. 7.10. Two diagonally-touching cells are not adjacent.

It should be noted here that in an n-variable Karnaugh map, an n-variable standard product, that is, a minterm, is mapped into a single cell. An $(n-1)$-variable product is mapped into the area of two adjacent cells, in a 2×1 cell arrangement. An $(n-2)$-variable product is mapped into a cluster of four adjacent cells in a 4×1 or a 2×2 cell arrangement. An $(n-3)$-variable product is mapped in a 2×4 cluster of adjacent cells. These observations, which hold true for maps with up to four variables, are illustrated in Fig. 7.16 on a four-variable map.

Notice that mapping the product $xy\bar{z}$ in the 2×1 cluster (Fig. 7.16) corresponds to the operation of mapping $xy\bar{z}w$ OR $xy\bar{z}\bar{w}$, according to the equation

$$xy\bar{z}w + xy\bar{z}\bar{w} = xy\bar{z}.$$

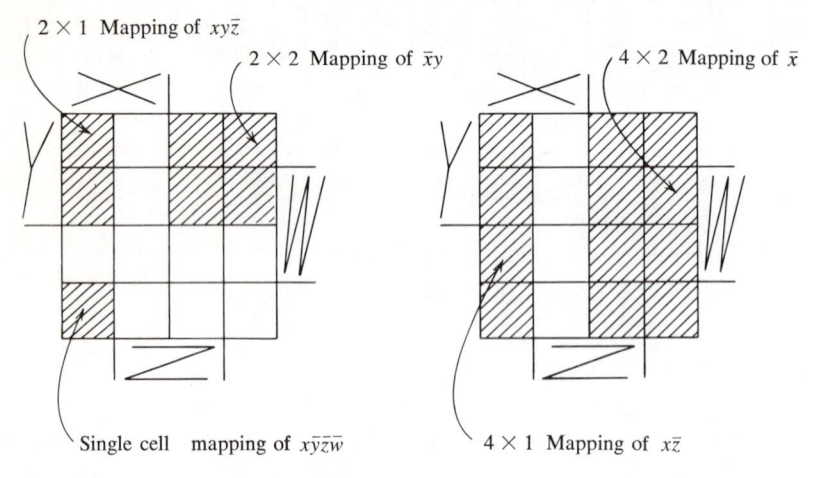

Fig. 7.16 Illustrations on a four-variable map of the following: single cell mapping of a four-variable product, 2×1 cluster mapping of a three-variable product, 2×2 and 4×1 cluster mapping of two-variable products, and 4×2 cluster mapping of a single variable.

Similarly, mapping $\bar{x}y$ in a 2×2 cluster (Fig. 7.16) corresponds to the operation of mapping $\bar{x}yzw$ OR $\bar{x}yz\bar{w}$ OR $\bar{x}y\bar{z}w$ OR $\bar{x}y\bar{z}\bar{w}$ in four single adjacent cells, or alternatively, it corresponds to the operation of mapping the three-variable products $\bar{x}yz$ OR $\bar{x}y\bar{z}$ in two 2×1 adjacent clusters in accordance to the equations

$$\bar{x}yzw + \bar{x}yz\bar{w} + \bar{x}y\bar{z}w + \bar{x}y\bar{z}\bar{w}$$
$$= \bar{x}yz + \bar{x}y\bar{z}$$
$$= \bar{x}y.$$

It is important to observe that opposite edge-cells are also adjacent. If the map is folded to form a cylinder, then opposite edge-cells become adjacent. Adjacent edge-cells are illustrated in Fig. 7.17. Notice, for example, that a pair of two-edge cells forms a 2×1 cluster, which maps an $(n-1)$-variable product;

ADJACENT CELLS   169

Mapping of $y\bar{z}w$

Mapping of $x\bar{z}$

Mapping of $\bar{x}z\bar{w}$ Mapping of $\bar{y}z$

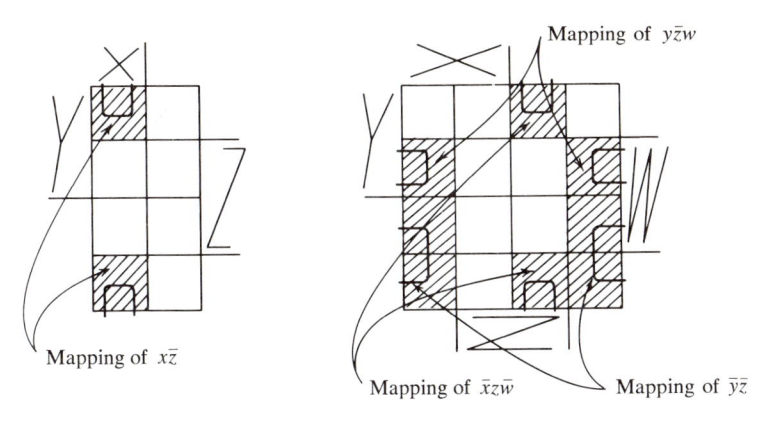

Fig. 7.17 Illustrations of edge adjacent cells.

that is, a two-variable product in the three-variable map on the left, and a three-variable product in the four-variable map on the right. The cluster of 2×2 edge-adjacent cells shown maps the two-variable product $\bar{y}\bar{z}$, that is, an $(n-2)$-variable product.

Corresponding single cells in the different levels of five- and six-variable maps may also be adjacent cells. Figure 7.18 shows a pair of adjacent cells in

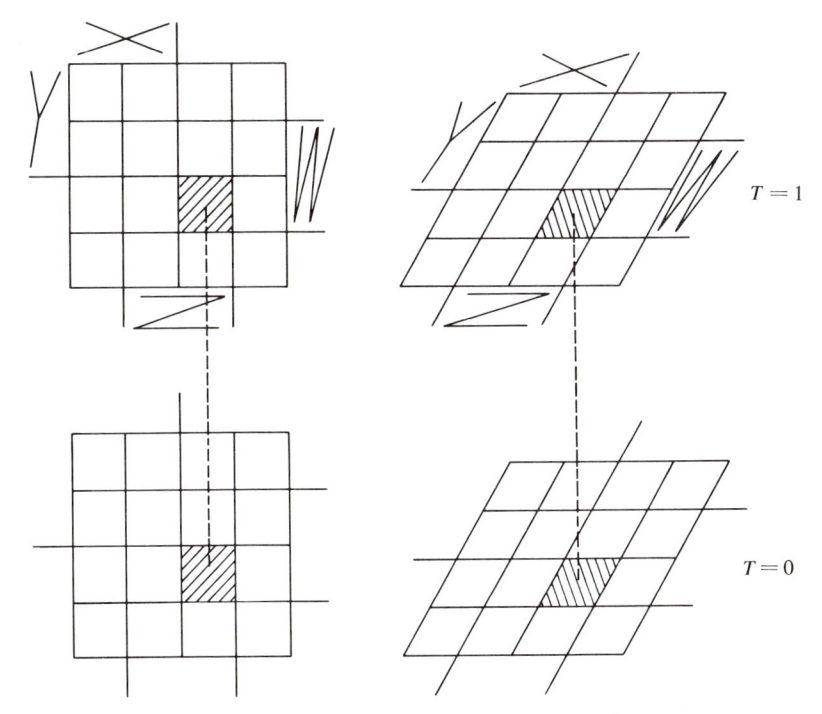

$T = 1$

$T = 0$

Fig. 7.18 Mapping of $\bar{x}\bar{y}zw$ in a five-variable map.

170 SWITCHING ALGEBRA

a five-variable map. The cluster of these two adjacent cells corresponds to the mapping of the $(n-1)$-variable product $\bar{x}\bar{y}zw$, in accordance to the equation

$$\bar{x}\bar{y}zwt + \bar{x}\bar{y}zw\bar{t} = \bar{x}\bar{y}zw.$$

A pair of 2×2 adjacent cell clusters mapping the $(n-2)$-variable products $y\bar{z}t$ and $\bar{y}z\bar{w}$ in a five-variable map are shown in Fig. 7.19.

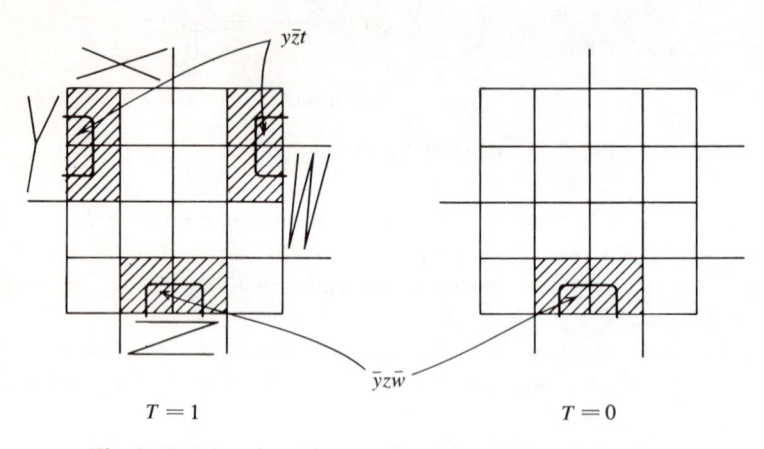

Fig. 7.19 Mapping of $y\bar{z}t$ and $\bar{y}z\bar{w}$ in a five-variable map.

Theoretically speaking this mapping method may be extended to any number of variables; however, the usefulness of the method in cases of more than five variables depends heavily on the ability of the individual to recognize patterns of adjacent cells. Considerable practice is required in order to obtain proficiency with maps of more than five variables. To facilitate the recognition of adjacent cells in a six-variable map, the map is arranged as shown in Fig. 7.20. Clusters of adjacent cells in any of the four-variable portions of the map are easily recognizable, as illustrated by the mapping of the product $\bar{z}w\bar{t}q$. However, clusters of corresponding cells in different four-variable portions of the map are more difficult to recognize. As we see in Fig. 7.20, diagonally-corresponding cells are nonadjacent, because they differ in two variables. Vertically- or horizontally-corresponding cells are adjacent.

7.12
DEMAPPING

The process of mapping a switching function by shading the appropriate cells was discussed in Section 7.10. A logical sum of two vari-

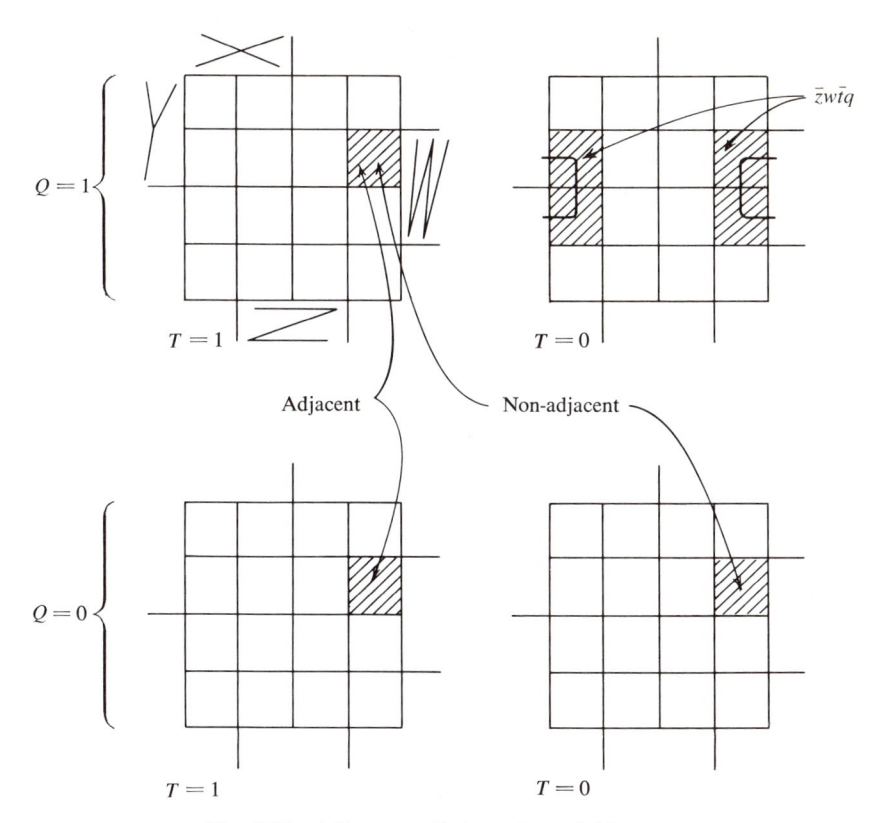

Fig. 7.20 Adjacent cells in a six-variable map.

either of the variables. A logical product is mapped by shading only the area which is common to the areas assigned to the components of the product. A switching function is mapped by mapping the variables and the operations as they appear in the function.

After a switching function has been mapped, the stage is set for the demapping process. If each shaded cell is realized in algebraic form (demapped) separately, then the canonic form of "sum of products" should result because each single cell is realized by a standard product (minterm). However, if the most simplified algebraic version of the function is sought, then the special patterns of clustered adjacent cells, such as the 2 × 1, 2 × 2, 4 × 1, 4 × 2 clusters, must be recognized and realized. Such patterns of adjacent cells are demapped into simplified products by the elimination of variables.

The process of selecting groupings of adjacent cells must continue until all shaded cells have been included *at least once*. Since we must always seek to realize the largest possible clusters of adjacent cells among the acceptable combinations (2 × 1, 2 × 2, 4 × 1, 4 × 2), we may expect to include any one of the shaded cells in any number of groupings. Algebraically the use of the same cell over and over in different groups corresponds to repetitively adding the same minterm in the corresponding process of algebraic simplification of the function. For example, in the demapping shown in Fig. 7.21, several cells have been realized twice. The double use of the cell marked with * corresponds to the addition

$$\bar{x}yzw + \bar{x}yzw = \bar{x}yzw$$

applied in the equivalent algebraic process of simplification, which corresponds to the graphical demapping procedure shown in Fig. 7.21.

If a cell does not fit in any appropriate grouping, then it must be demapped alone as a fundamental product (minterm). A switching function is demapped in the form of a logical sum of products, and it may be implemented by a set of parallel AND gates feeding a second-level OR-mixer. With experience gained with a little practice one may use Karnaugh maps to simplify switching functions with up to six variables, as will be illustrated with examples in the following section.

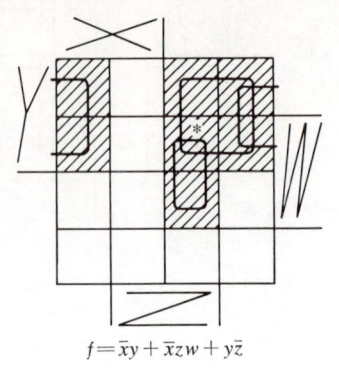

$$f = \bar{x}y + \bar{x}zw + y\bar{z}$$

Fig. 7.21 Repetitive realization of cells in demapping.

7.13
EXAMPLES OF KARNAUGH MAP SIMPLIFICATIONS

EXAMPLE ONE   Consider simplifying the function

$$f = \overline{(\bar{x} + \bar{y})(\bar{x} + y)}.$$

Figure 7.22 shows the steps of mapping and demapping the function. It is shown that $f = x$. This fact may also be verified by algebraic manipulations as follows:

$$\begin{aligned}f = \overline{(\bar{x} + \bar{y})(\bar{x} + y)} &= \overline{\bar{x} + \bar{x}\bar{y} + \bar{x}y + y\bar{y}} \\ &= \overline{\bar{x} + \bar{x}(\bar{y} + y) + 0} \\ &= \overline{\bar{x}} \\ &= x\end{aligned}$$

or, more simply, by applying DeMorgan's theorem:

$$f = \overline{(\bar{x} + \bar{y})(\bar{x} + y)} = xy + x\bar{y} = x$$

EXAMPLE TWO   A switching function is provided by its truth table in Fig. 7.23. The two canonical forms of realization are also shown.

EXAMPLES OF KARNAUGH MAP SIMPLIFICATIONS   173

$\bar{x}+\bar{y}$ $\bar{x}+y$ $(\bar{x}+\bar{y})(\bar{x}+y)$ $\overline{(\bar{x}+\bar{y})(\bar{x}+y)}=x$

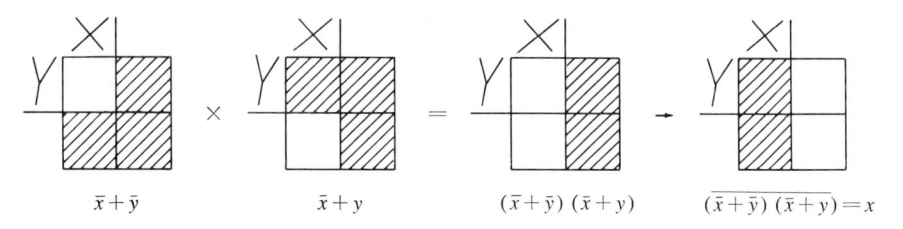

Fig. 7.22 Simplification of $f=\overline{(\bar{x}+\bar{y})(\bar{x}+y)}$ (Example one).

x	y	z	f
0	0	0	1
0	0	1	0
0	1	0	1
0	1	1	1
1	0	0	1
1	0	1	0
1	1	0	1
1	1	1	1

Truth table

$f = xyz + xy\bar{z} + \bar{x}yz + \bar{x}y\bar{z} + x\bar{y}\bar{z} + \bar{x}\bar{y}\bar{z}$

(Canonical form realization)

$f = (x+y+\bar{z})(\bar{x}+y+\bar{z})$

(Canonical form realization)

$f = y + \bar{z}$ (Minimal realization)

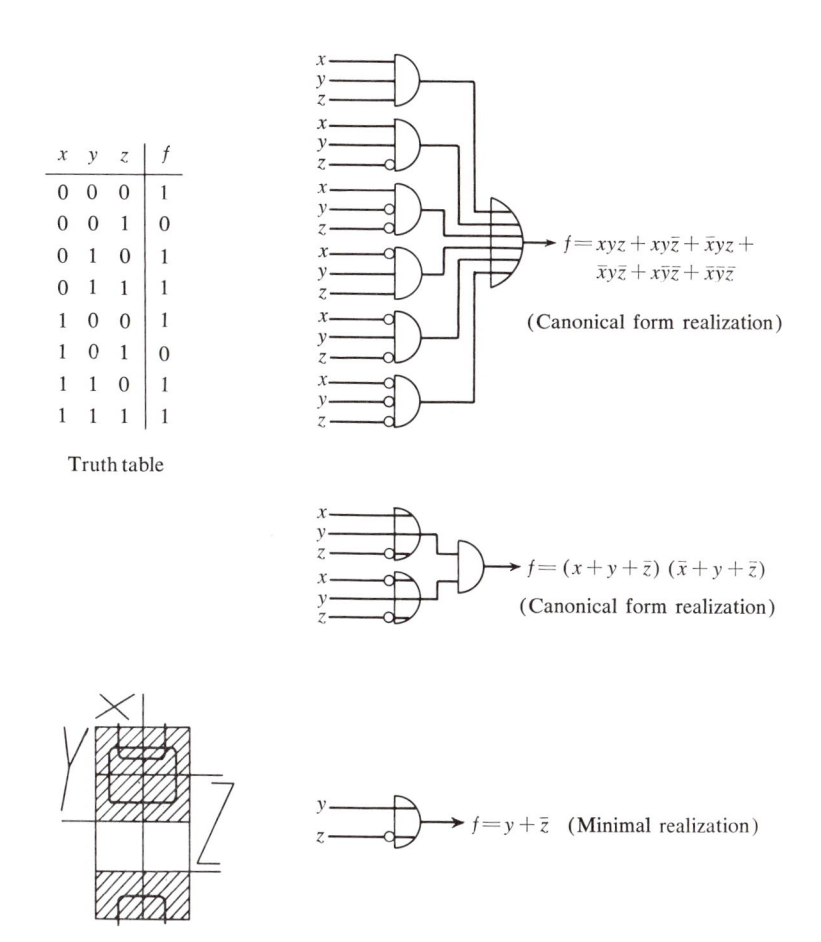

Fig. 7.23 Canonical forms and minimal realization using a Karnaugh map (Example two).

The function is then mapped and simplified with the help of a Karnaugh map. The minimal realization, which uses only a two-input OR gate and an inverter, is also shown in Fig. 7.23. Notice that in its simplified form f is not any more a function of the variable x. The simplification is also demonstrated below using algebraic manipulations:

$$\begin{aligned} f &= xyz + xy\bar{z} + \bar{x}yz + \bar{x}y\bar{z} + x\bar{y}\bar{z} + \bar{x}\bar{y}\bar{z} \\ &= xy(z + \bar{z}) + \bar{x}y(z + \bar{z}) + \bar{y}\bar{z}(x + \bar{x}) \\ &= y(x + \bar{x}) + \bar{y}\bar{z} \\ &= y + \bar{y}\bar{z} \\ &= y(1 + \bar{z}) + \bar{y}\bar{z} \\ &= y + \bar{z}(y + \bar{y}) \\ &= y + \bar{z}. \end{aligned}$$

EXAMPLE THREE   Assume that you are the quality engineer in a company that manufactures transistors. A quality test is devised so that a random sample made of three transistors is tested at periodic intervals of time. Four tests, A, B, C, and D, are performed simultaneously on each transistor in the sample. If three or more test-failures are found in any one transistor, the unit is rejected. If two or more defective units are found in any one sample, the production line becomes suspected and it is to be stopped until a larger sample is tested and the production faults are located. You are asked to automate the quality test by designing the appropriate logical circuits. It is assumed that all tests are made simultaneously.

Let x, y, z, and w represent the binary electrical signals, which are generated, respectively, by the tests A, B, C, and D. A test failure is signified by a 1, so that if $x = 1$ appears at the output of the A-test, it will signify a test failure in A.

Since three or more test-failures should result in rejecting the unit, the truth table for testing each unit in the sample is as shown in Fig. 7.24(a). The output functions f_i ($i = 1, 2, 3$) from testing the individual units (that is f_1, f_2, f_3) become 1 if three or more of the inputs are 1. The map minimization and circuit realization of f_i, are shown in Fig. 7.24(b).

The outputs from testing the individual units, that is, f_1, f_2, f_3, are now applied as inputs to another logical circuit, whose output F should be 1 if two or more of the inputs are 1. The value $F = 1$ will issue the alarm signal which will stop the production line.

x	y	z	w	f_i
0	0	0	0	0
0	0	0	1	0
0	0	1	0	0
0	0	1	1	0
0	1	0	0	0
0	1	0	1	0
0	1	1	0	0
0	1	1	1	1
1	0	0	0	0
1	0	0	1	0
1	0	1	0	0
1	0	1	1	1
1	1	0	0	0
1	1	0	1	1
1	1	1	0	1
1	1	1	1	1

(a)

(b) $f_i = xyz + yzw + zwx + wxy$

f_1	f_2	f_3	F
0	0	0	0
0	0	1	0
0	1	0	0
0	1	1	1
1	0	0	0
1	0	1	1
1	1	0	1
1	1	1	1

(c)

(d) $F = f_1 f_2 + f_2 f_3 + f_1 f_3 = f_1 f_2 + f_3 (f_1 + f_2)$

(e)

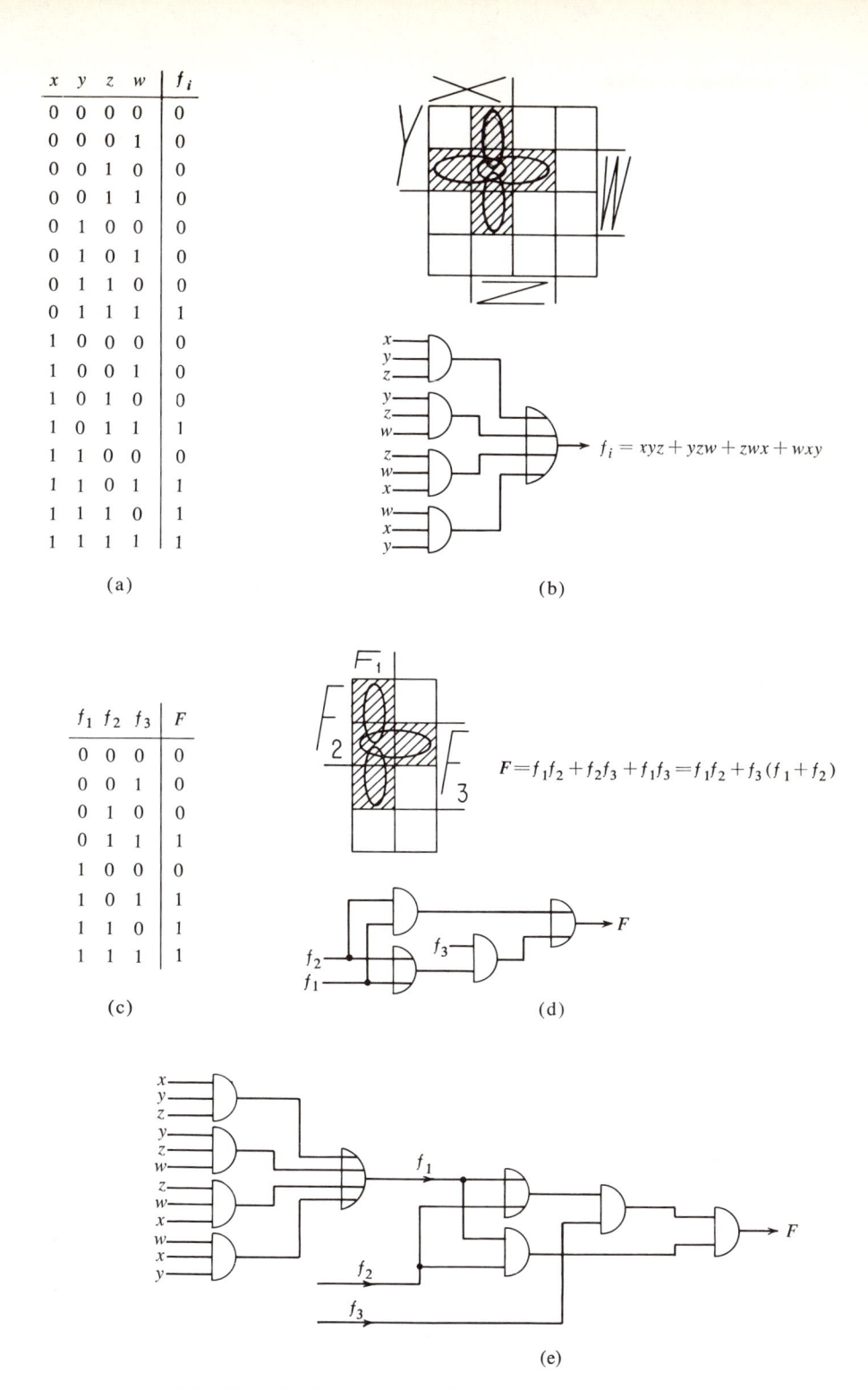

Fig. 7.24 Quality control logic circuit design for Example three.

The truth table for F is shown in Fig. 7.24(c), while the circuit realization is shown in Fig. 7.24(d). The overall quality test logical circuit is composed of three parallel circuits like the one shown in Fig. 7.24(b), all feeding the output circuit of Fig. 7.24(d), as shown in Fig. 7.24(e).

Notice that in deriving the optimum function from the Karnaugh map of Fig. 7.24(d), the expression $F = f_1 f_2 + f_3(f_1 + f_2)$ was preferred over the expression $F = f_1 f_2 + f_2 f_3 + f_1 f_3$. This was done because we have assumed previously (Section 7.8) that the cost of realizing OR gates is lower than that of AND gates. Recalling that we have assumed a cost of \$3 per input for an AND gate and \$2 per input for an OR gate, we find that the $F = f_1 f_2 + f_2 f_3 + f_1 f_3$ realization (three parallel two-input AND gates feeding a three-input OR gate), would cost $3(2)(\$3) + 1(3)(\$2) = \$24$. The $F = f_1 f_2 + f_3(f_1 + f_2)$ realization, which is shown in Fig. 7.24(d), would cost $2(2)(\$3) + 2(2)(\$2) = \$20$.

Karnaugh maps may also be used to verify algebraic equations. It will be left up to the student as an exercise to construct the two-variable map which verifies the equation

$$\overline{x + \bar{y}} = \bar{x} y,$$

thus demonstrating the truth of DeMorgan's theorem for two variables.

7.14
BOOLEAN VERSUS ORDINARY ALGEBRA

A few brief comments may be timely here in order to delineate some of the most characteristic dissimilarities between the ordinary high-school algebra which is used in a great variety of problems and the rather specialized brand of algebra we have discussed in this chapter, which is used in conjunction with binary symbolic logic. The name "ordinary" is coined here for the high-school algebra only to distinguish this algebra from all other "modern" algebras, such as the "algebra of rings," the "algebra of lattices," and so forth.

To begin with, in its most elementary operations, ordinary algebra uses *numbers*, including fractions, roots, powers, and other similar concepts, all of which are based on the fundamental concept of "number." Symbols such as letters, which also stand for numbers, are used widely. In addition, ordinary algebra also deals with mathematical operations, such as the arithmetic ones, operations of differential calculus, and so on. Concepts like "variable," "unknown," "equality between two quantities," and so forth, are common in ordinary algebra.

Boolean algebra is quite similar to ordinary algebra in many ways, since it also represents another form of algebra. However, it also differs in several ways. Boolean algebra applies to classes of objects, to statements, and to conditions. It deals with "logical operations" on classes or statements, such as the logical transformations which were defined previously in the text. The elementary logical transformations OR, AND, NOT, and others like these, may be applied to the calculus of "classes" or they may be used as logical connectives when operating on "propositions" or "statements." Boolean algebra also uses the concept of "variable," together with the concept of the "value" of a variable, but only to denote the *condition* in which the variable exists rather than its quantitative properties.

A substantial difference to be noted is also found in the distributivity properties. In Boolean algebra the operations " + " and " · " are distributive over each other, that is,

$$x + (y \cdot z) = (x + y) \cdot (x + z) \quad \text{distributivity of " + " over " · "}$$
$$x \cdot (y + z) = x \cdot y + x \cdot z \quad \text{distributivity of " · " over " + "}$$

not true in ordinary algebra.

Care should be exercised when identical symbols are used for arithmetic operations in ordinary algebra and for logical transformations in Boolean algebra, in order to avoid confusion. Thus, in ordinary algebra, we have $a + a = 2a$, whatever number the symbol a may represent. In Boolean algebra we have $a + a = a$, whatever "class" or "statement" the binary "variable" a may represent. Also, in ordinary algebra we have $a \cdot a = a^2$, while in Boolean algebra $a \cdot a = a$. Notice also that binary *arithmetic* is a form of ordinary arithmetic, based on the radix "two," so that $1 + 1 = 10$, while in Boolean algebra $1 + 1 = 1$.

To illustrate a case where the use of the switching algebra provides a simple solution while ordinary algebra does not, consider the following case: "We are requested to construct the electrical circuit for a three-way switch to operate the light bulb in the hallway of a house." The light may be switched on or off from either end of the hall. The four possible combinations of values for the two light switches are related to the values of the light bulb, L, in the following manner:

S_1	S_2	L
0	0	0
0	1	1
1	0	1
1	1	0

178 SWITCHING ALBEGRA

The light bulb will light ($L = 1$) only when either one, but *not* both, of the switches is "up," that is, has a value of 1 ($S_1 = 1$, $S_2 = 0$, *or* $S_1 = 0$, $S_2 = 1$). The above table shows that L is an EXCLUSIVE-OR function of S_1 and S_2. Symbolically,

$$L = S_1 \bar{S}_2 + \bar{S}_2 S_2$$

(see function f_6 in Table 7.3).

The exclusive-OR relation suggests that two pairs of series switches (AND gates) be connected in parallel(OR-ed). However, corresponding switches in the two parallel pairs should operate interdependently, so that when one switch is closed its parallel will be open, as shown in Fig. 7.25(a) for the pairs S_1, $\bar{S}_1$ and S_2, $\bar{S}_2$. The simplest circuit which would implement this switching operation consists of two single-pole double-throw switches connected in parallel, as shown in Fig. 7.25(b).

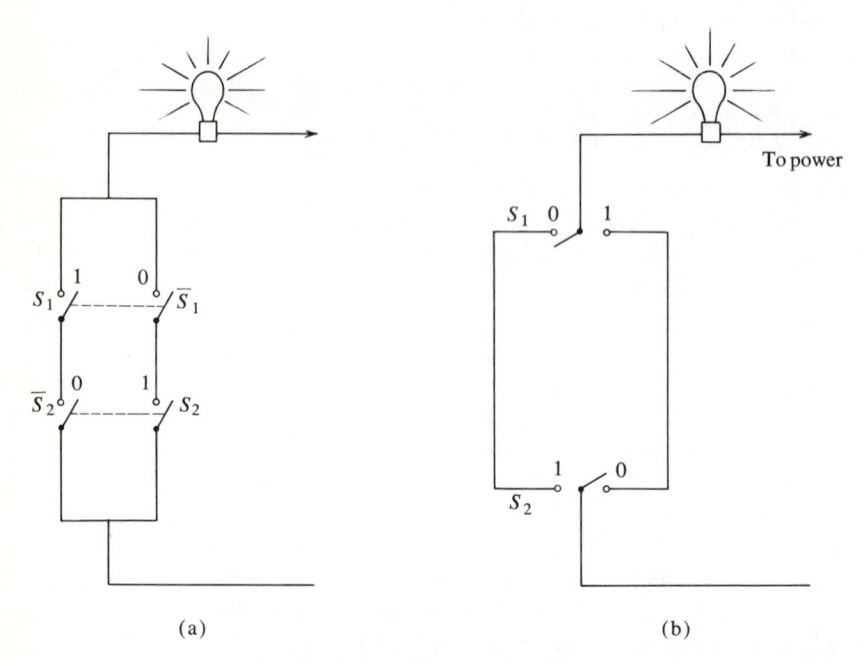

Fig. 7.25 Three-way switch.

7.15
TIME CONSIDERATIONS—BOOLEAN CALCULUS

If time is considered as an additional dimension of problems which involve binary variables, then one is led from the domain of the algebra of "states" (Boolean algebra) to the domain of an algebra of "states and events." The variables in Boolean algebra hold a single determinable "value" during any one problem, even if that value may be unknown. Time was completely disregarded in the illustrative examples which were used in this chapter. When time is accounted for, the conditions of a particular problem may change as time advances, and the variables may alter their values back and forth. Time is counted or measured then in the ordinary way.

Switching algebra applied to the design of circuits using all-or-none (binary) circuit elements will have to be extended appropriately to incorporate time considerations; the symbolic logic from which the switching algebra is derived will also have to be extended from such connectives as AND, OR, NOT, and so forth, to include connectives like BEFORE, AFTER, WHEN, BEGIN, FINISH, DURING, HAPPEN, and other similar ones, properly identified and defined.

A new algebraic structure which incorporates time considerations in problems of binary variables, is analogous to a calculus something like a "Boolean calculus." It is more appropriately referred to as the algebra of "states and events."[8] It constitutes essentially a natural extension of Boolean algebra. Without really delving into the difficulties of such a Boolean calculus, we shall deal with problems involving time considerations in the chapters devoted to the discussion of sequential machines later in the textbook.

Although time is counted in the ordinary manner, its introduction may bring about some interesting considerations. A variable may go through an "event" by changing its value 0 to 1 or from 1 to 0. If the change is not instantaneous, then during a change the variable goes through a "third" state since its value is neither 0 nor 1 but "undefined." Accordingly we may say for example that at time t, or during the time interval Δt, a two-state circuit element "is on," "is off," or "is undefined."

The definition of new operators and symbols to go with the ideas introduced by connectives such as WHEN, BEFORE, AFTER, DELAY, and so on, and their synonyms is in many ways rather evident. The enrichment of Boolean algebra by these new time-dependent operators becomes essential in the design of information-processing machines. Time-involving problems are discussed later in the chapters on sequential machines.

BIBLIOGRAPHY

Asser, G., *Einfuhrung in die Mathematische Logik*. Leipzig: Teubner, 1959.
Caldwell, Samuel H., *Switching Circuits and Logical Design*. New York: John Wiley & Sons, Inc., 1958.
Curtis, Herbert A., *Design of Switching Circuits*. Princeton, N.J.: D. Van Nostrand Co., Inc., 1962.

[8] Berkeley, E. C., *Symbolic Logic and Intelligent Machines*. New York: Reinhold Publ. Corp., 1959.

Davis, M., *Computability and Unsolvability*. New York: McGraw-Hill, Inc., 1958.
Flores, I., *Computer Design*. Englewood Cliffs, N.J.: Prentice-Hall, Inc., 1967.
Halmos, P. R., *Lectures on Boolean Algebras*. Princeton, N.J.: D. Van Nostrand Co., Inc., 1963 (a good abstract discussion of Boolean algebras for the more mathematically inclined reader).
Hohn, Franz. *Applied Boolean Algebra*. New York: The Macmillan Co., 1960.
McCluskey, E. J. and Bartee, T. C., *A Survey of Switching Theory*. New York: McGraw-Hill, Inc., 1962.
Mendelson, E., *Introduction to Mathematical Logic*. Princeton: D. Van Nostrand, Inc., 1964.
Meschkowski, H., *Einfuhrung in die Moderne Mathematik*. Mannheim: Bibliographisches Institut, 1964.
Phister, M., Jr., *Logical Design of Digital Computers*. New York: Wiley & Sons, Inc., 1963.
Schmidt, H. A., *Mathematische Gesetze der Logik*. Berlin: Springer, 1960.
Shannon, C. E. and Weaver, W., *The Mathematical Theory of Communication*. Urbana, Ill.: Univ. of Illinois Press, 1949.
Shannon, C. E., "A Symbolic Analysis of Relay and Switching Circuits," *Trans. of the Am. Institute of Electrical Engineering*, vol. 57, 1938.

Problems

7.1 If $x = 1$, $y = 0$, and $z = 0$, what is the value (0 or 1) of each of the following Boolean expressions?

$$f_1 = x + y\,\bar{z},$$
$$f_2 = (x + y)\,\bar{z},$$
$$f_3 = x\,y\,\bar{z},$$
$$f_4 = x\,(y + \bar{z}),$$
$$f_5 = \bar{y} + x\,z,$$
$$f_6 = \overline{\bar{x}\,y + x\,\bar{z}}.$$

7.2 Prove that

$$a + \bar{a}\,b = a + b.$$

7.3 Construct the appropriate truth tables to verify the following rules for negating logical sums and products:

(a) $\overline{a + b} = \bar{a}\bar{b}$,
(b) $\overline{ab} = \bar{a} + \bar{b}$.

7.4 Justify the following relationships:

(a) $a + a\,b = a$,
(b) $a(a + b) = a$,
(c) $a + \bar{a}\,b = a + b$,
(d) $a(\bar{a} + b) = a\,b$.

7.5 Show that

$$ab + \bar{a}c + bc = ab + \bar{a}c$$

(a) algebraically,
(b) using a truth table.

7.6 Show that

(a) $a + \bar{a}b = a + b$,
(b) $a(\bar{a} + b) = ab$.

7.7 One wishes to design a three-input logical circuit whose output is a 1 when an odd number of inputs bear a 1. Using a Karnaugh map, verify that either canonical form of the output function is a minimal (fewer number of gates).

7.8 (a) Assuming the cost of \$3 per input in AND gates, \$2 per input in OR gates, and \$4 for a NOT circuit, determine the cost of the minimal (least cost) circuit which you will obtain if you simplify the function

$$f = \overline{w + \bar{z}(\bar{x} + y)}.$$

(b) Repeat for the function

$$f = \overline{x\bar{y}w} + yz.$$

(c) Find the cost of the canonical realizations of the above two functions.

7.9 Use either algebra or a truth table to prove that

$$ab + bc + ca = (a+b)(b+c)(c+a).$$

7.10 Simplify

$$f = \bar{x}(y + x\bar{y}z) + \bar{z}y$$

using
(a) De Morgan's Theorem and algebraic manipulations,
(b) the map method; then
(c) determine the cost of the canonical and of the minimal realizations of this function using
  \$3 per input AND gate,
  \$2 per input OR gate,
  \$4 for a NOT circuit.

7.11 Demap the function mapped as shown in Fig. 7.11.1.

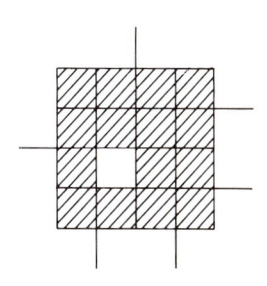

7.11.1

7.12 For the switching circuit shown in Fig. 7.12.1, obtain an algebraic expression for f, then simplify to the minimum sum of products. Draw the minimal circuit block diagram.

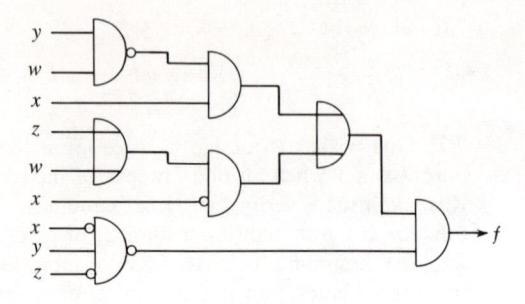

7.12.1

7.13 Examine the following logical circuits (Fig. 7.13.1). Find the Boolean functions they perform, that is, $z = f(a, b \ldots)$. Construct a truth table for each and see if you can simplify the z functions. If yes, construct the simplified logical circuit that will produce the same function.

7.14 Express each of the following functions in sum-of-products and product-of-sums standard forms:
(a) $f = x_1x_2 + x_1x_3 + x_1x_4 + x_2\bar{x}_4$
(b) $f = (\bar{x}_1 + x_2x_3)(x_2 + \bar{x}_3)(x_1 + \bar{x}_4)$
(c) $f = x_1\bar{x}_2x_3 + \bar{x}_1x_2x_3 + \bar{x}_1x_3\bar{x}_4 + \bar{x}_2\bar{x}_4$
(d) $f = x_1\bar{x}_2 + x_2\bar{x}_3(\bar{x}_1 + \bar{x}_3)(x_1 + \bar{x}_4)$
(e) $f = x_1x_2x_3\bar{x}_4 + x_1\bar{x}_3\bar{x}_4 + \bar{x}_1\bar{x}_2x_3x_4 + \bar{x}_1\bar{x}_2\bar{x}_4 + x_2x_3x_4$

7.15 The Boolean functions f are given mapped as shown in Fig. 7.15.1 (a), (b), (c). Demap and express each function f in the minimum sum of-products and the minimum product-of-sums forms.

7.16 Given the truth table below, determine the two canonical forms of the function f.

x_1	x_2	x_3	f
0	0	0	1
0	0	1	1
0	1	0	1
0	1	1	1
1	0	0	1
1	0	1	0
1	1	0	0
1	1	1	1

PROBLEMS 183

(a)

(b)

(c)

(d)

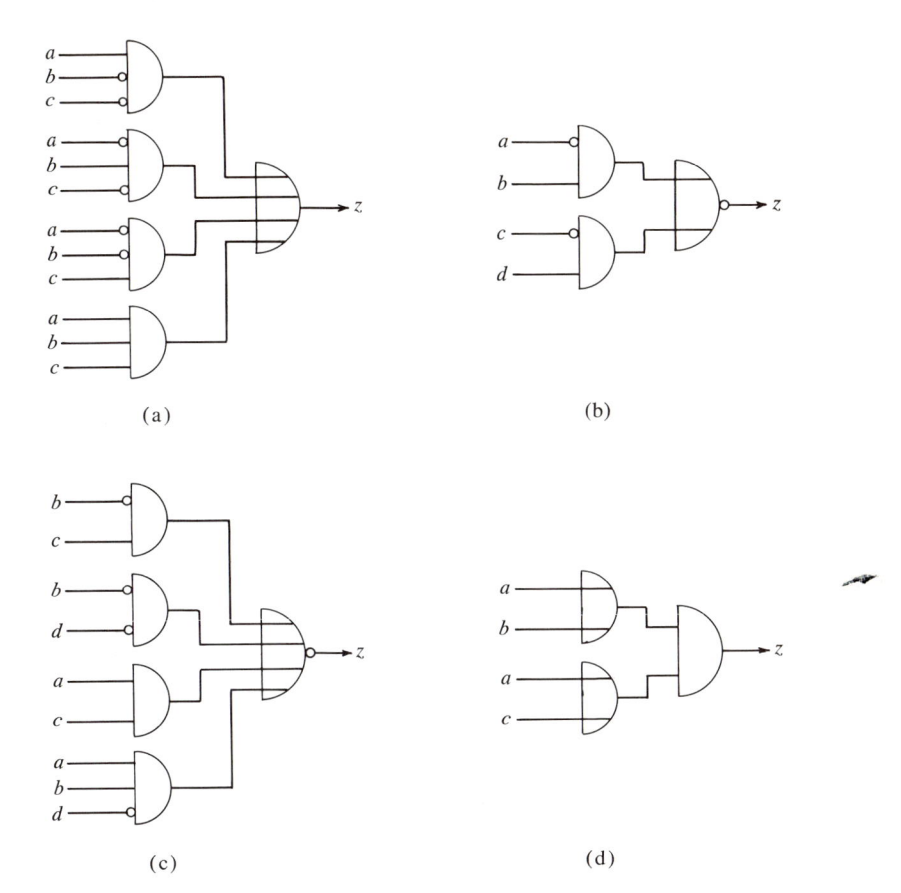

7.13.1

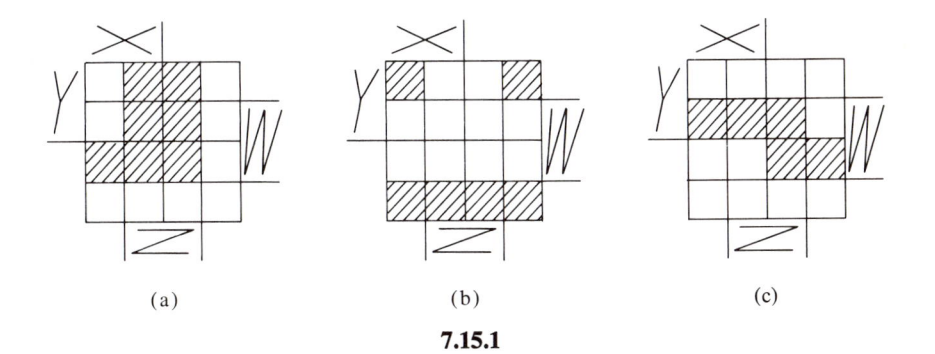

(a)

(b)

(c)

7.15.1

8
Combinational Circuits

A *combinational*[1] circuit is one whose output values at a given time t depend exclusively on the input values at time t. In contrast, a *sequential* circuit is one whose output values at a given time t depend on the sequence of input values at times before t (see Chapter 11). Both inputs and outputs are *binary* variables, that is, quantities which can acquire only either one of two possible values. Each binary variable may represent a physical quantity (a "signal"), as, for example, a voltage impulse which may be either positive or negative.

The design of combinational circuits begins with an examination of the problem specifications as discussed in Chapter 5. The problem specifications are then translated by way of various methods into switching functions. A switching function may be implemented by various circuit configurations. The one which is selected in a particular design depends on the definition of "optimum performance," which is dictated by the existing design constraints, such as available hardware elements, reliability considerations, speed demands, and so forth.

The design of the "optimum" logical circuit is often a most difficult job, because many factors have to be considered simultaneously. For example, the kind of

---

[1] Also referred to as "logical" or "switching" circuit.

available hardware would determine the class of logical circuits from which the optimum is to be found. In the early days of switching circuit designs, the AND, OR, and NOT gates were used almost exclusively, because they were the best realizable elementary logical blocks with the then-available relay and diode circuits. However, the development of sophisticated transistor switching circuits has dictated the design of combinational circuits using NAND and NOR transistor blocks.

In certain cases the designer may aim at producing a "minimal" circuit, that is, the one with fewer gates. However, the minimal circuit does not always correspond to minimum cost, and often cost may be traded for increased reliability, speed, or other desirable features. While a minimal circuit is often desirable, if integrated circuits are to be used one would prefer a circuit configuration with fewer interconnections rather than fewer gates. A practical method for logic circuit design would take into consideration all the essential constraints of cost, reliability, speed, available hardware, and so on, and produce the "optimum" logic circuit by making the most appropriate compromises among the various constraints. Unfortunately, a universal method for combinational circuit design is not available today. Special methods, satisfying narrower objectives, such as delay time minimization alone or minimization of the number of components ("minimal" circuit design), are available. Most practical designs begin by satisfying some narrower objectives, such as producing a minimal circuit, and then proceed to meet additional performance demands by techniques largely dependent on experience and intuition.

Since the details of the design constraints are often peculiar to the particular class of elementary logic blocks or physical switching devices used, it appears unnecessary to discuss in great detail any one particular component. Only the important characteristics of the physical switching devices most commonly used in logic circuit and memory designs, together with the related elementary logic blocks, are discussed in this and the next chapter. In this respect, it is demonstrated that existing hardware constraints may modify the methodology of logic design. For example, if the transistor is the available elementary switching device, the kind of elementary logic blocks normally constructed using transistors would generate functions of the type: $\bar{x}_1 + \bar{x}_2 + \bar{x}_3, \ldots$, or, $\bar{x}_1 \bar{x}_2 \bar{x}_3 \ldots$. It is then found convenient to develop an off-shoot of the switching algebra, which is particularly suited for working with such functions (Section 8.4).

8.1
COMBINATIONAL OPERATIONS—
"DON'T CARES"

Let us consider that a number n of binary physical signals, such as two-valued voltages or currents, are available. Each signal may be represented by a binary variable x_i, where $i = 1, 2, \ldots, n$. One may wish to

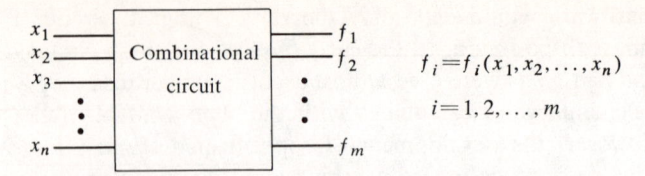

Fig. 8.1 Black-box representation of an *n*-input *m*-output combinational circuit.

$$f_i = f_i(x_1, x_2, \ldots, x_n)$$
$$i = 1, 2, \ldots, m$$

perform certain operations on the available variables, in accordance with certain specified relationships between the variables and the desired outcome. All legitimate operations, that is, those we deal with in this text, are restricted to the ones which yield as outputs only *binary* variables (signals).

In the functional block illustrated in Fig. 8.1, both the inputs $(x_1, x_2, \ldots, x_n)$ and the outputs $(f_1, f_2, \ldots, f_m)$ are binary variables that can only take the values 0 and 1. The values of the outputs $(f_1, f_2, \ldots, f_m)$ at a given time t are determined exclusively by the values of the inputs $(x_1, x_2, \ldots, x_n)$ at the time t. An example involving *n*-input *m*-output combinational circuit analysis will be discussed in a later section, in connection with the design of an *NBCD* to 2-out-of-5 code translator.

As discussed in Section 7.4, there exist only a finite number of possible unique operations to be performed on a given set of *n* input binary variables. The number of all possible unique operations is that of all possible functions generated by *n* binary-input variables, and it is equal to 2^{2^n}. Thus there are 16 possible operations (functions) for two binary input-variables x_1, x_2 (Table 7.3).

Any one operation (function) may be expressed either *tabularly* by a truth table or *algebraically* by an algebraic expression or *graphically* by Karnaugh maps. In order to have a complete description, the binary value of the output function for each of the 2^n possible combinations[2] of values for the *n* binary input variables must be specified.

In frequent combinational design problems there may be certain constraints upon the input combinations. It may be that not all conceivable input combinations can occur in normal operation. If this is the case one may wish to arbitrarily assign output values for the input combinations which are not expected to occur normally. These values, which are commonly referred to as "don't cares," may be either 1's or 0's.

One might think that it matters very little whether the output values corresponding to nonoccurring input combinations are specified. It must be noted, though, that whether these input combinations appear or not, the circuit will have the capability to respond to these input combinations *in some way* if they should appear. Therefore, if there is any design advantage in requiring special

---

[2] Also referred to as "states," by analogy to points in binary space, where each point represents a state of the input.

COMBINATIONAL OPERATIONS—"DON'T CARES" 187

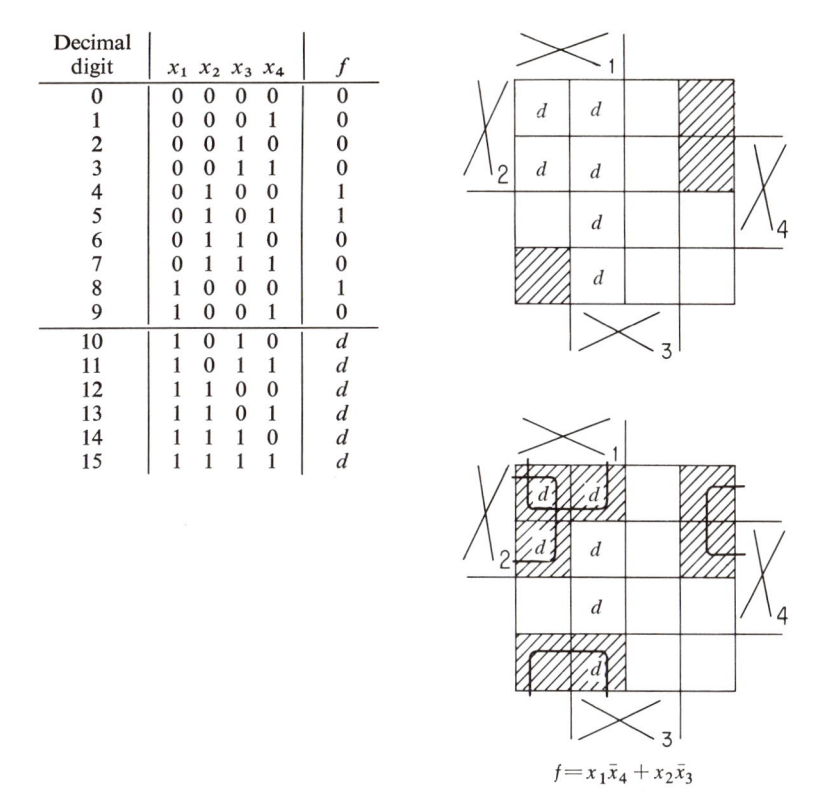

Decimal digit	$x_1\ x_2\ x_3\ x_4$	f
0	0 0 0 0	0
1	0 0 0 1	0
2	0 0 1 0	0
3	0 0 1 1	0
4	0 1 0 0	1
5	0 1 0 1	1
6	0 1 1 0	0
7	0 1 1 1	0
8	1 0 0 0	1
9	1 0 0 1	0
10	1 0 1 0	d
11	1 0 1 1	d
12	1 1 0 0	d
13	1 1 0 1	d
14	1 1 1 0	d
15	1 1 1 1	d

$$f = x_1 \bar{x}_4 + x_2 \bar{x}_3$$

Fig. 8.2 Example of the use of "don't cares" in combinational circuit design.

output values for these input combinations, the designer may desire to do so. As it turns out, if used judiciously, the "don't care" conditions may facilitate the optimum logic design and lead to simpler circuit realizations.

To illustrate the use of the "don't care" conditions, let us suppose that the inputs to a combinational circuit are composed of four binary lines (X_1, X_2, X_3, X_4) upon which coded representations of the decimal digits from 0 to 9 may appear. Of the 16 possible input combinations, only ten are assigned to the ten decimal digits, leaving six unassigned (don't care) conditions, as shown in the table of Fig. 8.2.

Let us also assume that the combinational circuit has a single output, whose value must be 1 when the combinations corresponding to the decimal digits 4, 5, and 8 appear at the input. The output function is mapped as shown in Fig. 8.2., and the "don't cares," marked by "d," are then selected so that the minimal realization is obtained. Notice that of the six "don't care" conditions, four were made 1's and two were made 0's. More examples on the use of "don't cares" in minimal realizations will be offered later in this chapter.

8.2 COMPLETE SETS OF ELEMENTARY BLOCKS

The design of a combinational circuit with n inputs involves the realization of combinational logic functions with n variables. Of the 2^{2^n} possible unique functions of n variables (see Section 7.4) there exist certain ones that are of special practical importance; they are the ones most suitable for hardware implementations using the available switching devices and for realizing all other switching functions. Such functions, referred to as elementary logic functions, are the AND, OR, and NOT functions defined in Table 7.3. They are most suitable for constructing AND, OR and NOT elementary functional blocks using relays and diodes, which, in turn, are used to realize any other combinational switching function.

That any given switching function can be realized using exclusively AND, OR, and NOT elementary functional blocks is apparent from the fact that any switching function may be written in either one of its two canonical forms, that is, as a logical sum-of-products or as a logical product-of-sums. Minimal realizations obtained from Karnaugh maps are also realizable with AND, OR, and NOT blocks exclusively.[3]

Any set of elementary functional blocks which is sufficient for realizing any given switching function is a "complete set." It turns out that the sets of (AND, NOT) and (OR, NOT) are complete sets. As shown in Fig. 8.3, it is possible to form an OR operation using only the blocks of (AND, NOT) and to form an AND operation using only the blocks of (OR, NOT). This fact establishes that either one of the sets (AND, NOT) or (OR, NOT) is a complete set.

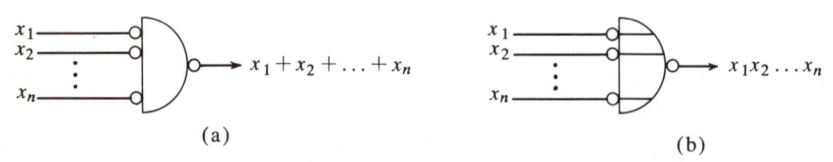

Fig. 8.3 (a) OR operation formed using the blocks of (AND, NOT) complete set; (b) AND operation formed using the blocks of (OR, NOT) complete set. Notice the use of De Morgan's theorem in forming the output functions.

As mentioned previously, there exist also other logical functions that are of practical importance. More specifically, the NAND function forms by itself a complete set, in that it is the only operation required to realize all combinational switching functions. The NAND, otherwise referred to as STROKE operation, is formed by a succession of AND and NOT operations in the configuration shown in Fig. 8.4(a). To see that the NAND block is itself a complete set, notice that if

---

[3] It is also anticipated since these functional blocks implement the only essential operations set forth in the axiomatic definition of this Boolean algebra (Section 7.2).

COMPLETE SETS OF ELEMENTARY BLOCKS   189

$$f = \overline{x_1 x_2 \ldots x_n} = \bar{x}_1 + \bar{x}_2 + \ldots + \bar{x}_n$$

(a)

$$f = \overline{x_1 + x_2 + \ldots + x_n} = \bar{x}_1 \bar{x}_2 \ldots \bar{x}_n$$

(b)

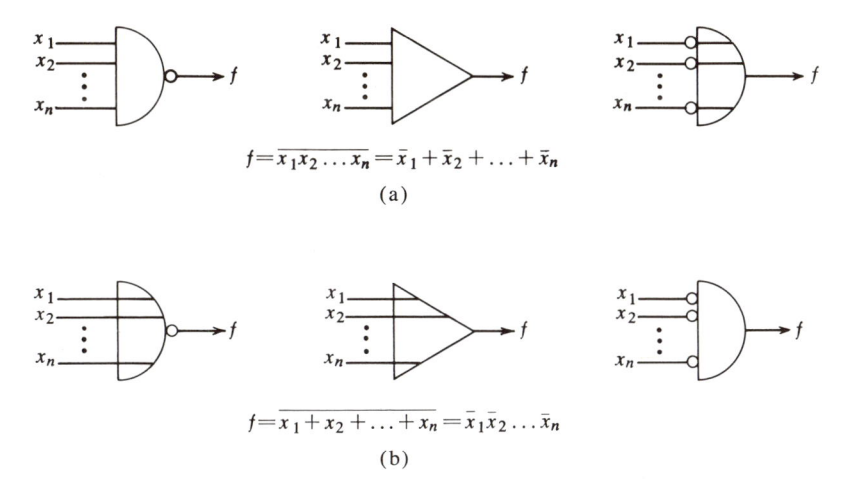

Fig. 8.4 NAND and NOR functions and symbols. (a) Equivalent symbol representation of the NAND (STROKE) function. (b) Equivalent symbol representation of the NOR (DAGGER) function.

one sets $x_1 = x_2 = \ldots = x_n$, the result is the NOT operation. Also, since the negation (NOT) is possible, the OR operation is possible by using $\bar{x}_1, \bar{x}_2, \ldots, \bar{x}_n$, as inputs. [See Fig. 8. 4(a).] Having realized OR and NOT, we have again a complete set.

The dual element to the NAND, that is, the NOR, also referred to as the DAGGER element, which forms the function $\bar{x}_1 \bar{x}_2 \ldots \bar{x}_n$, as shown in Fig. 8.4(b), also forms a complete set by itself. Again, the NOT operation is formed by letting $x_1 = x_2 = \ldots = x_n$, while the AND operation is obtained by using $\bar{x}_1, \bar{x}_2, \ldots \bar{x}_n$, as inputs.

The use of the NAND (STROKE) and the NOR (DAGGER) as elementary logical blocks in combinational circuits is discussed further in Section 8.4. Here we may only add that another elementary logical block, called the MINORITY element, also forms a complete set. The MIN block requires an *odd* number of inputs, and it yields a 1 output if, and only if, a minority of the inputs is 1. The three-input MIN block, shown in Fig. 8.5, is sufficient to perform all combinational logic. A NOT is formed by making all three inputs the same. An AND can be constructed by supplying $\bar{x}_1$ and $\bar{x}_2$ to the two inputs and a constant 1 to the third input. Having realized AND and NOT, we have again a complete set.

The MAJORITY element, which also requires an odd number of inputs and yields a 1 output if and only if the majority of the inputs are 1, does not form by itself a complete set because obviously it is not possible to obtain the NOT function. The pair (MIN, MAJ) form, of course, a complete set. A three-input MAJ element is shown in Fig. 8.5.

There are various other complete sets possible, as for example the set (EXCL OR, OR). For each complete set there exist equivalent canonical forms of realization

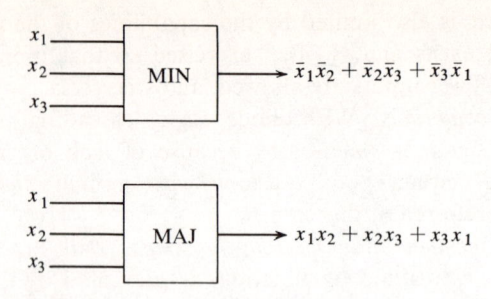

```
x1 ─┐
x2 ─┤ MIN ├─→ $\bar{x}_1\bar{x}_2 + \bar{x}_2\bar{x}_3 + \bar{x}_3\bar{x}_1$
x3 ─┘

x1 ─┐
x2 ─┤ MAJ ├─→ $x_1 x_2 + x_2 x_3 + x_3 x_1$
x3 ─┘
```

Fig. 8.5 Three-element minority and majority blocks.

for a given switching function. NAND and NOR realizations of switching functions are discussed further in Section 8.4.

8.3 HARDWARE CONSIDERATIONS

Any physical device which is capable of displaying either of two discrete well-defined stable-equilibrium conditions may be used as a switching component for constructing combinational circuits. The operation of such devices is based on their ability to switch the value of some physical parameter (scalar or vector), such as the index of refraction, the electrical conductivity, or the magnetic moment between two discrete conditions. The change must take place reliably, repeatedly, and in response to an external command. Switching devices which serve as components of logical circuits are discussed in more detail in Chapter 9.

In the design of combinational circuits several constraints may be imposed by the nature of the physical switching devices used. Four typical types of such constraints are illustrated in the following with illustrative examples.

1. *"Fan-in"* and *"Fan-out"* *constraints.* Computer circuits are commonly required to drive at a predetermined speed as many other similar units in parallel as possible, each driven unit being in turn capable of driving others, and so on. The number of units that a computer circuit is capable of driving in parallel from its output at a given speed is usually referred to as the fan-out factor. The fan-out capabilities of an electronic logical circuit depend on the current gain of the active device used in the circuit, and are limited by the energy losses in the circuit. Since the current gain of an active device, such as a transistor, is finite and is actually quite limited, there is always a restricted drive capability—there is a certain maximum fan-out at any given speed. The fan-in factor, which is the maximum number of logical inputs that may be designed with a given

active device, is also limited by the capabilities of the active device used. A trade-off exists and is often exercised so that more outputs may be obtained if fewer inputs are allowed, and vice versa.
2. *Cascading conditions.* With diode logic, cascading of stages typically causes the signal to deteriorate because of lack of amplification. Two-level logical circuit configurations with amplification between stages should be preferred in this case.
3. *Suitability for generating certain functions.* With transistors again, the generation of AND and OR functions is prevented by the phase inversion which is encountered on the input signal. This signal phase inversion makes the transistor more suitable for generating such functions as NAND, NOR, and NOT, that is, $\bar{x}_1 + \bar{x}_2$, $\bar{x}_1 \bar{x}_2$, $\bar{x}$, to be discussed later in this chapter.
4. *Time-delay restrictions.* When combinational circuits are used as parts of larger sequential circuits, the time delays of signals passing through the combinational circuits may become of critical importance. The combinational circuits themselves may be designed without regard to time delay. However, in sequential logic the time delays suffered by a signal passing through a chain of combinational circuits may become intolerable if they exceed some permissible value and, accordingly, restrictions may have to be imposed on acceptable time delays. Two-level logic circuits, where the input signal needs to go through only two gates before arriving at the output, may become desirable if time delays are to be minimized.

Further hardware considerations, with discussions of a small but representative number of switching devices and elementary functional circuits, are left for Chapter 9 where the most prominent of the switching devices used in modern information-processing machines are examined strictly from the point of view of the logic operations they can perform in combinational and memory circuits.

8.4
NOR AND NAND LOGIC

Combinational circuits are frequently designed using transistor NOR and NAND elementary functional blocks. These circuits are superior to the diode OR and AND circuits, in that many of them can be cascaded without the loss of signal (voltage) level.

Table 8.1 shows the truth tables of the two-variable NOR and NAND functions, together with those of the corresponding OR and AND functions.

The symbols $\downarrow$ and $/$ are commonly used for the NOR and the NAND function, respectively, to signify the following easily verifiable algebraic relationships:

$$x \downarrow y = \overline{x + y} = \bar{x}\bar{y} = \bar{x}/\bar{y},$$
$$x/y = \overline{\bar{x}\bar{y}} = \bar{x} + \bar{y} = \overline{\bar{x} \downarrow \bar{y}}.$$

Notice a form of De Morgan's theorem for NOR *and* NAND *operations*

Table 8.1. Truth tables for NOR, NAND, OR, and AND functions.

x	y	OR $x+y$	NOR $x \downarrow y$	AND xy	NAND x/y
0	0	0	1	0	1
0	1	1	0	0	1
1	0	1	0	0	1
1	1	1	0	1	0

The terms DAGGER and STROKE may be used interchangeably with NOR and NAND, respectively.

The above algebraic definitions of the NOR and NAND functions may be extended to many variables:

$$x_1 \downarrow x_2 \downarrow \ldots \downarrow x_n = \overline{x_1 + x_2 + \ldots + x_n} = \bar{x}_1 \bar{x}_2 \ldots \bar{x}_n,$$
$$x_1 / x_2 / \ldots / x_n = \overline{x_1 x_2 \ldots x_n} = \bar{x}_1 + \bar{x}_2 + \ldots + \bar{x}_n.$$

The electronic circuits performing these functions are readily extended from two to many inputs. The block diagram symbols of these functions are shown in Fig. 8.4.

With respect to the algebraic properties of these functions, it is important to note that the NOR and NAND functions are commutative

$$x \downarrow y = y \downarrow x,$$
$$x/y = y/x;$$

but they are *not* associative or distributive,

$$x \downarrow (y \downarrow z) \neq (x \downarrow y) \downarrow z,$$
$$x/(y/z) \neq (x/y)/z,$$
$$x \downarrow (y/z) \neq (x \downarrow y)/(x \downarrow z),$$
$$x/(y \downarrow z) \neq (x/y) \downarrow (x/z).$$

The reader may easily verify the above statements.

It should also be noted that a cascading of NOR and NAND functions using symbolic notation without parentheses is meaningless, because the functions are not associative. Thus, for example, it is meaningless to write $x \downarrow y / z$ without parentheses, because $(x \downarrow y)/z \neq x \downarrow (y/z)$. In writing cascaded NAND and NOR operations parentheses must be used to signify the proper sequence of these operations.

Because of the prominence of the NAND and NOR circuits in the design of combinational logic, simple rules and procedures have been developed for the conversion of canonical (AND and OR) expressions and circuits into equivalent

$$f = x_1\bar{x}_2\bar{x}_3 x_4 = \overline{\bar{x}_1 + x_2 + x_3 + \bar{x}_4}$$

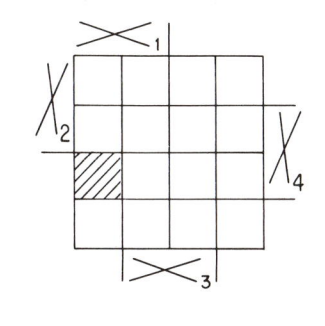

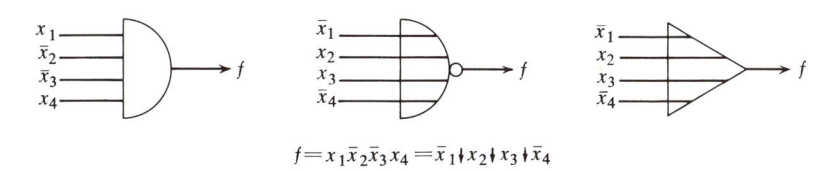

Fig. 8.6 Conversion of a standard product to a single NOR function.

NAND and NOR expressions and circuits. These procedures will be discussed in the following with illustrative examples of conversion from logical products and sums to NAND and NOR expressions.

An important rule to remember is that any function of n variables can be realized by at most a three-level cascade of NAND or NOR circuits, where the inputs to the first level are the variables in their uncomplemented or complemented forms. To show this, consider first a standard product of n variables (minterm), which is mapped into a single square. The product can be expressed as a single NOR function among the complements of the variables of the product. The conversion is illustrated in Fig. 8.6 with the four-variable product conversion,

$$f = x_1\bar{x}_2\bar{x}_3 x_4 = \overline{\bar{x}_1 + x_2 + x_3 + \bar{x}_4} = \bar{x}_1 \downarrow x_2 \downarrow x_3 \downarrow \bar{x}_4 \quad \text{(DeMorgan's theorem was used here).}$$

This conversion is extended to nonminterms, that is, to any logical product. Figure 8.7 illustrates the conversion of such a product into a single NOR function.

In order to convert a switching function which is given in the form of sum-of-products, first convert all products in the sum into equivalent NOR functions and then sum up the outputs of the NOR blocks by use of the identity,

$$a + b + \ldots + z = \overline{a \downarrow b \downarrow \ldots \downarrow z}.$$

This can be verified easily.

In order to illustrate the conversion of a given function into a NOR circuit, let us consider the function $f = x_2\bar{x}_3 + \bar{x}_1 x_4$, which has been expressed in a

194 COMBINATIONAL CIRCUITS

$$f = x_2 \bar{x}_3 = \bar{\bar{x}}_2 \downarrow x_3$$

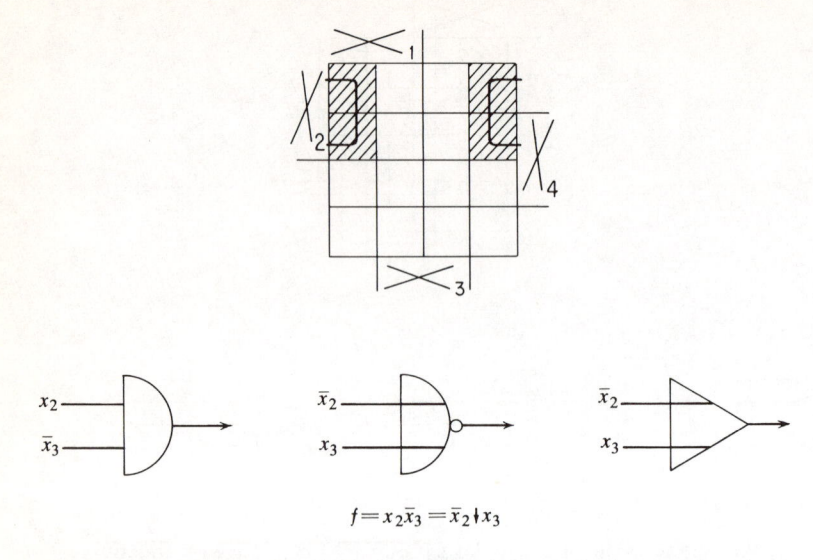

Fig. 8.7 Conversions of a logical product into a single NOR function. Notice that the inputs to the NOR block are the complements of the variables in the logical product.

sum-of-products minimal form by the mapping method. In accordance with the above discussion, the conversion equation will be

$$f = x_2\bar{x}_3 + \bar{x}_1 x_4 = \bar{\bar{x}}_2 \downarrow x_3 + x_1 \downarrow \bar{\bar{x}}_4 = \overline{(\bar{\bar{x}}_2 \downarrow x_3)} \overline{(x_1 \downarrow \bar{\bar{x}}_4)} = \overline{(\bar{\bar{x}}_2 \downarrow x_3) \downarrow (x_1 \downarrow \bar{\bar{x}}_4)}.$$

Since there are three conversions in cascade (product, sum, and negation), it becomes apparent now that any given function can be expressed by at most a NOR-NOR-NOR circuit, as shown in Fig. 8.8 for the above function. The equivalence of the illustrated NOR circuit to the corresponding AND, OR circuit, also shown in Fig. 8.8, is easily proven and is left as an exercise for the student (Problem 8.4).

Following a similar procedure one can convert any given function of n variables alternatively to a cascade of at most three levels of NAND circuits. The function is now expressed as a product-of-sums; each sum is then converted into a single NAND function among the complements of the variables appearing in the sum; finally, the product is realized by two more levels of NAND operations. The conversion is illustrated in Fig. 8.9. The function is expressed in a product-of-sums form, and converted as follows:

$$f = \overline{(\bar{x}_1 + \bar{x}_3)(x_2 + x_4)(\bar{x}_1 + x_2)(\bar{x}_3 + x_4)} \\ = (x_1/x_3)/(\bar{x}_2/\bar{x}_4)/(x_1/\bar{x}_2)/(x_3/\bar{x}_4).$$

The verification of the equivalence between the two circuit implementations shown in Fig. 8.9 is again left as an exercise for the student.

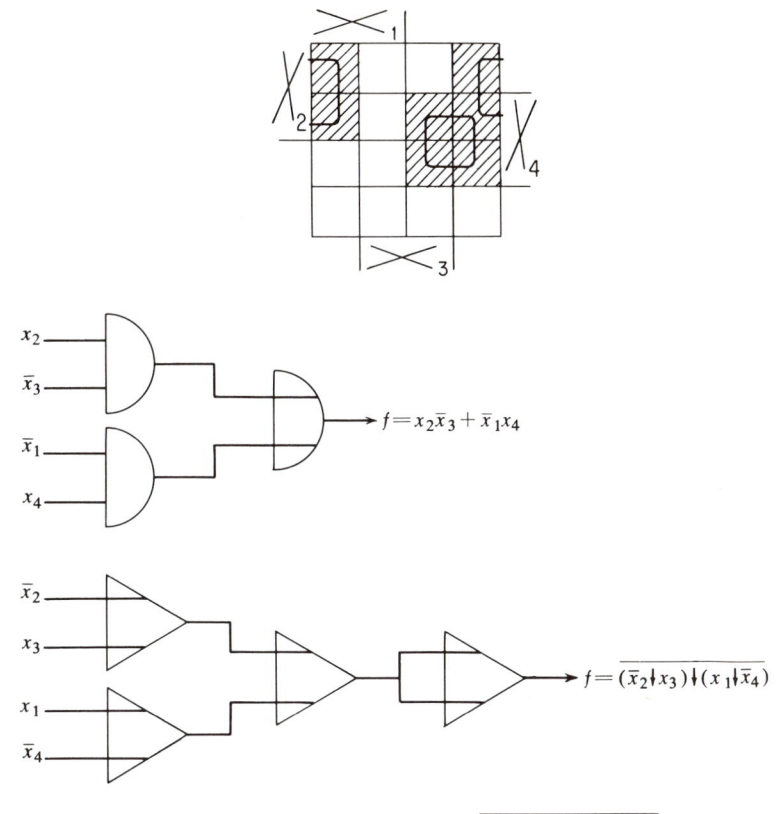

Fig. 8.8 Conversion $f = x_2\bar{x}_3 + \bar{x}_1 x_4 = \overline{(\bar{x}_2 \downarrow x_3) \downarrow (x_1 \downarrow \bar{x}_4)}$.

The conversion of products or sums into NOR or NAND functions is based on the following generalized conversion equations:

$$x_1 x_2 x_3 \ldots x_n = \overline{\bar{x}_1 \downarrow \bar{x}_2 \downarrow \bar{x}_3 \downarrow \ldots \downarrow \bar{x}_n}$$
$$= x_1/x_2/x_3/\ldots/x_n$$

and

$$x_1 + x_2 + x_3 + \ldots + x_n = \overline{x_1 \downarrow x_2 \downarrow x_3 \downarrow \ldots \downarrow x_n}$$
$$= \bar{x}_1/\bar{x}_2/\bar{x}_3/\ldots/\bar{x}_n$$

where the x's represent either single variables or functions of variables.

We may generalize the above conversion procedures by noting that any function put in its sum-of-products or product-of-sums form is readily converted to a NOR or NAND expression. In each case, we first secure the form of the function by the use of proper parentheses; then, we replace all the $+$ and $\cdot$ symbols either by $\downarrow$ (NOR), or by $/$ (NAND) symbols, respectively. Finally, in the cases of

196 COMBINATIONAL CIRCUITS

$$f = \overline{x_1 x_3 + \bar{x}_2 \bar{x}_4 + x_1 \bar{x}_2 + x_3 \bar{x}_4} = (\bar{x}_1 + \bar{x}_3)(x_2 + x_4)(\bar{x}_1 + x_2)(\bar{x}_3 + x_4)$$

$$f = (\bar{x}_1 + \bar{x}_3)(x_2 + x_4)(\bar{x}_1 + x_2)(\bar{x}_3 + x_4)$$

$$f = \overline{(x_1/x_3)/(\bar{x}_2/\bar{x}_4)/(x_1/\bar{x}_2)/(x_3/\bar{x}_4)}$$

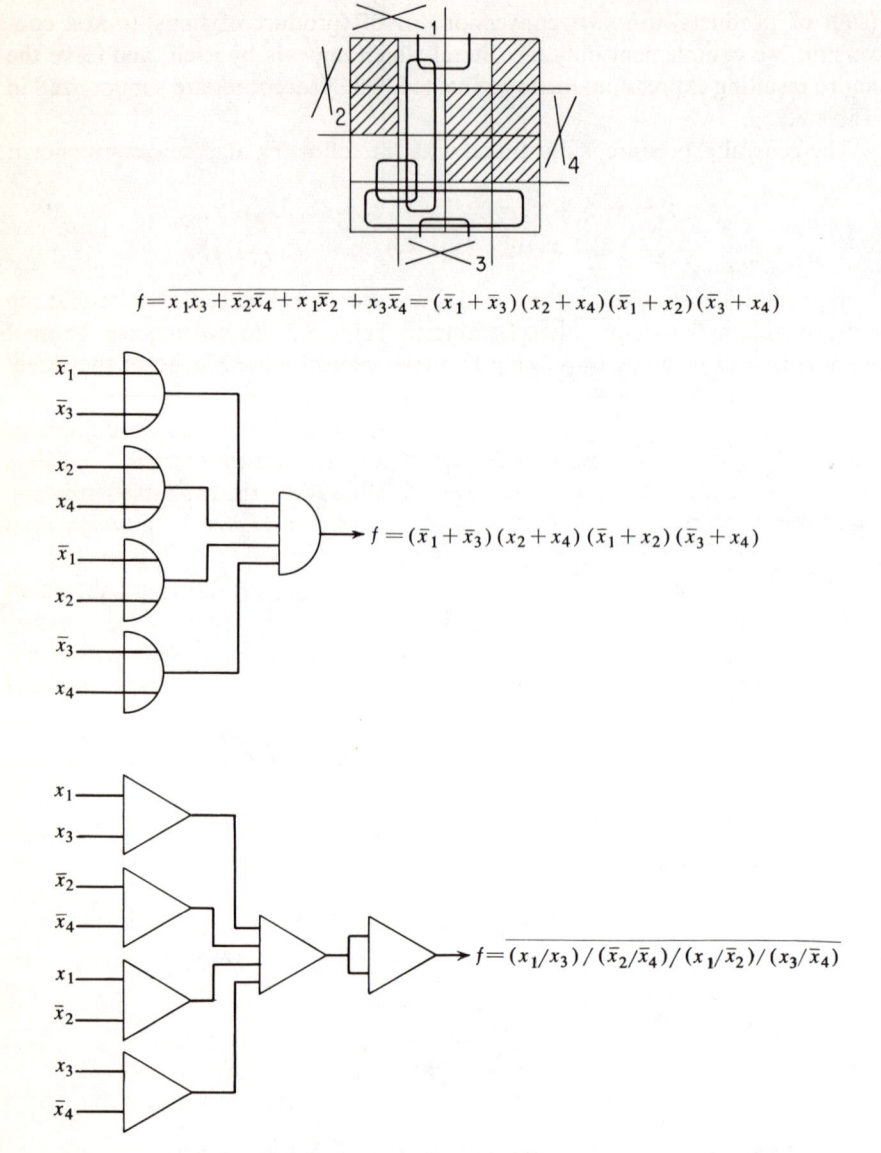

Fig. 8.9 Conversion $f = (\bar{x}_1 + \bar{x}_3)(x_2 + x_4)(\bar{x}_1 + x_2)(\bar{x}_3 + x_4)$
$= \overline{(x_1/x_3)/(\bar{x}_2/\bar{x}_4)/(x_1/\bar{x}_2)/(x_3/\bar{x}_4)}.$

The reading of the function f in a product-of-sums form is first illustrated at the top of the figure; then the implementations using standard AND, OR gates and NAND blocks are demonstrated.

(sum of products)-to-NOR conversion, or of (product of sums)-to-NAND conversion, we complement all literals of each term except those which appear alone, and take the complement of the entire resulting expression. In the cases of (sum of products)-to-NAND conversion, or of (product of sums)-to-NOR conversion, we complement only any literal which appears by itself, and leave the entire resulting expression uncomplemented. The procedures are summarized in Table 8.2.

The general procedure is illustrated with the following algebraic conversions:

$$f_a = x_1\bar{x}_2 + x_3 = \overline{(\bar{x}_1 \downarrow x_2) \downarrow x_3} = (x_1/\bar{x}_2)/\bar{x}_3,$$
$$f_b = (\bar{x}_4 + x_5)x_6 = \overline{(x_4/\bar{x}_5)/x_6} = (\bar{x}_4 \downarrow x_5) \downarrow \bar{x}_6.$$

The circuit implementations for f_a and f_b are shown in Fig. 8.10. Notice that the "diagonal" implementations, according to Table 8.2, do not require the final complementation stage, thus being two-level realizations, instead of the three-level "vertical" realizations.

In NOR or NAND logic design, one may look first for the minimal form of the function with fewer literals since that form would use fewer inputs and thus would require fewer input resistors. Then, if allowed by the available hardware, the designer should pursue a "diagonal" conversion (Table 8.2), which does not require the final complementation stage, thus using fewer circuit levels.

These observations are illustrated in the realization of the function f, shown on the Karnaugh map of Fig. 8.11(a). In its product-of-sums form the function uses a total of twelve inputs, while in its sum-of-products form it uses only seven. Thus the sum-of-products form is preferred, and converted into a two-level

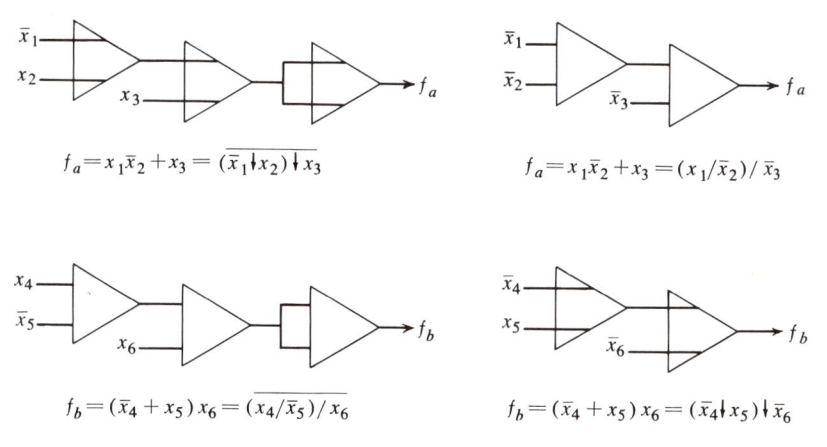

Fig. 8.10 NAND and NOR circuit implementations of a sum-of-products function, f_a, and of a product-of-sums function f_b. The left-hand implementations are "vertical" conversions, while the right-hand implementations are "diagonal" conversions according to Table 8.2.

Table 8.2. NOR, NAND conversions.

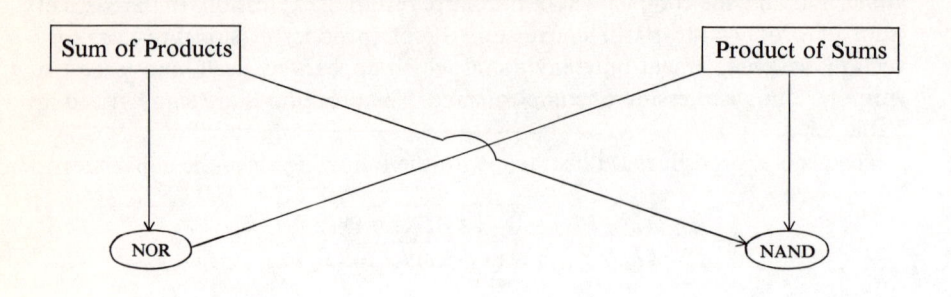

```
        Sum of Products                    Product of Sums
              |        \          /              |
              |          \      /                |
              |            \  /                  |
              |            /  \                  |
              |          /      \                |
              ↓        ↙          ↘              ↓
            (NOR)                              (NAND)
```

Conversion Procedure.
1. Use parentheses to secure the form as a sum of products, or a product of sums.
2. Replace $+$ and $\cdot$ with $\downarrow$ or $/$, whichever is appropriate.
3. (a) In case of a *vertical* conversion: complement all literals of each term, except those which appear alone; take the complement of the final expression.
   (b) In case of a *diagonal* conversion: complement only any literal which appears by itself; leave the final expression uncomplemented.

NAND circuit, as shown in Fig. 8.11(b). It may be noted that the function f can be factored and then converted into a NAND expression, as follows:

$$f = (x_1 + \bar{x}_2)x_3 + x_4$$
$$= (Ax_3) + x_4$$
$$= (A/x_3)/\bar{x}_4,$$

where

$$A = x_1 + \bar{x}_2 = \overline{\bar{x}_1 x_2} = \bar{x}_1/x_2.$$

Thus:

$$f = [(\bar{x}_1/x_2)/x_3]/\bar{x}_4.$$

This realization requires only a total of six inputs, but also three levels, as shown in Fig. 8.11(c).

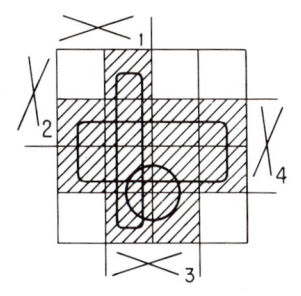

Sum-of-products form: $f = x_1 x_3 + \bar{x}_2 x_3 + x_4$
Product-of-sums form: $f = (x_1 + \bar{x}_2 + x_4)(\bar{x}_1 + x_3 + x_4)(x_2 + x_3 + x_4)$

(a)

$f = (x_1 x_3) + (\bar{x}_2 x_3) + (x_4) = (x_1/x_3) / (\bar{x}_2/x_3) / \bar{x}_4$

(b)

$f = (x_1 + \bar{x}_2) x_3 + x_4 = \left[(\bar{x}_1/x_2)/x_3\right]/\bar{x}_4$

(c)

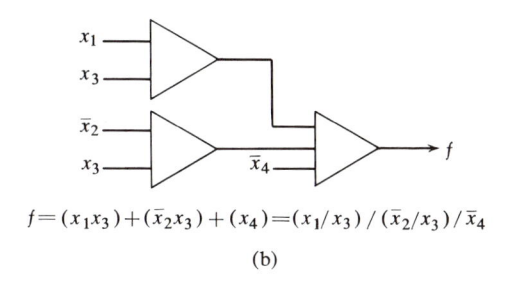

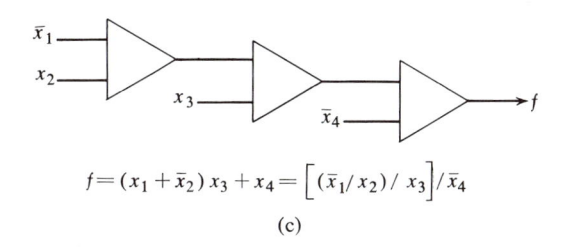

Fig. 8.11 NAND realizations of mapped function f using fewer inputs and stages.

The design of combinational circuits is illustrated with selected practical design problems. The complete design of multioutput combinational circuits (Fig. 8.1) requires the application of Boolean matrix techniques of combining single-output functions, which demands skill, experience, and the best judgment from the designer. The design of a code translator (Section 8.6) illustrates only the essentials of the problem. The complete details of the design are beyond the scope of this textbook.

8.5
"INTERROGATION" CIRCUIT FOR A FOUR-BIT REGISTER

Consider that a four-bit register has been built, composed of a sequence of four binary storage "flip-flop" devices (see Chapter 9). The register is used to store one decimal digit, according to the XS3 code (Table 3.7) shown in the table of Fig. 8.12(a). The register is not expected to contain any

Decimal digits	XS3 Code $x_1\ x_2\ x_3\ x_4$	Output f
0	0 0 1 1	0
1	0 1 0 0	0
2	0 1 0 1	0
3	0 1 1 0	0
4	0 1 1 1	0
5	1 0 0 0	1
6	1 0 0 1	0
7	1 0 1 0	1
8	1 0 1 1	0
9	1 1 0 0	1
10	1 1 0 1	d
11	1 1 1 0	d
12	1 1 1 1	d
13	0 0 0 0	d
14	0 0 0 1	d
15	0 0 1 0	d

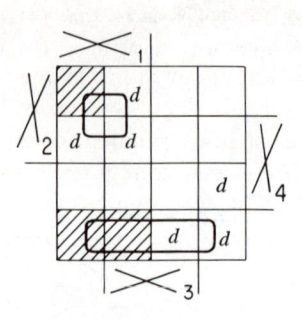

Sum-of-products form: $f = x_1 x_2 + \bar{x}_2 \bar{x}_4$
Product-of-sums form: $f = (x_1 + \bar{x}_2)(x_2 + \bar{x}_4)$

(a) Truth table

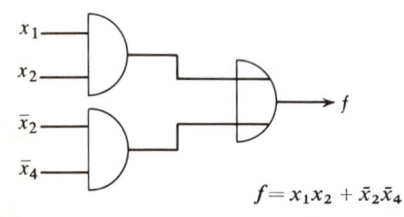

(b) Karnaugh map minimization

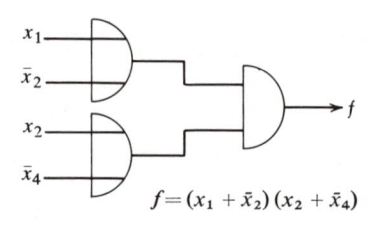

$f = x_1 x_2 + \bar{x}_2 \bar{x}_4$

$f = (x_1 + \bar{x}_2)(x_2 + \bar{x}_4)$

(c) Diode (AND, OR) realization.

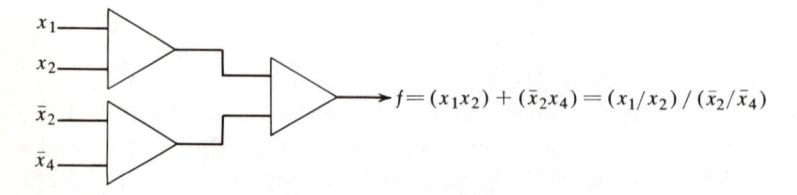

$f = (x_1 x_2) + (\bar{x}_2 x_4) = (x_1/x_2)/(\bar{x}_2/\bar{x}_4)$

(d) Minimal NAND realization.

Fig. 8.12 "Inquire" circuit design for a four-bit register.

other code, or any of the last six combinations shown in the table, which thus correspond to "don't care" conditions. We wish to design a combinational circuit which will put out a 1 if the register contains the decimal digit 5, or 7, or 9.

The inputs to the logic circuit are four binary interrogation lines, coming from the four binary flip-flop cells of the register. On command, the lines are activated to bear a "one" or a "zero" signal each, corresponding to the contents of the register.

The design problem is similar to the one illustrated in Fig. 8.2. Here we shall discuss the circuit design procedure, using either diode gates or transistor NAND blocks.

The output function f is mapped and minimized as shown in Fig. 8.12(b), making use of the "don't care" conditions. Both the sum-of-products and the product-of-sums minimal forms have the same number of literals (see Problem 8.5). The circuit configurations using diode AND and OR gates,[4] are shown in Fig. 8.12(c). The sum-of-products form is converted into a minimal NAND circuit, as shown in Fig. 8.12(d). In all circuits shown, only the three interrogation lines x_1, x_2, and x_4 are used.

8.6
NBCD TO 2-OUT-OF-5 CODE CONVERTER

Consider the problem of designing a four-input and five-output combinational circuit which will convert from the NBCD "8421" code (Table 3.4) into the 2-out-of-5 "74210" code (Table 4.1). The input-output truth tables are shown in Fig. 8.13(a). No illegitimate combinations are expected at the inputs, so that the tables will include six "don't care" conditions for each of the five output functions.

To facilitate the mapping of four-variable functions, such as the output functions of this design problem, we prepare the map shown in Fig. 8.13(b). Its cells are marked with the decimal digits corresponding to the input combinations for easier recognition. The map shown in Fig. 8.13(c) illustrates the positions of the "don't care" input combinations, which correspond to the decimals 10, 11, 12, 13, 14, and 15, for all the output functions.

The output functions, f_7, f_4, f_2, f_1, and f_0, are mapped and minimized individually as shown in Fig. 8.13(d). The AND/OR and NAND circuit configurations for the output function f_7 are shown in Fig. 8.13(e) as examples of individual single-output function realizations.

To produce the code converter circuit, one may decide to combine in parallel the individual single-output realizations, like the ones shown in Fig. 8.13(e). However, this would result in a nonminimal, and certainly costly, circuit design.

---
[4] Circuit diagrams of diode and transistor gates are discussed in Chapter 9.

202 COMBINATIONAL CIRCUITS

The method of combining the minimal output functions, $f_7, f_4, f_2, f_1,$ and f_0, in order to design the minimal four-input and five-output code converter, could not be explained in the limited space allotted in this text and is beyond the scope of an introductory discussion on information-processing machines. It is therefore left up to the interested student to pursue further information on this subject.[5]

Multioutput combinational networks are designed by a mixed graphical and tabular method, involving repeated iterations and exercise of judgment. In the process, the single-output minimized functions are combined in products of various kinds. Because of the rather general importance of the subject of combining and decomposing of given functions and maps, we discuss this subject in the closing section of this chapter.

8.7
THE DESIGN OF A HALF-ADDER

The addition of two binary numbers is done by the ordinary column method, in direct analogy to the familiar decimal addition. The first addition step involves the two "lowest order," that is, right-end, bits of the two numbers. The second addition-step involves the next to lowest order pair of bits and a carry from the previous addition, and so on until all bits have been processed. The first addition-step is a two-input problem and its realization in hardware requires a "half-adder" logical circuit. The second addition-step

Decimal Digit	NBCD-Code $x_8\ x_4\ x_2\ x_1$	2-out-of-5-Code $f_7\ f_4\ f_2\ f_1\ f_0$
0	0 0 0 0	1 1 0 0 0
1	0 0 0 1	0 0 0 1 1
2	0 0 1 0	0 0 1 0 1
3	0 0 1 1	0 0 1 1 0
4	0 1 0 0	0 1 0 0 1
5	0 1 0 1	0 1 0 1 0
6	0 1 1 0	0 1 1 0 0
7	0 1 1 1	1 0 0 0 1
8	1 0 0 0	1 0 0 1 0
9	1 0 0 1	1 0 1 0 0
10	1 0 1 0	d d d d d
11	1 0 1 1	d d d d d
12	1 1 0 0	d d d d d
13	1 1 0 1	d d d d d
14	1 1 1 0	d d d d d
15	1 1 1 1	d d d d d

Fig. 8.13 (a) Truth tables.

---

[5] For bibliography on this subject see: Bartee, T. C., "Computer Design of Multiple-Output Logical Networks," *IRE Trans. Elect. Comp.*, March 1961, pp. 21–30.

(b)

(c)

$f_7 = x_8 + x_4 x_2 x_1 + \bar{x}_4 \bar{x}_2 \bar{x}_1$

$f_4 = \bar{x}_8 \bar{x}_2 \bar{x}_1 + x_4 x_2 + x_4 \bar{x}_1$

$f_2 = x_8 x_1 + \bar{x}_4 x_2 + x_2 \bar{x}_1$

$f_1 = x_8 \bar{x}_1 + \bar{x}_8 \bar{x}_4 x_1 + \bar{x}_8 \bar{x}_2 x_1$

$f_0 = \bar{x}_8 \bar{x}_4 \bar{x}_2 x_1 + \bar{x}_4 x_2 \bar{x}_1 + x_4 x_2 x_1 + x_4 \bar{x}_2 \bar{x}_1$

(d) Mapping and minimization

$f_7 = \bar{x}_3 / (x_4 / x_2 / x_1) / (\bar{x}_4 / \bar{x}_2 / \bar{x}_1)$

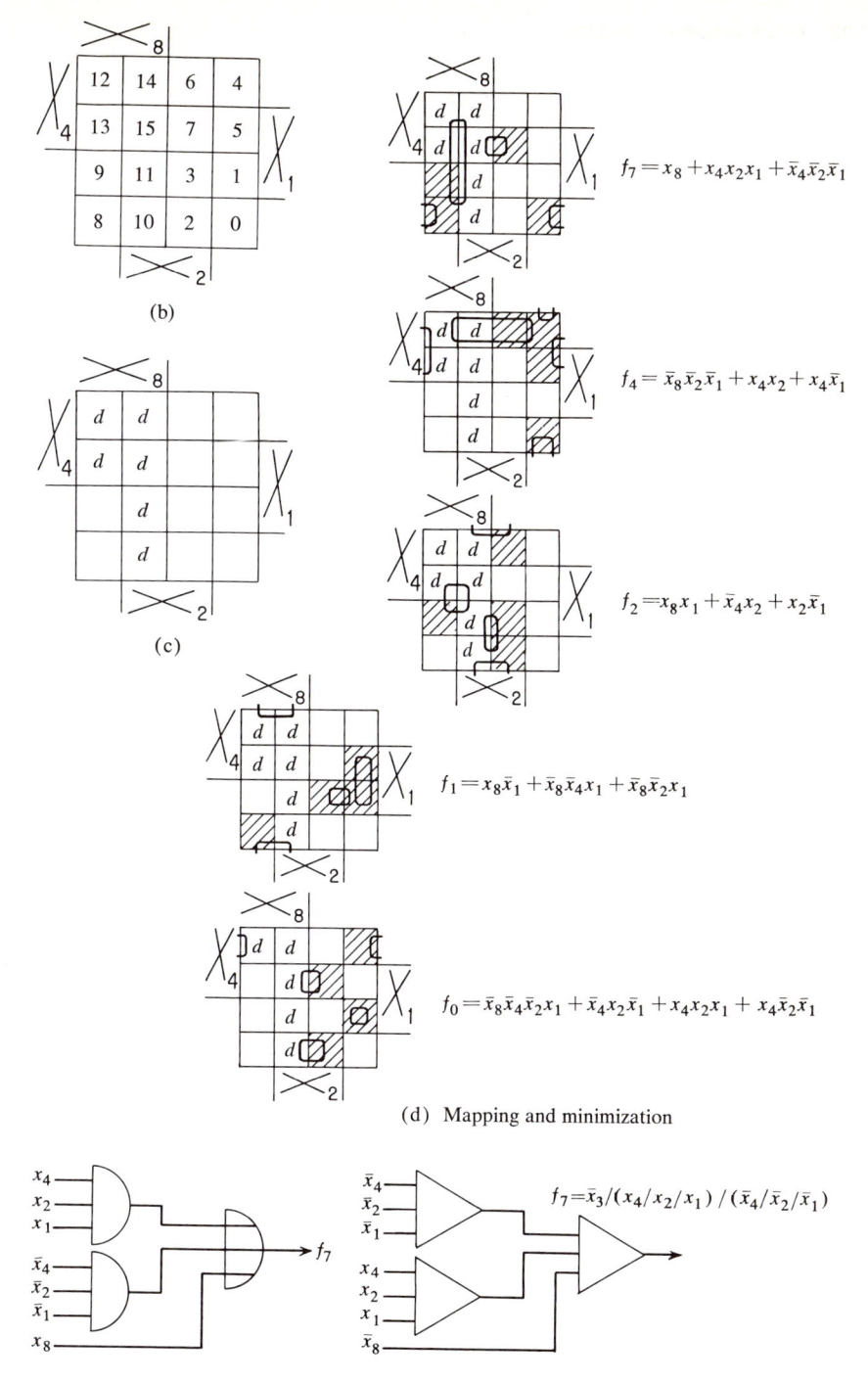

(e) Example of AND/OR and
NAND implementation of f_7

Fig. 8.13 NBCD to 2-out-of-5 code Converter design.

(and all others) is a three-input problem, requiring a "full-adder" circuit, to be discussed later in the text.

A half-adder is then an essential part of a full-adder, which is the circuit for the implementation of the arithmetic addition of two binary numbers. The half-adder itself is a logical (combinational) circuit, whose operation may be defined as follows: "The half-adder circuit accepts two input bits x_1 and x_2 and yields two output bits s and c, in accordance with the relations,

$$s = x_1 \bar{x}_2 + \bar{x}_1 x_2,$$
$$c = x_1 x_2,$$

which are derived from the rules of binary arithmetic, as illustrated in the truth table of Fig. 8.14." The quantity "cs" is the binary sum of the inputs x_1 and x_2.

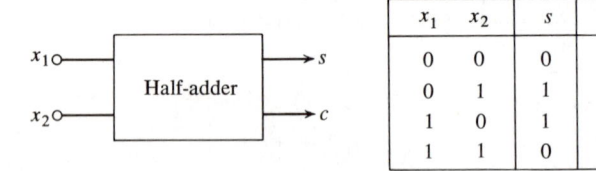

x_1	x_2	s	c
0	0	0	0
0	1	1	0
1	0	1	0
1	1	0	1

Fig. 8.14 Truth table for a half-adder.

The bit s is known as the partial "sum bit," while c is known as the partial "carry bit." The term "partial" is used in the operation of the half-adder because the sum and carry bits must be processed further in order to yield the bits in the final sum.[6]

The AND/OR circuit configurations for the above functions s and c are shown in Fig. 8.15, connected in parallel to form a half-adder combinational circuit.

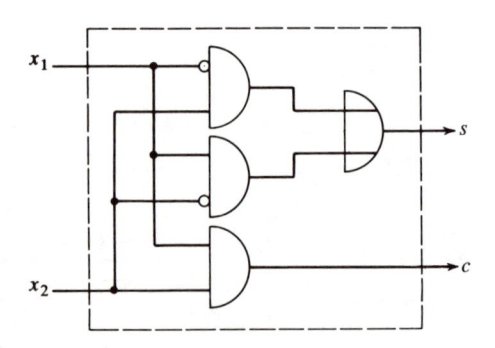

Fig. 8.15 AND/OR half-adder combinational circuit.

[6] Chapter 10.

THE DESIGN OF A HALF-ADDER

An economical realization of a half-adder circuit may be afforded by rewriting the function s as follows:

$$\begin{aligned} s &= x_1\bar{x}_2 + \bar{x}_1 x_2 \\ &= (x_1 + x_2)\overline{(\bar{x}_1 + \bar{x}_2)} \\ &= (x_1 + x_2)\overline{x_1 x_2}. \end{aligned}$$

The AND/OR minimal circuit configuration is shown in Fig. 8.16.

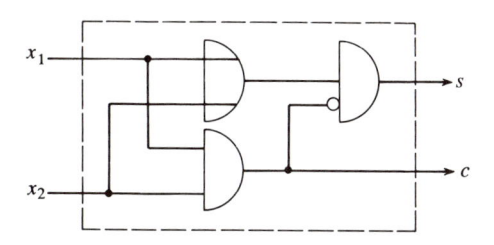

Fig. 8.16 Minimal AND/OR half-adder circuit.

It should be remembered that the design of combinational circuits makes no allowance for time delay. Therefore, the physical time delays caused in passing signals through logic circuits may have to be accounted for separately in consideration of synchronization of circuit operations. In cases where time delay is an integral part of the circuit action, a sequential circuit design method should be used. The subject of sequential circuit design is discussed later.

If a time delay is available, one may obtain the half-adder circuit shown in Fig. 8.17 by the following transformation of the function c:

$$\begin{aligned} c &= x_1 x_2 \\ &= x_1(x_1 x_2 + \bar{x}_1\bar{x}_2) \\ &= x_1\overline{(\bar{x}_1 + x_2)(x_1 + \bar{x}_2)} \\ &= x_1\overline{(x_1\bar{x}_2 + \bar{x}_1 x_2)} \\ &= x_1\bar{s}. \end{aligned}$$

We also have

$$s = x_1\bar{x}_2 + \bar{x}_1 x_2 = x_1 \oplus x_2.$$

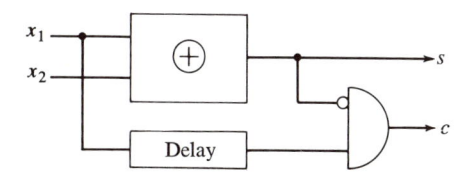

Fig. 8.17 Half-adder circuit with time delay.

206 COMBINATIONAL CIRCUITS

The symbol ⊕ is used for the EXCLUSIVE-OR operation otherwise known as "sum modulo two" (see Section 4.8, Footnote p. 97).

8.8 COMPOSITION AND DECOMPOSITION OF MAPS AND FUNCTIONS

The product of two n-variable maps is also an n-variable map, which contains only those shaded cells found in common in both the component maps. Map multiplication is illustrated in Fig. 8.18, with the mapping of the operation $f = f_1 f_2$. The justification of the above rule becomes apparent if one expresses the functions represented by the maps in their sum-of-products canonical form, and proceeds to perform the algebraic multiplication. Then only the minterms common to the two factors will survive in the resulting function.

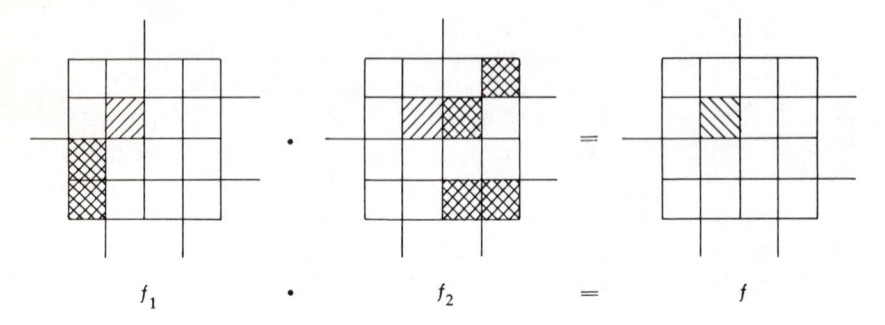

Fig. 8.18 Map multiplication (composition).

The inverse operation, that is, the decomposition of a mapped function into a product of two other functions, is often found useful in the design of logical circuits. A case in point is: "Given the function, $f = x_1 x_2 \bar{x}_3 + \bar{x}_1 \bar{x}_2 x_4$, design a logical circuit realization using only two-input AND/OR gates." The function f includes only three-input products, so that its map consists of only 2×1 shaded cell clusters. It must be decomposed so that $f = f_1 f_2$, where the f_1-map and the f_2-map are selected to include added shaded cells to form only 2×2, 4×1, and 4×2 shaded cell clusters, which correspond to two-variable products or single-variable terms.

One such selection is demonstrated in Fig. 8.19, together with the two-input AND/OR realization. Notice that the added shaded cells (cross-hatched in the drawing) have been selected to form 4×1, 2×2, and 4×2 clusters with the originally given shaded cells.

the circuit by making use of the optional "don't care" terms to obtain the minimal sum-of-products form.

8.8 Given the inputs x_1, x_2, x_3, x_4 and their complements, realize the function

$$f = x_1 \bar{x}_2 x_3 x_4 + \bar{x}_1 x_2 \bar{x}_4$$

using only two-input (a) AND/OR gates,
(b) NOR blocks,
(c) NAND blocks.

8.9 The circuit shown below in Fig. 8.9.1 is a portion of a binary (NBCD)-to-decimal code converter. Explain its operation and demonstrate it using an example.

8.10 To illustrate the fact that circuits with feedback loops

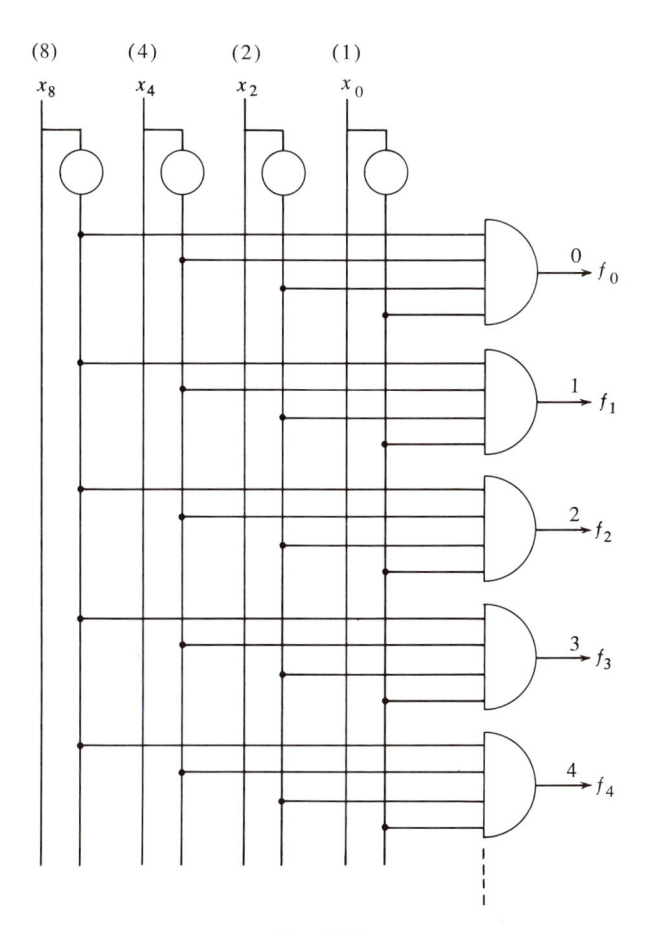

Fig. 8.9.1

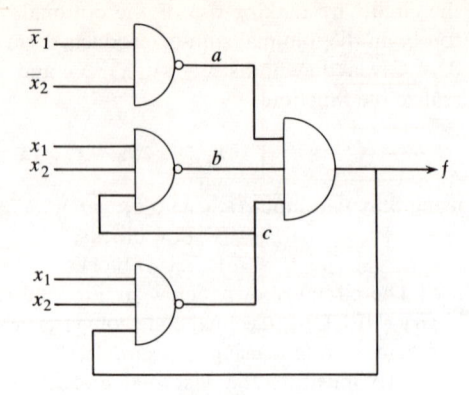

Fig. 8.10.1

do not necessarily involve regenerative action (as is the case with flip-flops, [Chapter 9]) consider the combinational circuit given in Fig. 8.10.1, and derive expressions for f, a, b, c, in terms only of x_1, x_2, and their complements. Draw a simplified circuit to obtain f.

8.11 Simplify each of the following functions making best use of the "don't care" conditions:

(a) $f = x_1 x_2 x_3 + x_1 \bar{x}_2 \bar{x}_3 + \bar{x}_1 x_2 \bar{x}_3$
"Don't cares" $= x_1 x_2 \bar{x}_3 + \bar{x}_1 \bar{x}_2 x_3$

(b) $f = x_1 x_2 x_3 + x_1 x_2 \bar{x}_3 x_4 + \bar{x}_1 x_2 \bar{x}_3 \bar{x}_4 + x_2 x_3 \bar{x}_4$
"Don't care" $= \bar{x}_1 x_2 x_3 x_4$

(c) $f = \bar{x}_1 x_3 \bar{x}_4 + x_2 \bar{x}_3 x_4$ "Don't care" $= \bar{x}_1 x_2 x_3 x_4 + x_1 x_2 x_3 \bar{x}_4$

8.12 Transform the function $f = x_1 x_2 + \bar{x}_1 x_3$ into an all NOR function, and implement it in block diagram form (a) as a three-level NOR circuit, and (b) as a two-level NOR circuit.

8.13 Analyze the logical network shown in Fig. 8.13.1 to obtain the minimum sum-of-products and the minimum product-of-sums forms of the function f.

8.14 Equations (10.1) and (10.2) of Chapter 10 provide the carry and the sum of a full-adder. These equations may be written as follows:

(a) $c_{i+1} = a_i b_i + a_i c_i + b_i c_i$

(b) $S_i = a_i \bar{b}_i \bar{c}_i + \bar{a}_i b_i \bar{c}_i + a_i b_i c_i + \bar{a}_i \bar{b}_i c_i$

Provided that the variables a_i, b_i, c_i, are available in both their

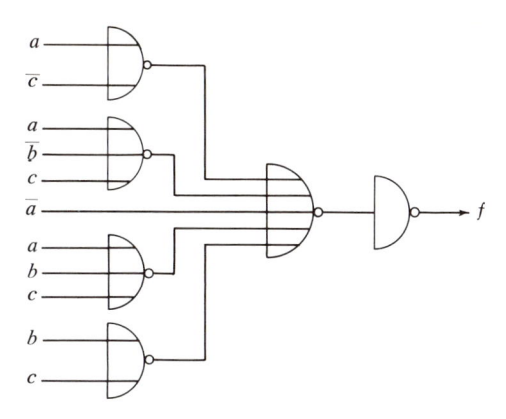

Fig. 8.13.1

uncomplemented and complemented forms, design (a) a two-level NOR circuit for the carry c_{i+1}, and (b) a three-level NOR circuit for the sum S_i.

8.15 A four-bit register has been built to store one decimal digit according to the code shown below. No other combination except the ones shown in the code below may be contained in the register. Design a minimal logical circuit which, when inputted by the register's cells, will recognize and put out a 1 if the register contains a: 0, 2, 6, 7, or 8.
(a) Use only AND/OR/NOT blocks
(b) Use only NAND blocks
(c) Use only NOR blocks.

Decimal digit	$x_1 x_2 x_3 x_4$
0	0 0 1 1
1	0 1 0 0
2	0 1 0 1
3	0 1 1 1
4	1 0 0 0
5	1 0 0 1
6	1 0 1 1
7	1 1 0 0
8	1 1 0 1
9	1 1 1 1

8.16 Determine and minimize the function f of the logic circuit shown in Fig. 8.16.1.

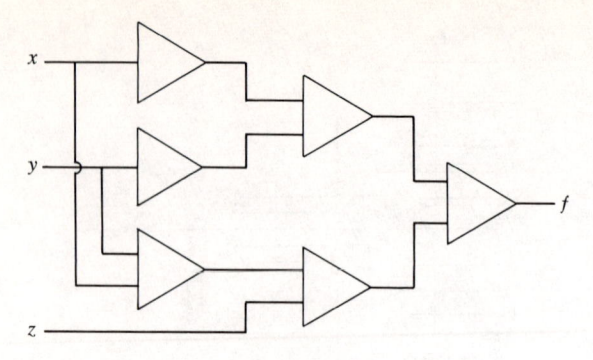

Fig. 8.16.1

8.17 Design minimum logical circuits for problems 5.5 and 5.6 using AND/OR/NOT gates.

8.18 Design minimum logical circuits for problems 5.5 and 5.6 using only NOR blocks.

9

Elementary Components: Switching Devices and Circuits

In previous chapters we developed the principles and the methodology of designing logical circuits which would perform various information-processing tasks. The desirability of using a binary logic, which is implemented in hardware by the switching algebra, leads us to a well-pointed definition of the characteristic types of devices and elementary circuits necessary for constructing these essential information-handling logical circuits.

In designing information-processing machines there is a basic need to implement logical, amplifying, and memory functions, in addition to a host of other auxiliary functions. We have already shown that a "complete" set of logical blocks, such as the AND/NOT set, would suffice for the realization of all logical functions. In this chapter, the logical and amplification functions are often combined in hardware implementations, as for example the case with transistor logic circuits. If a "delay" element or a flip-flop is added to a complete logical set, then memory functions are also implemented along with the logical ones.

The necessary elementary components that are needed in order to build the various logical and memory units of an information-processing machine are various types of binary switching devices and elementary circuits. A large modern digital computer may contain nearly half a million switches of various kinds for performing logical functions and several million elementary memory components. Elementary components include switching devices, such as the electromagnetic relay, the magnetic core, transistors, and diodes, in a variety of elementary circuits, such as multivibrators, inverters, clippers, amplifiers, and so forth. There is a wide variety of electronic, electromagnetic, electro-optical, and other devices and circuits. Within the scope of this book, we discuss only the most representative and commonly used elementary components in order to underline only the important features of such devices and circuits.

9.1
ENGINEERING SPECIFICATIONS OF SWITCHING DEVICES

Devices which distinctly possess either of two discrete well-defined stable-equilibrium conditions are called "switching" devices. They are used as components in constructing binary logic and memory circuits, like the ones discussed in previous chapters.

The operation of two-state switching devices is based on their ability to switch the "value" of some physical parameter (scalar or vector), such as the index of refraction, the electrical conductivity, the magnetic moment, and so on, between two discrete conditions. The action is exerted either on the magnitude of the parameter if the parameter is a scalar quantity, or it may also be exerted on the direction of its equilibrium position if it is a vector quantity. The change of condition must occur reliably and in response to an external command. It must also be repeatable bilaterally between the two conditions, so that it may be effected over and over at will.

A great variety of natural phenomena can be put to use in building switching devices. Under proper control the devices must be capable of ascertaining either of two discrete and stable conditions. To be useful in a logic circuit, the switching device must be able to pass *reversibly* from one state to the other under the influence of an appropriate external control. This requirement prevents an ordinary fuse in an electric circuit, for example, from being used as a switching component; once a fuse has been used, it cannot be used again. On the other hand, a piece of litmus paper dipped in a chemical solution makes a switching device which functions reversibly. Its color turns to red if the solution is made acid and blue if the solution is made basic. The condition of this device may be detected by a light sensor with different sensitivity to blue or red light reflected from the litmus paper.

To be acceptable as a component of logic circuits, the switching device must satisfy a number of engineering requirements. The most important are:

1. easy and reliable engineering means of detecting the change of condition;

2. ample margin of stability in either condition, reliable transition between conditions, relative insensitivity to environmental influences (temperature variations, and so forth), and good signal-to-noise ratio;
3. economy of fabricating the device and compliance to mass production;
4. physical size of the device and adaptability to micropackaging ("time-of-flight" problems in high-speed computers[1]);
5. speed of switching between the two discrete conditions, compatible with the task requirements;
6. communication, that is, uniformity of the type of signals (electric voltage or current, light, and so on) used between the device and the other computer components;
7. low switching power requirements. The power stimulus necessary for switching the device between its two discrete conditions must be tolerably low. In two-state memory devices, the requirement of "holding power," that is, the power required to sustain either of the two equilibrium conditions, is also an important consideration;
8. in the case of switching memory ("flip-flop") devices, the principle of addressing an array of such similar devices is an additional engineering consideration, in that it determines the reliability of addressing the computer memory;
9. devices are sought which by their switching action deliver at their output enough energy to drive to switching at least two or three other similar devices. This property is often referred to as the "fan-out gain" of the switching device.

According to their operational features, switching devices may be classified as *switches*, *relays*, and *flip-flops*.

9.2 SWITCHES

A diode *switch* is a two-terminal "valve" which, under proper external control, may permit or disrupt the flow of some physical quantity such as air, liquid, or electric current. The switching mechanism is built into the device and is stimulated by external means. Of the great variety of existing switches, we are interested in the electronic type. When inserted in an electric circuit, this type of switch allows current flow only in one direction, as illustrated with the ideal diode symbol in Fig. 9.1(a). The solid-state semiconductor

---

[1] In very fast modern computers, the time it takes the signal to travel between switching devices, "time-of-flight," may become comparable to the switching time of the device itself. In such a case, microminiaturization is sought in order to combat the "time-of-flight" restriction on the speed of modern machines.

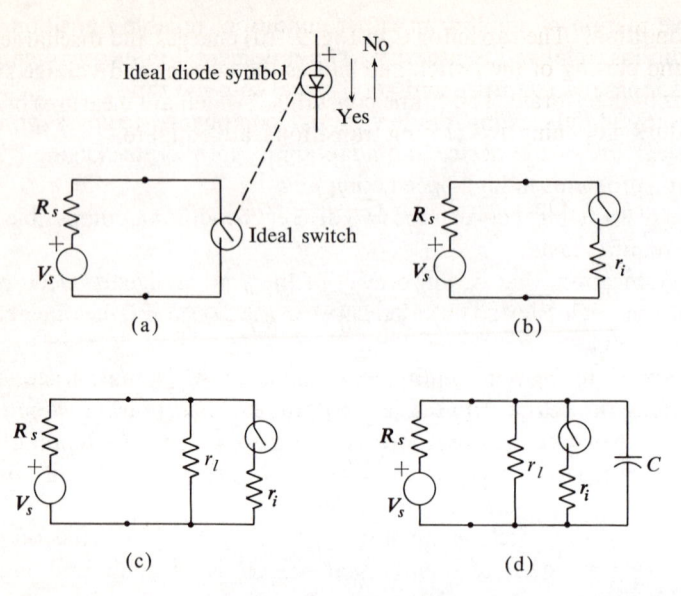

Fig. 9.1 (a) Ideal switch, (b) switch with internal resistance, (c) switch with internal and leakage resistance, and (d) real switch with inertia and losses.

diode is typical of such switches. It is used as a rectifier or as a clamper, and so forth, in more conventional electronic applications.

Ideal Diode Switch. Figure 9.1(a) shows an ideal switch, independently operated, and connected to a voltage source V_s of an internal resistance R_s. The switch opens and closes instantaneously (no inertia) and presents zero resistance to the current when "closed" and infinite resistance when "open."

Diode Switch with Internal Resistance. When a real switch is in the "closed" condition, the voltage across the switch does not drop to zero because of some internal resistance r_i present in the switch as shown in Fig. 9.1(b). This causes a slight decrease in the "closed" current and a nonzero voltage.

Diode Switch with Internal and Leakage Resistance. When open, a real switch may leak, thus presenting a high but not infinite resistance path to the electric current. This situation is shown in Fig. 9.1(c).

Diode Switch with Internal and Leakage Resistance, and Possessing Inertia. In the previous cases, no "closing" or "opening" time is required when operating the switch. The transitions are instantaneous, and the slopes of the current and voltage waveforms during the transitions are infinite. Real switches, however, always have some capacitance between their terminals, manifested by the presence of inertia during the transition of the device between the two conditions. The capacitor C in Fig. 9.1(d) charges and discharges during opening and closing of the switch and causes the current and voltage switching waveforms to deteriorate. The "time constants," which are measures of the time required during opening and closing transitions, are equal to

$$\text{Opening} = \tau_{op} = C\frac{R_s r_l}{R_s + r_l} \cong R_s C,$$

$$\text{Closing} = \tau_{cl} = C\frac{R_s r_l r_i}{R_s r_l + R_s r_i + r_l r_i} \cong r_i C.$$

It is worthwhile noting that, under any circumstances, the switch always opens more slowly than it closes; that is,

$$\tau_{cl} < \tau_{op}.$$

The capacitance C together with the surrounding resistances determine the speed with which the switch changes condition; that is, their values become of importance in transient analysis. The power dissipation of a real switch is larger during the closing phase, and special care should be taken in designing the switch to keep the dissipated power lower than the maximum permissible dissipation at any instant.

The inertia effects described above and displayed by the capacitance C may be due to parasitic or load capacitances, as well as to inherent properties of the switch. Inertia effects are present in all practical devices in all three classes of switches, relays, and flip-flops.

Sudden closing or opening of a switch in several electronic pulse circuits may be used to obtain jumps or discontinuities in currents or voltages. Pulse shaping is accomplished by superimposing such discontinuities in time to obtain the desired pulse waveforms. In calculating voltage or current transient waveforms, the switch is replaced by a current source when open and by a voltage source when closed. The value of the source is equal but opposite in sign to the existing current through the switch or the voltage across the switch just before switching.

Mechanical switches approach the ideal switching operation as far as "fast break" and "fast make," that is, transitions between "open" and "closed" states, is concerned. They may present practically zero internal resistance, almost infinite leakage resistance, and a small capacitance; therefore, mechanical switches are capable of very short τ_{op} and τ_{cl} times. However, because of slow-moving mechanical parts, mechanical switches operate at comparatively slow rates of repetition. Mainly because of this limitation, electronic switches are used exclusively where fast repetitive switching becomes of prime importance. Electronic switches such as the common semiconductor diode, as well as optical, magnetic, and other special types, are normally less nearly perfect than mechanical switches as far as sudden transitions are concerned, but offer great speed of cyclic operation, higher reliability, and lower cost.

9.3
ELECTRONIC DIODE AND/OR GATES

The solid-state semiconductor diode is a most attractive switch in logic circuit design because of fast switching, low cost, small size, and low power consumption. However, because of inability to amplify and to invert input signals, most diode logic circuits also use active devices, such as transistors and vacuum tubes. The lack of amplification in cascaded diode circuits causes eventual deterioration of the input signals, and thus limits the "fan-out," that is, the maximum number of outputs obtainable. Signal inversion is also required in order to realize the complementation operation.

Assuming "positive logic,"[2] where 1 is represented by a more *positive* voltage level V_1 and 0 is represented by a less positive voltage level V_2, we may use the diode circuits shown in Fig. 9.2 to implement the two-input AND and OR operations. The circuits may be extended to any practical number of inputs.

In the AND diode gate, shown in Fig. 9.2(a), the output voltage f is raised to $+5$ V if, and only if, *both* input voltages x_1 and x_2 are raised to $+5$ V. If at least one of the input voltages is at zero voltage level, then the output voltage is also at zero voltage level. More generally, the output always equals the *smallest* of the input voltages as long as $V_1 < V_2 < E_1$. As illustrated with the voltage pulses in Fig. 9.2(c), one can say that the AND circuits shown will provide a positive output voltage, that is, a one-bit, when both inputs are positive, that is, ones; it will provide a zero output voltage, that is, a zero-bit, when at least one of the input voltages is at zero voltage level, that is, a zero-bit. We should, then, easily recognize the AND operation by constructing the truth table shown in Fig. 9.2(a).

The OR diode gate is analyzed similarly. If any one, or more, of the input voltages is raised to $+5$ V, then the corresponding diodes would conduct (be "closed") to make the output voltage also rise to $+5$ V and back-bias ("open") the rest of the diodes. Only if all input voltages were at the zero level, would the output voltage also be at the zero level. Generally, again, the output equals the *largest* of the input voltages as long as $E_2 < V_1 < V_2$. The operation is again illustrated with voltage pulses and with the corresponding truth table, as shown in Fig. 9.2.

Ideal diode gates will operate with any input voltage levels in the way described above as long as $E_1 \geq V_2 \geq V_1 \geq E_2$. With real diodes, however, one should see to it that the input voltage swing $(V_2 - V_1)$ is much larger than the forward voltage drop of the diode. The value of the resistor R is limited by the finite reverse resistance of the practical diode used in the circuits. The value R is selected large to minimize power consumption, but not too large in comparison with the reverse resistance of the diode. The junction and the diffusion capacitances, which depend on the applied voltages, together with the parasitic capacitances, limit the switching speed of the diode and thus the speed of the

---

[2] Also referred to as "positive true."

ELECTRONIC DIODE AND/OR GATES 221

AND

Circuit

(AND circuit diagram with diodes D_1, D_2, resistor R, supply $E_1 = 10$ volts, inputs x_1, x_2, output $f = x_1 x_2$)

Truth table

x_1	x_2		
0	0	0	$D_1 =$ ON, $D_2 =$ ON
0	+5	0	$D_1 =$ ON, $D_2 =$ OFF
+5	0	0	$D_1 =$ OFF, $D_2 =$ ON
+5	+5	+5	$D_1 =$ ON, $D_2 =$ ON

Symbol

(a) $x_1, x_2 \to f = x_1 x_2$

OR

(OR circuit diagram with diodes D_1, D_2, resistor R, supply $E_2 = -10$ volts, inputs x_1, x_2, output $f = x_1 + x_2$)

x_1	x_2		
0	0	0	$D_1 =$ ON, $D_2 =$ ON
0	+5	+5	$D_1 =$ OFF, $D_2 =$ ON
+5	0	+5	$D_1 =$ ON, $D_2 =$ OFF
+5	+5	+5	$D_1 =$ ON, $D_2 =$ ON

(b) $x_1, x_2 \to f = x_1 + x_2$

Input voltage waveforms

x_1: 1 1 0 1 — $V_2 = +5$ volts, $V_1 = 0$ volts
x_2: 1 0 0 1 — $V_2 = +5$ volts, $V_1 = 0$ volts

Output voltage waveforms

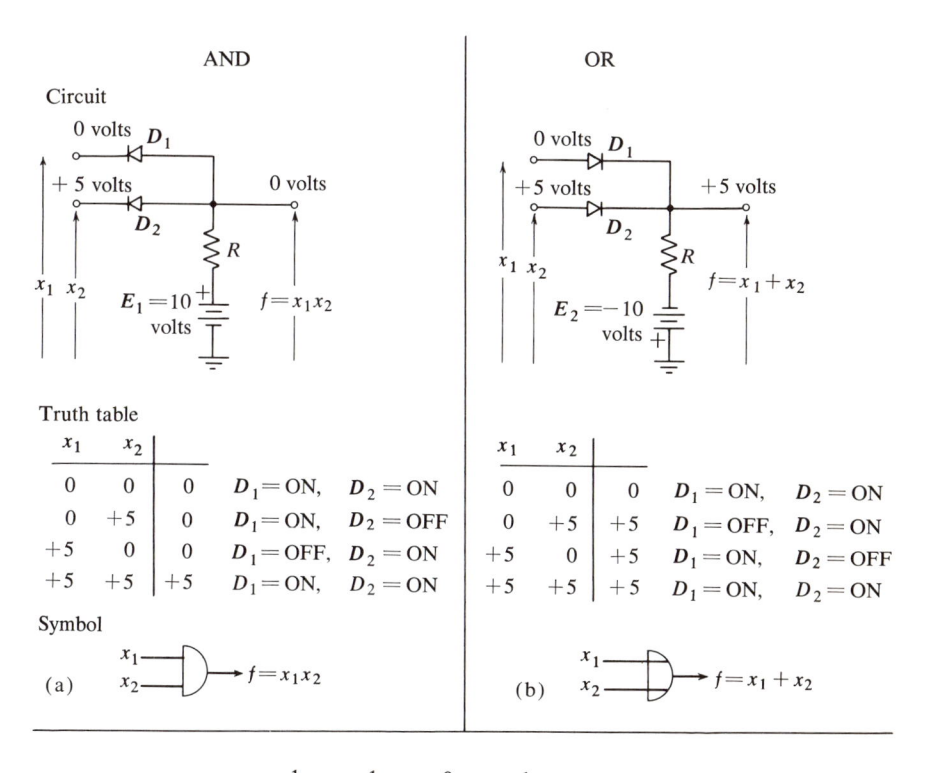

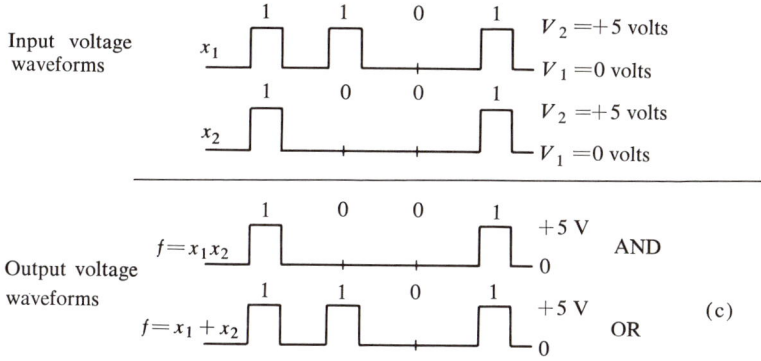

$f = x_1 x_2$: 1 0 0 1 — +5 V / 0 AND
$f = x_1 + x_2$: 1 1 0 1 — +5 V / 0 OR (c)

Fig. 9.2 Single level, two-input AND and OR gates. The voltage levels and power supplies are such that $E_1 \geq V_2 \geq V_1 \geq E_2$. Under these conditions on voltages, the output voltage equals the *smallest* of the input voltages in the case of the AND circuit and equals the *largest* of the input voltages in the case of the OR circuit.

circuit. In ultra fast logic circuits, the physical separation between input and output terminals also becomes of importance, because of "time-of-flight" limitations in the propagation of signals.

222 ELEMENTARY COMPONENTS: SWITCHING DEVICES AND CIRCUITS

Logic diagram

Diode circuit

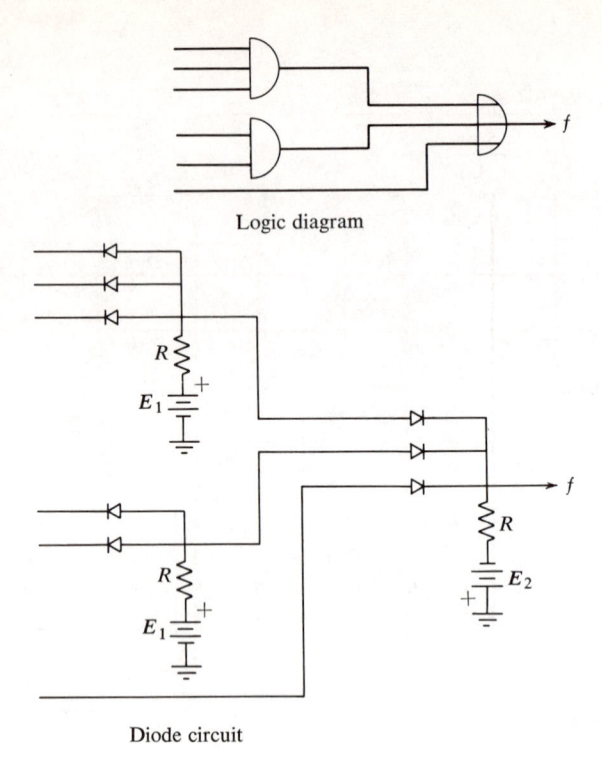

Fig. 9.3 Two-level diode logic circuit.

Multilevel Diode Logic Circuits. Diode gates may be cascaded to form multilevel AND/OR logic circuits. Figure 9.3 illustrates a two-level diode circuit. Two-level logic may be preferred in cases where the minimization of time required for the passage of the signal through the circuit is of special importance. The switching function is then expressed in a sum-of-products or product-of-sums form.

For reasons of economy, multilevel logic may be preferred in certain cases because considerable savings in the number of diodes may be effected. Diode multilevel circuits, however, suffer from attenuation and require more input power to be driven with. They should be designed to operate in conjunction with transistor amplifier stages, and in general they involve more design effort.

9.4 DIODE DECODING/ENCODING MATRICES

Diode AND circuits may be combined appropriately to form a "decoder." The matrix circuit shown in Fig. 9.4(a) decodes numbers in excess-3 code to decimal. It consists of ten AND circuits, each AND circuit having four diodes connected to a horizontal output line, as shown in Figure 9.4(b).

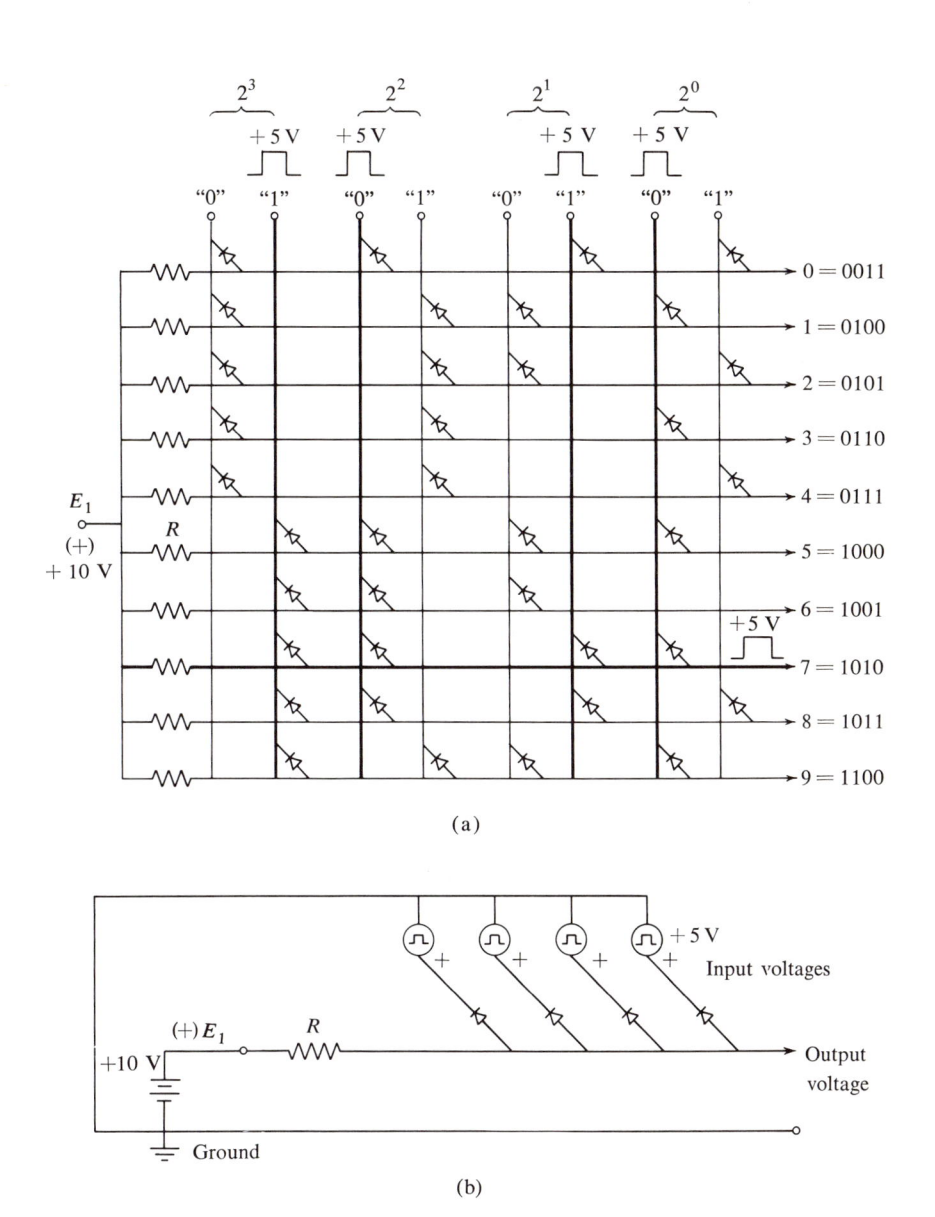

Fig. 9.4 (a) Diode decoding matrix: excess three to decimals. (b) A four-input AND circuit connected to a horizontal output line and showing connections to input voltage sources and to ground. Compare with the AND circuit of Fig. 9.2. Ten such circuits are used in the decoding diode matrix above. Only if all input voltages rise to +5V will the output voltage also rise to +5V.

A given four-bit excess-3 number is applied to the decoder by raising the potentials of the appropriate input vertical lines, that is, by applying simultaneously positive voltage pulses, for example, +5 V to either the (1) or the (0) line of each of the four pairs of vertical input lines. All other input lines are kept grounded. Then, only the one horizontal output line with all four diodes connected correspondingly to the four activated positive vertical lines will rise in potential to +5 V. All other horizontal output lines remain grounded because at least one diode per line remains grounded.

The decoding of 1010 into 7 is illustrated in Fig. 9.4(a) by drawing the activated positive lines bold. A positive voltage (+5 V) appears at the output of line 7 for the duration of the simultaneously applied input pulses. Since all diodes are assumed to be ideal, no voltage drop is experienced across any of them.

In a manner analogous to the one discussed above, diode OR circuits may be combined to form an "encoder." The matrix circuit shown in Fig. 9.5(a) encodes decimal numbers to NBCD. Each one of eight OR circuits has its diodes connected to a horizontal output line as illustrated in Fig. 9.5(b). A positive voltage pulse applied to one of the ten vertical input lines raises the potential of a certain combination of four output lines. For example, a positive voltage pulse (+5 V) on line 7 feeds with pulses (+5 V) the four output lines, drawn heavier in Fig. 9.5(a). The activated output lines, marked also with voltage pulses in Fig. 9.5(a), read 0111, that is, binary seven. One way to identify the activated output lines is by using "Set-Reset" flip-flops.[3] The flip-flops, which are bistable devices such as multivibrators or magnetic cores, are connected to the output lines as shown in the matrix configuration of Fig. 9.5(a). When a voltage pulse is applied to the set (S) input line, the flip-flop is "SET" in the one-condition, as a togle switch may be set in the ON-position. If a voltage pulse is applied to the reset (R) input line, the flip-flop is "RESET" to the zero-condition, as a togle switch may be reset to the OFF-position. Indicator lamps may be used to indicate the condition of the output flip-flops, so that one may read the encoded number visually, as it is illustrated with the number 0111 in the example of Fig. 9.5(a).

It should be noted here that the output lines of the decoder [Fig. 9.4(a)] may be connected to the input lines of the encoder [Fig. 9.5(a)] to form a code translator from excess-3 code to NBCD code. By means of logic design, the two circuits may be merged into a translator matrix using fewer diodes.

9.5
RELAYS AND THREE-TERMINAL ELECTRONIC SWITCHES

An electrical relay is generally a device by which a change of current or potential in one circuit can be made to produce a change in the

---

[3] See Section 9.7.

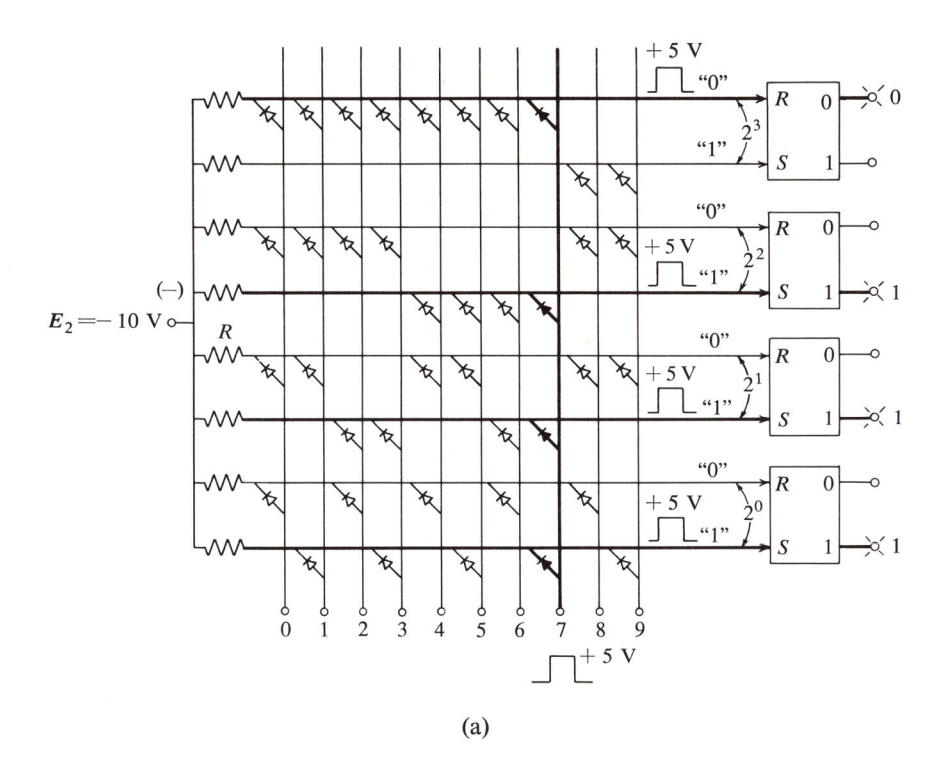

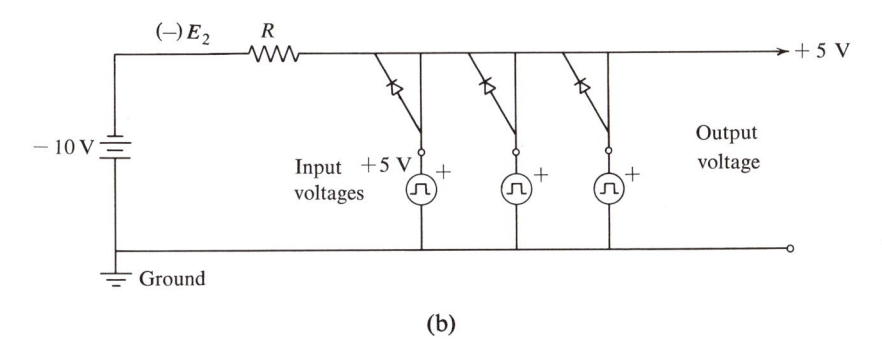

Fig. 9.5 (a) Diode encoding matrix: decimal to NBCD. (b) Illustrative configuration of a single OR circuit with a horizontal output line. Eight such multi-input OR circuits are used in the encoding diode matrix above. (If any one of input voltages rises to +5V, the output voltage also rises to +5V.)

226   ELEMENTARY COMPONENTS: SWITCHING DEVICES AND CIRCUITS

electrical condition of another circuit. Such devices often possess the ability to amplify electrical signals as, for example, is the case with all charge-controlled devices (for example, transistors).

Because the control may be extended from "full-conduction" (or "saturation") to "cut-off,"[4] three terminal electronic relay devices have been developed and used in electronic digital computers. Such devices as vacuum tubes, gas tubes, transistors, cryotrons,[5] and other optical and magnetic devices may ease or hinder the electrical contact between two terminals for as long as a controlling input signal is supplied to the third terminal. These devices are distinguished from flip-flop type devices in that they do not "latch on" to the acquired "closed" or "open" condition. They require the continuous presence of the control signal to maintain the acquired condition. Thus, magnetic cores and other similar devices are characterized, in contrast, as flip-flop devices (Section 9.7).

Although electromagnetic telephone relays (referred to from now on as "e/m relays") are not widely used at present in electronic computer logic circuits, they are, however, still of great interest, because a study of relay logic circuits may shed considerable light on the general principles of computer design. In general the e/m relays have two input (control) terminals and several pairs of controlled terminals.

The e/m Relays. Structurally, the e/m relays (Fig. 9.6) consist of three parts: (1) magnetic circuit which contains the "core," the pivoted armature, the frame, and an air gap, (2) winding for stimulating the magnetic circuit (control signal winding), and (3) several pairs of springs with contacts. An air gap is kept at all times between the moving armature, the core, and the frame, to keep the relay from "sticking." "Sticky" operation is corrected by adjusting a nonmagnetic screw.

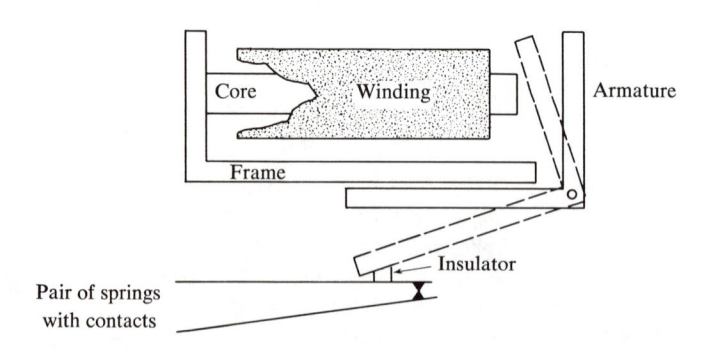

Fig. 9.6 Structural schematic of an e/m relay.

---

[4] These terms are used to represent the two states which the device may possess when used as a three-terminal switch.

[5] Flip-flop cryotron circuits are possible just as flip-flop transistor circuits are. However, cryotrons, like transistors, are considered as relay devices only when used properly.

A single e/m relay can often carry more than six pairs of changeover contacts in a variety of functional combinations. Because of this, quite elaborate relay operations can be performed with a small number of e/m relays. Figure 9.7 illustrates some typical e/m relay structures in principle.

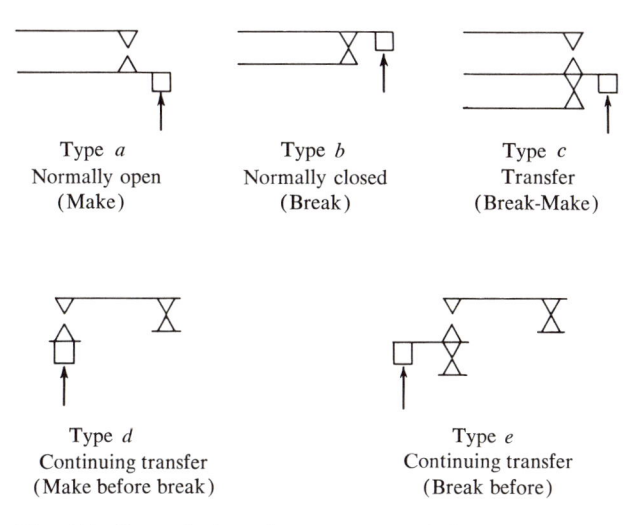

Type *a*
Normally open
(Make)

Type *b*
Normally closed
(Break)

Type *c*
Transfer
(Break-Make)

Type *d*
Continuing transfer
(Make before break)

Type *e*
Continuing transfer
(Break before)

Fig. 9.7 Several e/m relay structures (Schematic end views).

Although very versatile in circuit manipulations and suitable for many applications, e/m relays are constantly abandoned in areas where speed and reliability are important. Their operation becomes unreliable over long periods of time partly because the contact tips oxidize and partly because dust particles settle between the contact tips. Then too, they are not economical when used in massive information storage networks. For these reasons, e/m relay computers are very unlikely in the future, unless there is some unexpected breakthrough in their technology. Several e/m relay computers have been built. The best known of these are the Bell Telephone machines and the Harvard Mark I and Mark II. These machines are very similar, their main difference being that the circuits in the Bell machines were designed with self-checking provisions, while the Harvard machines were mainly program-checked.[6]

Three-Terminal Electronic Relay Devices (Three-Terminal Switches). Vacuum or gas tubes, transistors, and other three-terminal electronic devices may function as relays (or as three-terminal switches) under proper conditions. Vacuum tubes, for example, may change from full conduction to cut-off by proper control of their grid potential. Similarly, the emitter-to-collector

---

[6] The interested student is encouraged to study these computers. Relay arithmetic circuits are of great interest and much may be learned from such a study about the principles of computer design.

communication in a junction transistor may be regulated by proper regulation of the base current or potential. The two junctions in the transistor may be reverse biased, thereby reducing the emitter-collector current to very small values ("open" or "off" condition). Or, the junctions may be both biased forward, reducing the emitter-collector voltage to very small values ("closed" or "on" condition).

9.6 TRANSISTOR LOGIC

The transistor has proved to be a very useful and versatile device in the design of logical circuits. It makes possible the very important operation of complementation, and in addition to providing signal amplification, it offers high switching speed, low cost, small size, and reliability. From the great number of different systems of logic which have been developed, many are based on the use of transistors alone, while others refer to the use of transistors in combination with other elements, such as resistors or diodes.

A transistor in the common-emitter connection makes an excellent switch and provides the basis for several classes of transistor logical circuits. In comparison with the e/m relay, Fig. 9.8(a), the emitter and collector terminals correspond to a pair of relay contacts and the base current performs the control operation. The presence of a sizable base current will switch the transistor "ON," while a zero (or very small) base current will cause the transistor to switch "OFF."

The basic circuit of a transistor switch is shown in Fig. 9.8(b), together with the I_C–V_C characteristic curves. The switch is "open," that is, the transistor is turned OFF, by applying a zero or slightly negative base current, $I_B \lessgtr 0$. In this case the collector current is very small, $I_C \approx 0$, and the electrical resistance from collector-to-ground is very large (normally over a million ohms). This mode of operation corresponds to Region I on the I_C–V_C graph of transistor characteristic curves.

The transistor switch is "closed," that is, the transistor is turned ON,[7] by applying a positive base current $I_B > 0$. In this case the collector current can be very large, $I_C \gg 0$, and the electrical resistance from collector-to-ground drops to a small fraction of one ohm. The voltage difference from collector to emitter also drops to a small fraction of a volt. This mode of operation corresponds to Region III on the I_C–V_C graph of transistor characteristic curves.

For a full discussion of the characteristics of transistor operation and of the several modes of its application to logic circuits, the reader is referred to the literature. In this introductory treatment, we illustrate only the most characteristic configurations of transistor logic.

---

[7] The ON and OFF states of the transistor correspond to the "full conduction" or "saturation" and "cut-off" mentioned earlier (Section 9.5).

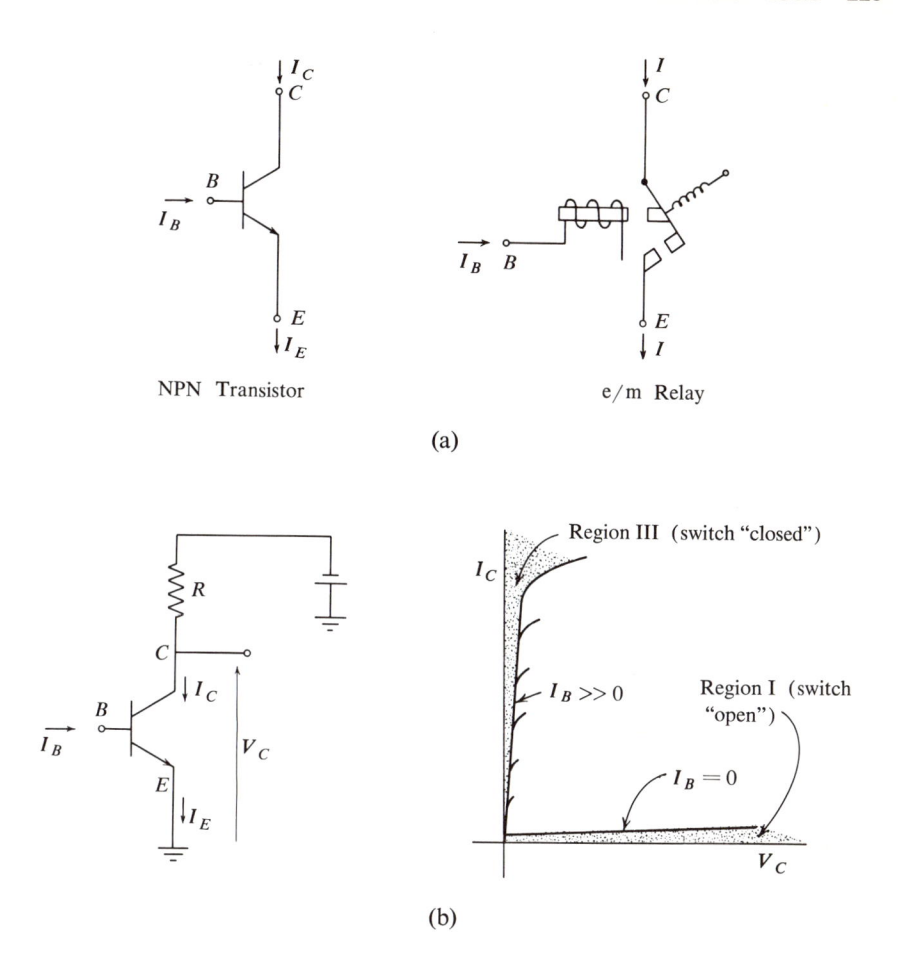

Fig. 9.8 (a) Transistor and e/m relay; (b) transistor switch and $I_c - V_c$ characteristics.

a. Transistor Inverter

A transistor inverter, which performs the complementation operation, is shown in Fig. 9.9. The inverter is designed so that with the single input x grounded (that is, "1") the transistor draws sufficient base current to be turned ON,[8] and the output is very nearly at the negative potential −6 V (that is, "0"). The required base-to-emitter forward voltage to turn the transistor ON, is found from the input characteristic curves which the transistor manufacturer provides. When the input is made negative, about −6 V (that is, "0"), the transistor is turned OFF, and the output rises to ground potential (that is, "1").

---

[8] When in the ON condition, the transistor operates in the "saturation" region with both the emitter and collector junctions being forward-biased.

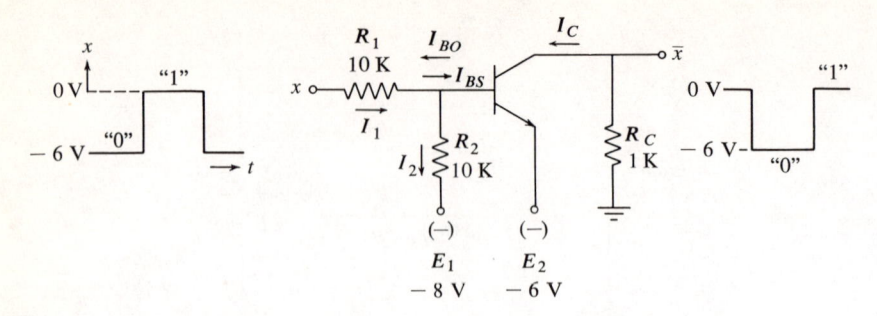

230 ELEMENTARY COMPONENTS: SWITCHING DEVICES AND CIRCUITS

Fig. 9.9 Transistor inverter (NOT) circuit. (Typical values are shown.)

The action of complementation is apparent in the above description. As with almost all other transistor applications, transistor inverters may also be constructed using PNP units,[9] if the polarities of the bias sources are reversed.

A transistor may be driven in the ON condition by forward biasing both junctions; this is known as the "saturation" condition. In order to drive a saturated transistor to the OFF condition (both junctions reverse-biased), excess charge accumulated in the base region during saturation must be removed. The accumulated excess charge increases with the "depth" of saturation preceding the switching. The operation of removing the accumulated base charge—that is, the ON-OFF switching operation—is comparatively slow, thus limiting the switching speed of the device.[10] In order to prevent the transistor from entering its saturation region but slightly, a diode-clamped load is used in the inverter circuit, shown in Fig. 9.10. The battery voltages E_3 and E_4 are adjusted so that

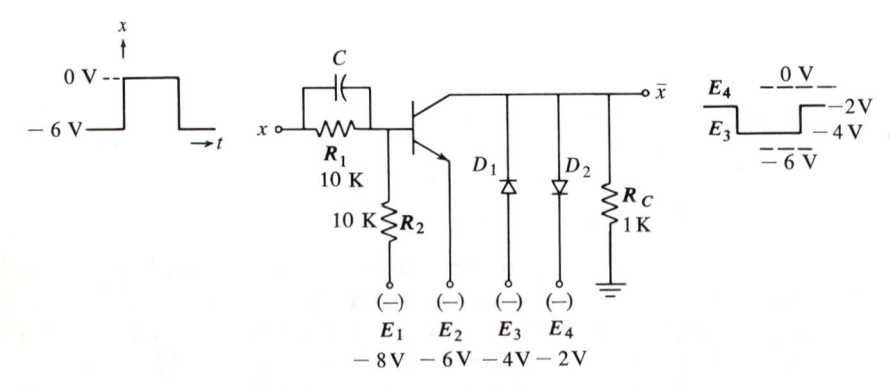

Fig. 9.10 Transistor inverter with diode-clamped load. (Typical values are shown. E_1 and E_2 may be equal. D_1 and D_2 are ideal.)

[9] The reader is referred to elementary books on transistor circuits for additional information on NPN and PNP transistor characteristics.

[10] The operation is analogous to discharging the capacitor of the diode switch discussed in Section 9.2, where it was observed that $\tau_{op} > \tau_{cl}$.

the collector voltage is not permitted to go below E_3 or above E_4. As the transistor conducts, the ideal diode D_1 conducts and clamps the collector voltage at E_3. When the collector voltage exceeds E_4, the ideal diode D_2 will conduct, clamping the collector at E_4 volts. The capacitance C also helps speed up the switching operation. One or more real semiconductor diodes connected in series are also commonly used to speed up the operation of the circuit.

b. Emitter-Follower Logic Circuits

The two circuits shown in Fig. 9.11 perform the AND and OR logic operations. Because of the emitter-follower configuration there is no phase inversion of the input signal. The OR gate employs NPN transistors in a positive logic operation, as follows: when any one or more of the inputs is at the high voltage level, (logical value "1"), then the output f is also at the same voltage except for a small voltage drop across the base-to-emitter junction. The transistors whose inputs remain at the low voltage level (logical value "0") are reverse

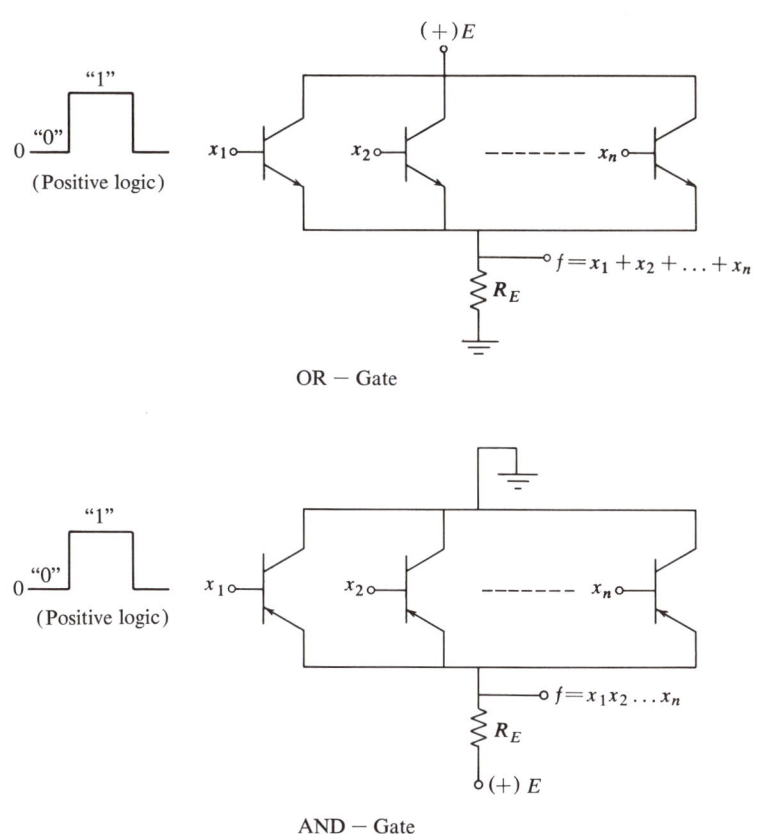

Fig. 9.11 Emitter-follower AND/OR circuits.

biased. If, and only if, all inputs are at the low voltage level, the output is also at low voltage level.

The AND gate employs PNP transistors, also in a positive logic operation. The output has the logical value "0" if one or more of the inputs is "0". If, and only if, all inputs have the logical value "1", that is, they are at sufficiently positive potentials, then the output is also "1".

c. Common-Emitter Logic

NAND and NOR circuits may be produced by connecting common-emitter stages in series or in parallel. A two-input NAND and a two-input NOR are illustrated in Fig. 9.12. The circuits employ positive logic. In the NAND circuit, if, and only if, both inputs are raised to sufficiently high positive potential (that is, "1"), then both transistors are turned ON, and the output drops to almost ground potential (that is, "0"). In the NOR circuit, if the base of either transistor (or both) is made positive enough (that is, "1"), the transistor

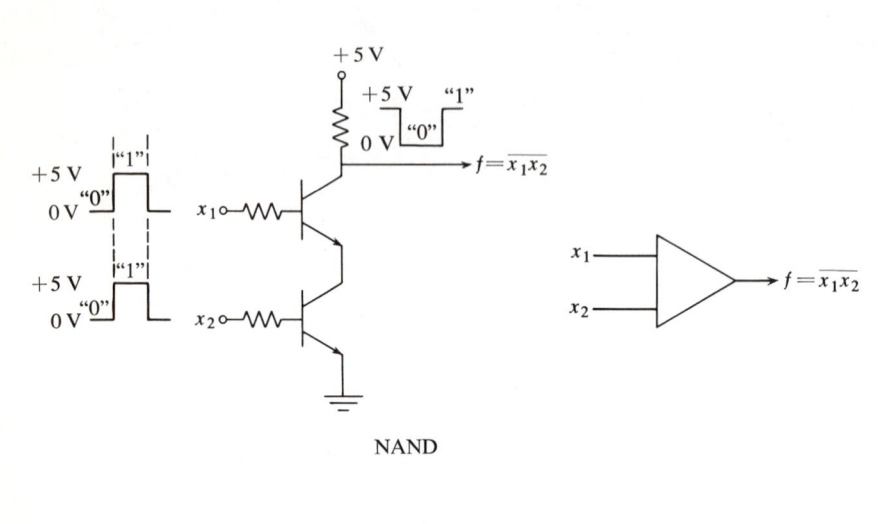

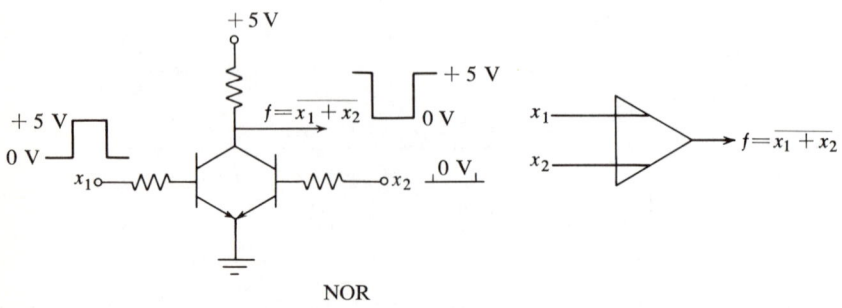

Fig. 9.12 Two-input NAND/NOR transistor circuits.

is turned ON and the output is almost grounded (that is, "0"). The NAND and NOR operations are apparent from the above description.

d. Resistor-Transistor Logic (RTL) Circuits

A summing network made of resistances and placed in series with an amplifying and phase inverting NPN or PNP transistor can perform the NOR or NAND operation, respectively. Two such circuits are illustrated in Fig. 9.13. Notice that the NOR circuit is similar to the inverter shown in Figs. 9.9 and 9.10, with more input resistances added. The NAND circuit is a similar extension of a PNP inverter.

Examining the NOR circuit shown in Fig. 9.13, we see that if *all* the input potentials are sufficiently negative, (that is, "0"), then the transistor is turned OFF, and the output rises to ground potential (that is, "1"). If any one, or more,

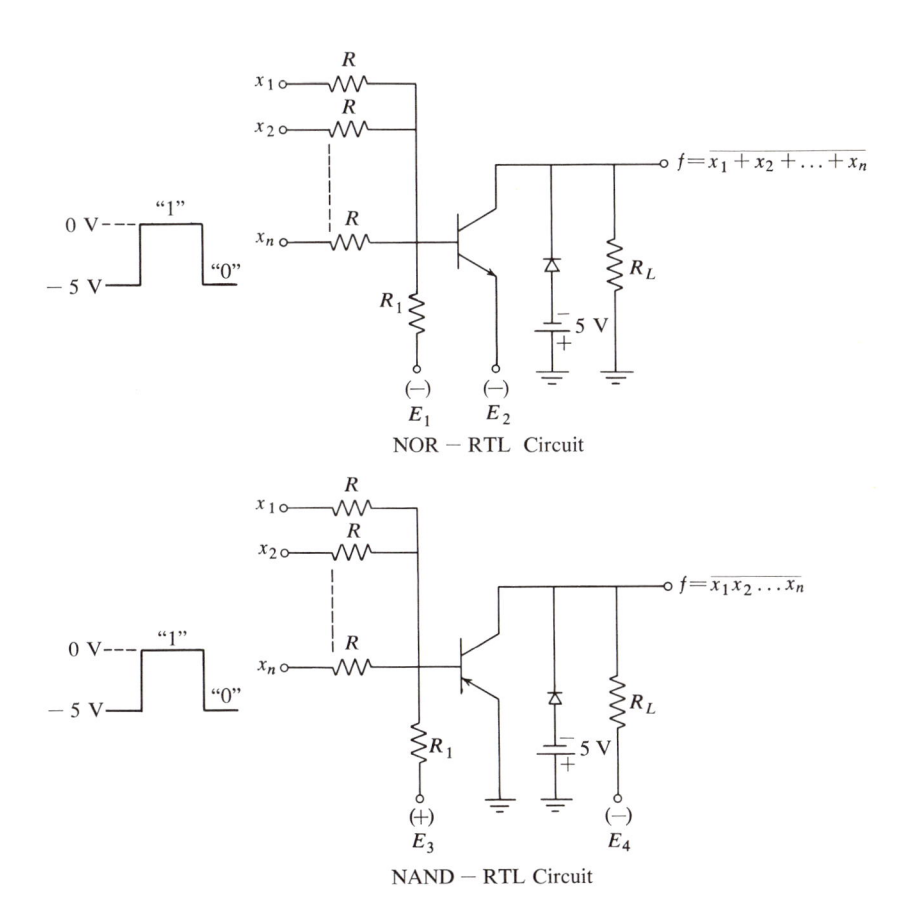

$f = \overline{x_1 + x_2 + \ldots + x_n}$

NOR — RTL Circuit

$f = \overline{x_1 x_2 \ldots x_n}$

NAND — RTL Circuit

Fig. 9.13 RTL positive logic circuits.

of the inputs is grounded (that is, "1"), then the transistor is turned ON, and the output drops to almost the negative supply voltage E_2 (that is, "0"). A truth table, constructed to describe the above operation, would verify the NOR function.

In the case of the NAND circuit, we may see that if *all* the inputs are grounded (that is, "1") the base of the transistor becomes positive, the transistor is turned OFF, and the output drops to the negative supply voltage E_4 (that is, "0"). If any one, or more, of the input potentials is made sufficiently negative (that is, "0"), the transistor is turned ON, and the output rises to near ground potential (that is, "1"). Again, a truth table tabulation of this operation should demonstrate that the logic function performed is that of a NAND function.

e. Diode-Transistor Logic (DTL) Circuits

Diode logic circuits, in conjunction with transistor amplifiers which restore the voltage levels and provide sufficient current gain for driving other logic circuits, form the transistor-diode logic. This type of logic circuit is used widely because it can implement all logic functions, is relatively inexpensive, fast, and may be microminiaturized. The transistor normally inverts the output from the diode circuit.

The NOR-DTL circuit shown in Fig. 9.14 is a positive logic circuit operating as follows: if *all* input potentials are sufficiently negative (that is, "0"), then all diodes are open; the base of the transistor becomes sufficiently negative; the transistor conducts heavily and the output rises to almost ground potential (that is, "1"). If any one, or more, of the inputs is grounded (that is, "1"), the point A in the circuit is also at ground potential; the base of the transistor becomes positive; the transistor is turned OFF; and the output drops to the negative supply voltage (that is, "0"). For a specified number of inputs the circuit designer must choose properly the values of the resistances and the power supplies. The capacitor C is used to improve the switching response of the circuit. One or more real semiconductor diodes connected in series are also commonly used to speed up the operation of the circuit.

Figure 9.14 also shows the NAND-DTL positive logic circuit. Similarly here we may notice the following operation: if *all* input potentials are sufficiently positive (that is, "1"), then all diodes are open; the base of the transistor becomes sufficiently positive[11]; the transistor conducts heavily; and the output drops to almost ground potential (that is, "0"). If any one, or more, of the inputs is grounded (that is, "0"), then the base of the transistor becomes negative; the transistor is turned OFF, and the output rises to the positive supply voltage (that is, "1"). That the above constitutes a NAND operation may be verified by constructing the corresponding truth table.

---

[11] With proper choice of bias resistances and supplies.

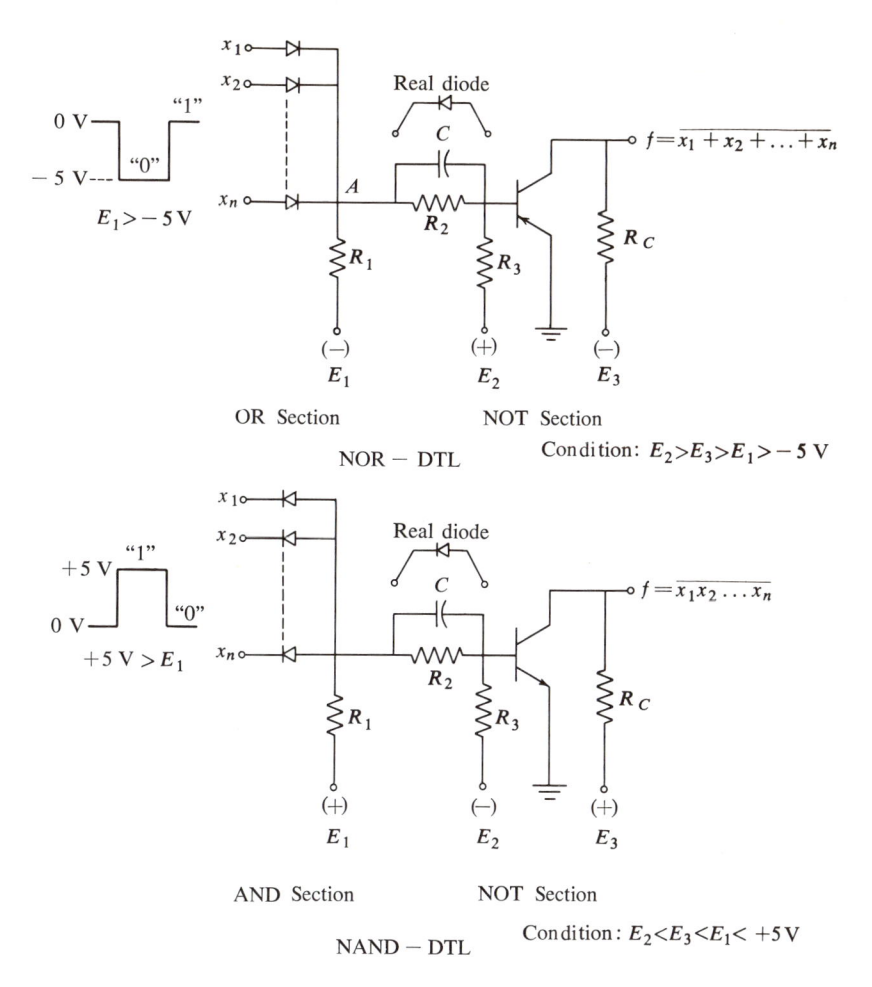

Fig. 9.14 DTL positive logic circuits.

9.7
FLIP-FLOPS

Elementary switching devices or circuits which, under proper control, may acquire and maintain either one of two stable conditions, labeled "one" and "zero", and thus may serve as single bit storage cells, are classified as "flip-flops." They are commonly found with one, two or three inputs and one or two outputs. The condition or "state"[12] of the *FF* (meaning "flip-flop") is identified by the logical values of its outputs. In electronic

---
[12] The term "state" is most frequently used to signify the bistable condition of the *FF*. It will therefore be used hereafter.

flip-flops the logical values of the outputs may correspond to "high" and "low" voltage levels. When two outputs are available, they have logical values which are always the logical complements of one another so that at any time, one output is at a "high" level (that is, a "1") and the other at a "low" level (that is, a "0"). The two outputs may be referred to as A and $\bar{A}$, where A is the logical variable which represents the state, so that the flip-flop is in the "one" state if $A = 1$ and $\bar{A} = 0$, and in the "zero" state if $\bar{A} = 1$ and $A = 0$. When only one output is available, the state of the flip-flop is identified by the logical value of the output, that is, $A = 1$ or $A = 0$.

Flip-flops are "sequential" bi-state elements whose state at any time[13] $t = n + 1$ is a function of its own state and its own inputs at time $t = n$ (Chapter 11, Section 11.1). Thus, the present state of a *FF* depends on its previous state and on the last event which occurred at its inputs. A *FF* remains in its present state until the appropriate input signal flips it to the opposite state, thus "remembering" which of the input lines was last pulsed.

The input lines are driven by proper signals, usually current pulses of appropriate polarity and sufficient magnitude and duration. Positive logic (Section 9.3) is employed most commonly, whereby a positive input current or voltage pulse signifies a one, while no pulse signifies a zero. A magnetic core, however, which may be looked upon as a *FF* element, ordinarily has a single input line and it is switched to either state by two types of input signals, usually electric current pulses of sufficient magnitude and duration but of opposite polarity (Section 9.12, Fig. 9.24).

The operation of a *FF* may be described, employing a truth table in a manner analogous to that used to describe the operation of logical circuits in Chapters 7 and 8. However, because time is of particular importance in the operation of a *FF*, the table should relate the "next state A_{n+1}" of the *FF*, that is, the state at $t = n + 1$, to the "present state A_n", that is, the state at $t = n$, and to the "present inputs." Such tables are better known as "transition tables" (Chapter 11, Section 11.5), because they describe all the possible transitions from one state to the next.

There are several types of flip-flops employed in the design of information-processing machines. In the following we describe the operation of the most prevalent ones using transition tables as illustrated in Fig. 9.15.

The *R-S* flip-flop, or "Set-Reset" flip-flop, [Fig. 9.15(a)] has two input and two output lines. If none of its input lines is pulsed, the *FF* remains in its previous state. If only the set-line is pulsed, $S_n = 1$, the next state will be a 1 (that is, $A_{n+1} = 1$). It may be observed that if $A_n = 1$, then we "don't care" if S_n is pulsed or not. If only the reset-line is pulsed, $R_n = 1$, the next state will be a 0 (that is, $A_{n+1} = 0$). Again, if $A_n = 0$, then we "don't care" if R_n is pulsed or not. Notice also that simultaneous pulsing of both input lines is "not allowed". The *R-S* flip-flop is the most commonly used one in the engineering design of

---

[13] In digital computers time is measured in discrete intervals.

FLIP-FLOPS 237

(a) $RS-FF$

R	S_n	A_{n+1}
0	0	A_n
0	1	1
1	0	0
1	1	Not allowed

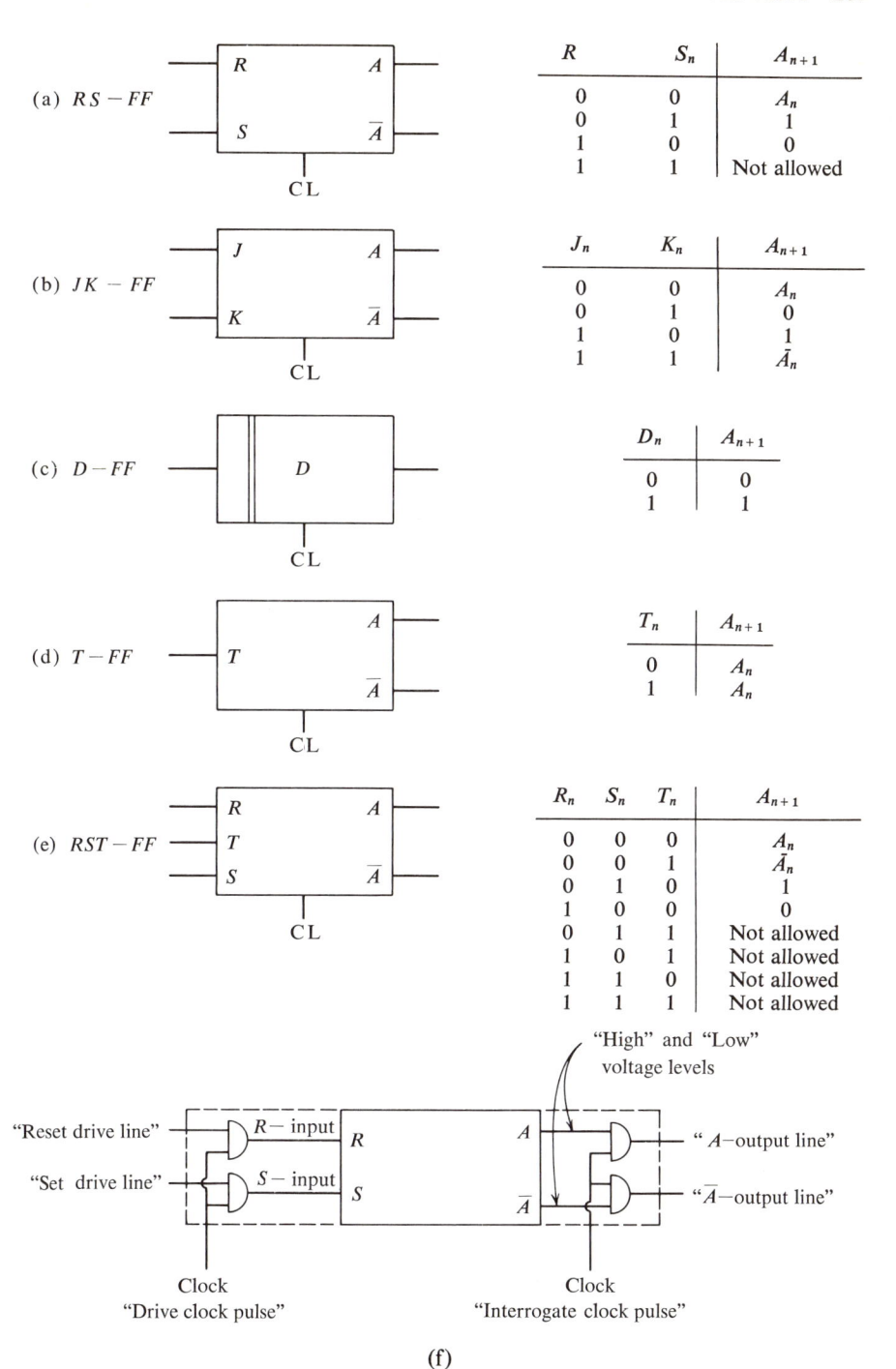

(b) $JK-FF$

J_n	K_n	A_{n+1}
0	0	A_n
0	1	0
1	0	1
1	1	$\bar{A}_n$

(c) $D-FF$

D_n	A_{n+1}
0	0
1	1

(d) $T-FF$

T_n	A_{n+1}
0	A_n
1	A_n

(e) $RST-FF$

R_n	S_n	T_n	A_{n+1}
0	0	0	A_n
0	0	1	$\bar{A}_n$
0	1	0	1
1	0	0	0
0	1	1	Not allowed
1	0	1	Not allowed
1	1	0	Not allowed
1	1	1	Not allowed

(f)
Fig. 9.15 (a) to (e) Various types of flip-flops; (f) clocking the operation of a flip-flop.

electronic circuits for information-processing machines. (See design example in Chapter 11.)

The *J-K* flip-flop is similar to the *R-S* flip-flop, except that simultaneous pulsing of its inputs reverses the state of the *FF*, as shown in the transition table of Fig. 9.15(b).

The *D* flip-flop is a one input-one output "delay" element [Fig. 9.15(c)]. Its state at time $t = n + 1$ is simply the input at time $t = n$. The *D* flip-flop is an essential part of a "full adder" for the performance of the carry operation (Chapter 10, Section 10.4a).

The *T* flip-flop, or "Trigger" flip-flop,[14] changes state each time its single input is pulsed, that is, $T_n = 1$, as illustrated in Fig. 9.15(d).

The *R-S-T* flip-flop is a combination of the *R-S* and the *T* flip-flop. It provides an output only when a single one of its three inputs is pulsed. If $T_n = 1$, the flip-flop reverses its state, that is, $A_{n+1} = \bar{A}_n$. If $S_n = 1$, then $A_{n+1} = 1$, and if $R_n = 1$, then $A_{n+1} = 0$. If none of the inputs is pulsed, the flip-flop maintains the same state as previously. Under no circumstances should there be more than one input pulsed at a time [Fig. 9.15(e)].

A few comments about the role of clocking (synchronization) pulses during the operation of flip-flops may be warranted at this point. Within computers which operate synchronously, action is timed by "clock" pulses generated by a "master timer" unit (Chapter 10, Section 10.6). Changes of state during the operation of flip-flops must also occur in the time interval between clock pulses. Therefore, clock pulses may be used to gate the "drive" signals to the input lines of the flip-flop. If, for example, the *R-S* flip-flop shown in Fig. 9.15(f) is to be driven to the RESET ("zero") state, the voltage on the "Reset drive line" should be raised to an appropriate magnitude. However, no action will take place until the "drive clock pulse" occurs. The clock pulse will gate the "Reset drive line" signal through the input AND gate which in turn will drive the "*R*-input" line to reset the flip-flop.

Similarly in interrogating the state of a flip-flop, a clock pulse may be used to gate the "high"-level output signal. If $\bar{A} = 1$, for example, the "interrogate clock pulse" will gate a positive pulse through the output AND gate to the $\bar{A}$-output line of the flip-flop [Fig. 9.15(f)]. If the flip-flop is in the SET state ($A = 1$), the "interrogate clock pulse" will cause a "high" voltage level to be gated through the output AND gate to the *A*-output line of the flip-flop [Fig. 9.15(f)].

Synchronized operation may also be accomplished using various other "clocking" techniques. The use of AND gates in clocking and gating input and output signals was preferred here [Fig. 9.15(f)], because it is easy to understand and it demonstrates clearly the clocked operation of flip-flops.

A flip-flop works in conjunction with combinational circuits that are used to drive it or to detect its state. Under proper conditions dictated by the design problem, its inputs are pulsed through logical circuits. To design such circuits

---

[14] Also known as complementing flip-flop (see Section 9.10).

the engineer must derive the "input equations," which are Boolean expressions for the input variables, such as R and S for the R-S flip-flop. Design techniques for determining the input equations resemble those discussed in the previous chapters concerning the design of combinational circuits. In Chapter 11 we examine the design of a small machine employing a R-S flip-flop.

In addition to the above discussed flip-flops, which are commonly implemented by electronic circuits such as bistable and monostable multivibrators, there exist many more types of single-bit storage elements and techniques. Since digital information-processing machines operate with discrete bits of information in a one-at-a-time fashion, single-bit storing elements and storage techniques are extremely important. All such single-bit storing elements may be looked upon as being flip-flops. For example, one may consider each magnetized spot in digital magnetic recording as being a flip-flop element. The recording and reading magnetic heads in this case serve as input and output transducers. Magnetic cores, to be discussed further in Section 9.12, are also flip-flop elements.

There is a similarity between the bistable multivibrator circuit and the magnetic core. Both provide memory for one bit of information; the core with its two extreme conditions of magnetic saturation, and the multivibrator with its two distinct stable conditions of electrical conduction. Both flip-flops are "set" and "reset" by electric current pulses properly coded and applied, and they both provide a lasting memory of the last event in their inputs. In comparison, the magnetic core is inexpensive and consumes no power to maintain its memory; a pulse of energy is required to switch the core from one condition to another, but no external power is required to hold it in that condition. The multivibrator, on the other hand, consumes power continuously to maintain its memory, and it may lose its memory if power is lost even momentarily. Therefore, it is natural that the magnetic core is preferred for large, long-term memories. One disadvantage of magnetic cores is that they require to be switched in order that they may be "sensed," that is, that their condition be detected; the multivibrator, on the other hand, provides a continuous indication of its condition by the level of voltage available at one of its outputs.

Flip-flops are typically used in computer applications as cells for the storage of information bits. Such applications include shift registers, counters, and memory systems. In the following we shall show some typical examples of transistorized flip-flop circuits and applications.

9.8
THE TRANSISTOR NOR FLIP-FLOP

A flip-flop may be formed by two NOR blocks connected as shown in Fig. 9.16. The circuit possesses two stable conditions, which may be described as follows: consider positive logic and assume that both inputs to

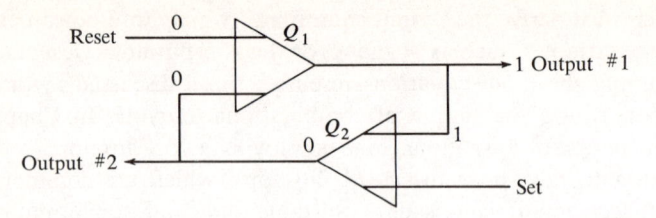

240 ELEMENTARY COMPONENTS: SWITCHING DEVICES AND CIRCUITS

Fig. 9.16 The NOR flip-flop.

Q_1 are grounded, that is, "0's"; Then the output of Q_1, is positive, that is, "1", thus making the output of Q_2 a "0" and confirming one of the inputs to Q_1. The FF is driven to this stable condition by pulsing the "SET" input positive, while keeping the "RESET" input grounded. When the set terminal is made positive, that is, a "1", both inputs to Q_1 are grounded, that is, "0's", thus making the output of Q_1 positive, that is, a "1". The positive pulse may now be removed from the "SET" input, and the circuit will maintain its condition. The condition of a FF may be termed as "1" or "0", depending on the convention used. Notice that in the condition described above, output #1 is positive, that is, a "1", while output #2 is grounded, that is, a "0". We may name this condition as the "1-condition" of the FF. Since the two outputs are always complementary, we need only recognize the value of one of the outputs. We may consider the value of the output #1 as representing the condition of the flip-flop as shown in the circuits of Fig. 9.17.

Once the FF has been pulsed and set in the "one" condition, as shown in Fig. 9.17(a), it is not affected by further pulsing the "set" terminal. The status-quo is maintained for either potential value present at the "set" terminal. If, however, the "reset" terminal is pulsed positive, as shown in Fig. 9.17(b), then the condition of the FF is reversed. The output of Q_1 is inverted to "0", and since the "set" terminal is already "0" the output of Q_2 becomes "1" and

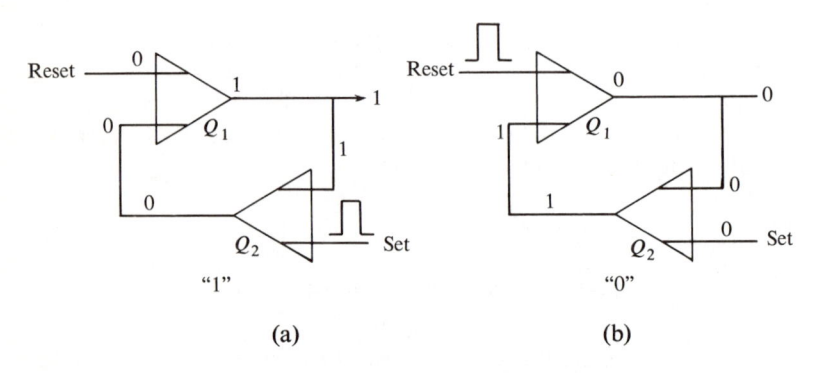

Fig. 9.17 (a) Set to "one" and (b) reset to "zero" by pulsing. Notice that the condition of the output indicates the condition of the FF.

maintains the new status-quo after the removal of the pulse. The *FF* is now at the "zero-condition," as may be recognized by the potential of the output. Further pulsing of the reset terminal would not alter the condition of the *FF*. The condition may be changed again by pulsing the set terminal positive.

The transistor NOR flip-flop is illustrated in Fig. 9.18, making use of the basic transistor NOR block shown in Fig. 9.12.[15] Two transistor NOR blocks are used in a cross-coupled configuration, in accordance with the block diagram of Fig. 9.16. Notice that the output of each NOR feeds one of the inputs of the other NOR. The output of the *FF* is taken at the output of the reset-NOR. The operation of this *FF* is identical to the one described above.

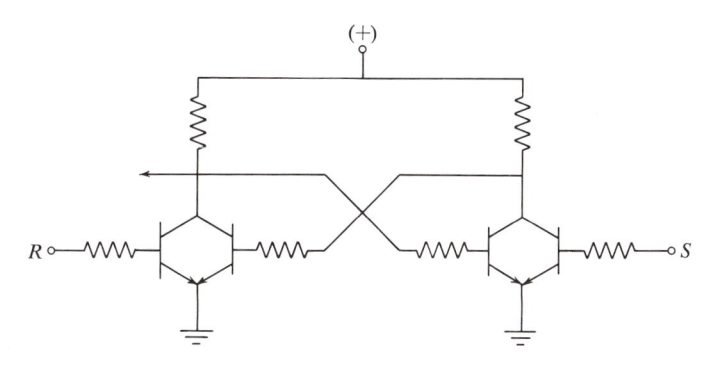

Fig. 9.18 Transistorized NOR flip-flop, making use of the NOR block shown in Fig. 9.12.

9.9
SHIFT REGISTER

A register is a temporary storage device consisting of a number of storage cells. A shift register is a register in which bits may be shifted from one cell to the next by means of shift pulses. Flip-flops and modified delay elements may be used as single-bit storage cells.

Several flip-flops combined with AND circuits and delay elements in the way illustrated in Fig. 9.19 may form a shift register. The information bits are entered in the register through the bit (top) input line. A positive voltage pulse is used to set *FF*1 to the one-condition. The pulse is gated through the input AND gate by simultaneously pulsing the bit-shifting line. If no pulse is applied to the bit-input line, *FF*1 remains in the zero-condition.

Each time the bit-shifting line is pulsed, a bit is transferred from one *FF* to the next. This pulse gates the positive potential of the one- or the zero-output of

---

[15] The vacuum tube version of this circuit is better known as the Eccles-Jordan flip-flop. Other names also applied to flip-flops are: latch, trigger, and toggle.

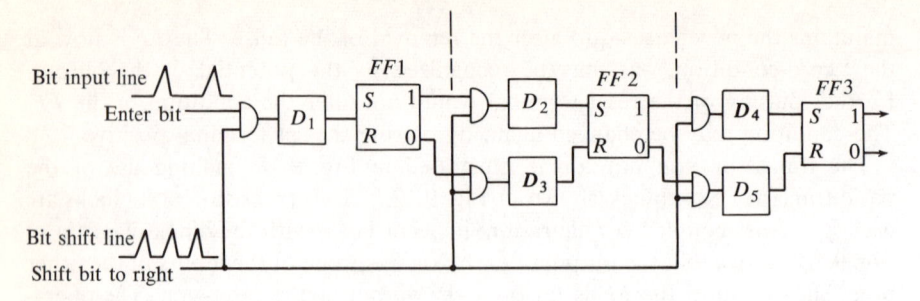

Fig. 9.19 Shift register block diagram.

the *FF* through one of the output AND gates. A positive pulse from the one-output of *FF*1, for example, sets *FF*2, and a positive pulse from the zero-output of *FF*1 resets *FF*2. Thus, the condition of *FF*1 (that is, the stored bit), is transferred to *FF*2. Each succeeding shift pulse advances this bit to the following flip-flop down the line, at the same time bringing a new bit to *FF*1. The delay elements between storage cells make sure that each *FF* does not flip before it has completed the transfer of its information to the following stage.

Only three stages of a "shift-to-right" shift register are shown in Fig. 9.19. More elaborate circuit designs would permit shifting either to-left or to-right. More reliable operation may be obtained at the expense of more *FF*'s by means of a *double-rank* shift register. In this type, two parallel registers are used with two rows of flip-flop cells with no delay elements in between the cells. During shifts, the bit being shifted is transferred from an upper-row cell to a lower-row cell and back up to the next upper-row cell down the line.

It may be noted here that a shift register may also be built with storage cells composed of delay elements and feedback alone, as shown in Fig. 9.20. A signal

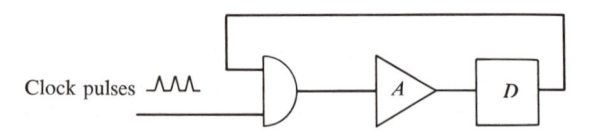

Fig. 9.20 Storage cell composed of a delay element and feedback.

bit is gated into the cell through the AND gate and then amplified. Each time the signal reaches the end of the delay element, it is fed back to the input gate. Proper clocking pulses allow the signal (bit) to enter the cell and recirculate once more. The signal is maintained in circulation as long as clock pulses are applied. The delay element itself may be an acoustic magnetostrictive delay line. The signal is induced on the line at the one end, picked up at the other end, amplified and then returned through a gate back to the entrance point on the line.

9.10
BINARY COUNTERS

A flip-flop, like the one in Fig. 9.18, may be easily converted to a "complementing flip-flop," that is, one which reverses its condition upon application of an input pulse.[16] The conversion of an *FF* to a *CFF* ("complementing flip-flop") may be accomplished with the addition of a pair of input diodes; with a differentiator added at the 0-output of the flip-flop, the *CFF* forms the building block for a binary counter.

As illustrated in Fig. 9.21, if the flip-flop is in the zero-condition with the left side ON and the right side OFF, then a positive input pulse will have no effect on the left-hand NOR, but it will turn ON the right-hand NOR. The condition of the flip-flop is reversed to the one-state and a negative potential step appears at the zero-output of the flip-flop (time t_1). The negative potential step is transformed to a negative narrow pulse by the output differentiator and is blocked by the output diodes.

The next input pulse will again reverse the flip-flop back to the zero-condition.

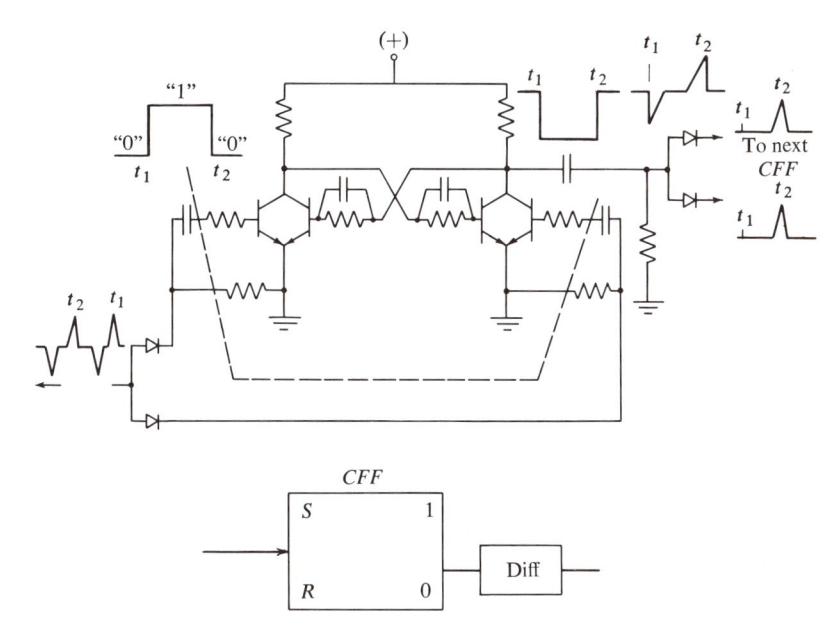

Fig. 9.21 A complementing flip-flop and symbol. The transistorized NOR flip-flop (Fig. 9.18) is contained within the dashed lines. The voltage waveforms at critical points for a 0-1-0 transitions cycle are also illustrated. The complementing flip-flop is also known as a "trigger" flip-flop (see Chapter 11).

---

[16] Also known as "trigger flop-flop."

Now the zero-output of the flip-flop goes through a positive potential step transition, which is transformed to a narrow positive pulse through the output differentiator and transmitted through the output diodes. If each time the flip-flop returns to the zero-condition a positive output pulse is fed to another *CFF*, then that second *CFF* will flip back and forth at half-rate compared to the first *CFF*. The second *CFF* may feed a third, flipping at one-fourth the rate of the first. A binary counter may thus be formed by connecting several *CFF*'s in tandem. A binary counter with n stages of *CFF*'s counts from 0 up to $2^n - 1$.

**9.11
RING COUNTERS**

Consider a string of connected *FF*'s, all reading 0. The flip-flops are numbered from left-to-right in sequence as *F*-0, *F*-1, *F*-2, ..., *F-N*. A one-bit is entered into the first register (*F*-0) and is used as a "marker." In the first position (*F*-0), the marker bit indicates the count 0. It is then shifted, step-at-a-time, to the right, to indicate the count 1, 2, and so on up to N, as shown in Fig. 9.22(a). This constitutes a "ring counter," capable of counting up to N. In actual circuit design, delay elements are used in between the storage

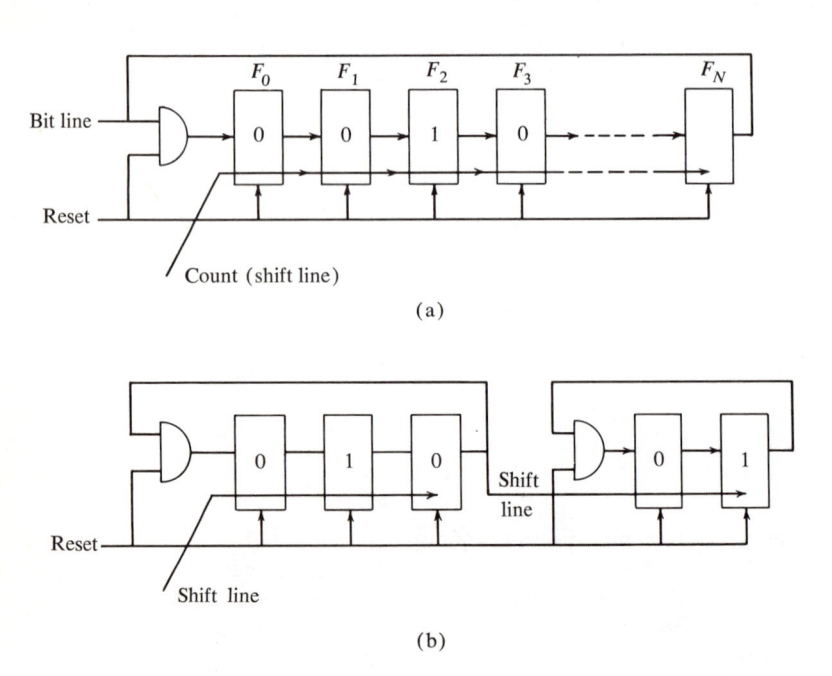

Fig. 9.22 (a) Ring counter. The present count is two; (b) Two stage (3 × 2) counter. The present count is four.

cells, that is, the flip-flops, just as is done in serial shift registers, discussed previously, in order to avoid erroneous bit transfers.

The ring counter is designed so that for each shift pulse each cell transfers its content to the next cell down the line. At any time all but one cell carry zero; only one cell carries the "marker" bit one. In the beginning of the count the ring counter is "cleared" by resetting all *FF*'s to zero except the one representing 0, which is set to one. The shift line is activated by pulses to effect the shifting of the "marker" one-step-per-shift pulse. The count may be read at any time by identifying the position of the "marker" bit one.

A ring counter is connected so that the count is repeated after the maximum count is reached. For that, the "marker" bit is transferred from the last cell back to the first cell; the counter is in the "clear" condition and ready to repeat another cycle of counting.

If a ring counter with *N* flip-flops is connected to a second ring counter with *M* flip-flops so that the second counts how many times the first counter has reached maximum count, then the two-stage ring counter can count up to $NM-1$. For example, if each stage has ten flip-flops, and the shift line of the second stage is energized by the pulse on the return line of the first stage, as illustrated in the example of Fig. 9.22(b), then the two-stage counter can count up to 99. More stages may be added, each stage counting the number of maximum counts of the previous one. Such multistage counters, decimal or otherwise, are widely used for a variety of tasks throughout the digital computer operation.

9.12
MAGNETIC CORE MEMORIES

The magnetic core is the most commonly used flip-flop device for binary information storage. It consists of a minute toroid of magnetic material, whose hysteresis loop is as nearly as possible rectangular, like the one shown in Fig. 9.23. In most cases, magnetic cores used in computer memories are made of ferrimagnetic ferrites. Small toroidal ferrite cores are manufactured by pressing powdered ferrite material mixed with ceramic "glue" and baking it in an oven.[17] Wound toroids are also used, made of thin, narrow molybdenum ribbon wound on a steatite bobbin.

The device possesses two discrete conditions of magnetization "remanence" marked by $-B_r$ and B_r, respectively, in Fig. 9.23. It may be "switched" from one remanent condition to the other on command, using properly shaped electric current pulses which are carried by a wire passing through the hole of the

---

[17] "Ferrites" are made of iron oxide, manganese, nonmetallic oxides, and an organic binder, all mixed, suspended, and baked in a ceramic base. Depending on the proportions of the mixture, the baking temperature, and time, the material exhibits ferrimagnetism in various degrees and kinds.

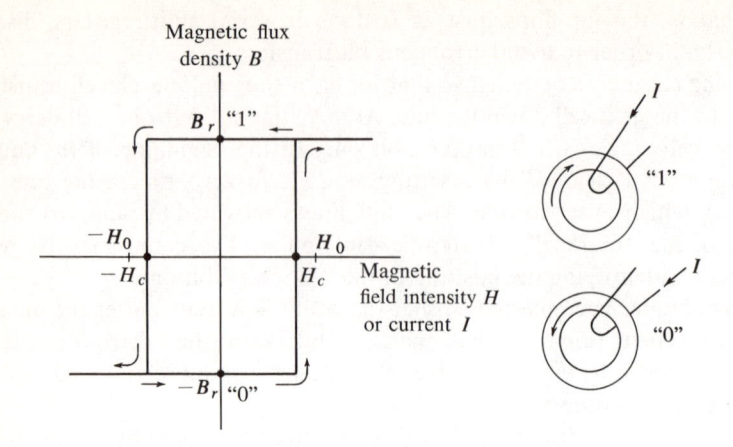

Fig. 9.23 Nearly rectangular hysteresis loop.

toroid. In switching, the magnetization of the material reverses between the clockwise and the counter-clockwise orientation around the hole of the toroid. The switching of the magnetization, and thus also of the magnetic flux, may also be traced around the hysteresis loop, as illustrated in Fig. 9.23.

Binary information is stored in the magnetic core by arbitrarily assigning the two remanent conditions to represent the bits "1" and "0," as shown in Fig. 9.23. Notice that it is the *direction* and not the magnitude of magnetization which is used to represent the stored bit. The information is inserted and retrieved from the magnetic core by electromagnetic induction of magnetization reversals. An electric current pulse of proper magnitude, duration and direction, carried through the toroid by a wire, generates the necessary[18] magnetic field intensity H_o, which establishes the desired orientation of the magnetization and of the magnetic flux around the toroidal core. The magnetization of the core is reversed by reversing the direction of the electric current pulse. Thus, by driving the proper current pulse through the wire, we may "write-in," that is, store, the desired information bit in a given magnetic core, as illustrated in Fig. 9.24. It may be added here that the toroidal shape of the magnetic cores is preferred over any other shape, because it best preserves the established magnetization condition, without the need for external holding power, that is, it contributes to the rectangular shape of the hysteresis loop.

Magnetic Core Matrices. Ferrite core memories are commonly realized in the form of planar matrices made of geometrical arrays of small ferrite toroids. Many such planar matrix frames are stacked in volume, as illustrated in the photograph of Fig. 9.25. For example, the IBM 7090 computer has a core memory of 1,179,648 cores, which are arranged in 72 planar frames, each

---

[18] Slightly larger than H_c.

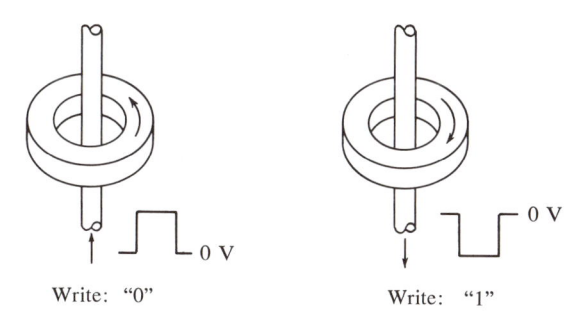

Fig. 9.24 Storing bits by current pulsing.

frame having 16,384 cores. Such a memory provides for the storage of 32,768 thirty-six-bit "words." In modern computer design, core memories with capacity of 16, 32, and even 64 thousand words are commonplace.

The magnetic core memories are organized and fabricated in different fashions. Because of practical problems, arising primarily from the fact that the cores do not exhibit the postulated ideal hysteresis characteristics, special fabrication techniques have been devised for the construction of the core memory matrices. The choice of the special matrix configuration of threading the magnetic cores, and of the methods used for inserting and retrieving information, depend primarily on the size and purpose of the core memory system being designed.

A "word-organized" memory, for example, is formed by planar rectangular arrays of cores, with each row containing one machine word. Each magnetic

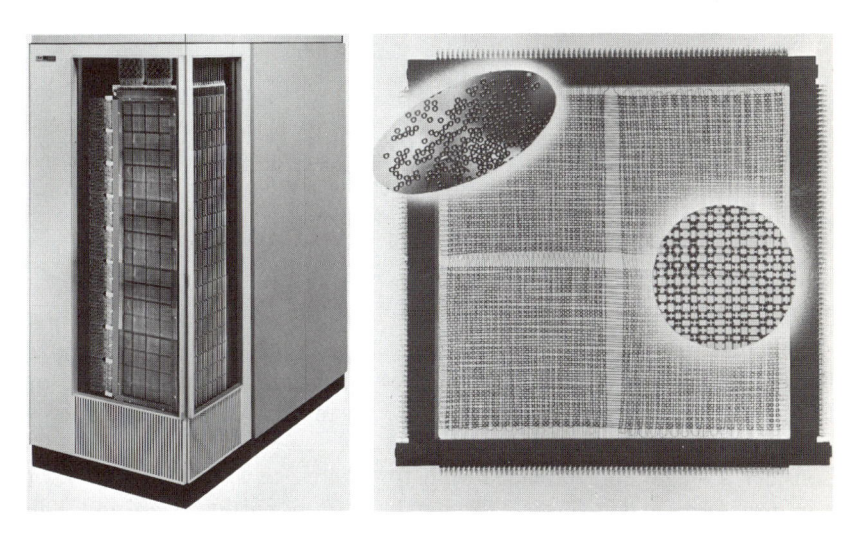

Fig. 9.25 Left: Magnetic core memory assemblies. Right: Core memory assembly without external connections. (*Courtesy of IBM.*)

core is threaded by three conductors, as illustrated in Fig. 9.26. In order to insert ("write-in") a word into the memory, one word-select line is chosen and "cleared" by pulsing it with a current pulse of sufficient magnitude and duration and of proper polarity, so that all the cores in that row are reset to the zero-condition. Then a current pulse is applied again, of opposite polarity to the one before, to produce a magnetic field H_o (Fig. 9.23) in the vicinity of each core in that row. This current drive would switch all cores in that row to the one-condition. Those cores, however, which must remain at the zero-condition are held from switching by proper currents which have been impressed on the corresponding "inhibit" lines. Each inhibit-bias current produces a magnetic field of value between zero and $-H_o$ in the vicinity of each core in its column. The inhibit current must be present before and after the word-select line current, in order that not all of the magnetic cores in that row are switched to the one-condition. In order to read the contents of a row, all cores are driven to the zero-condition, that is, the row is "cleared" by driving the word-select line with a current pulse sufficient to produce a magnetic field $-H_o$ in the vicinity of each core. Only those cores which were one would produce large voltages, induced by a $2B_r$-switching which is detected at the "sense" line terminals. Notice that each row must be cleared before inserting a new word,

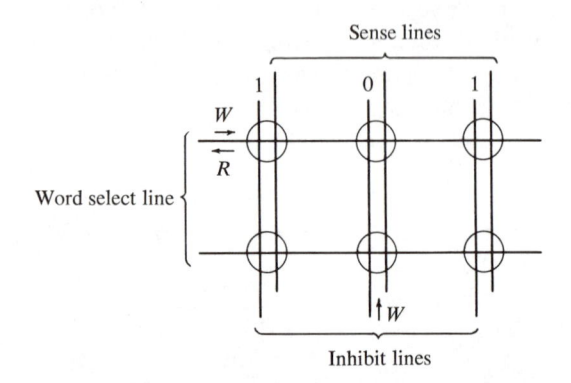

Fig. 9.26 Section of a word organized memory.

and it must be reset after each "reading" operation. This is characteristic of the "destructive read" memory.

The cost of large magnetic core memories is substantially reduced by using a basic technique of a three-dimensional core organization. We shall discuss this memory organization only in principle.

Consider a single core threaded with four wires, as shown in Fig. 9.27(a). The information content of the core may be read by pulsing the X and Y lines simultaneously, each with current which produces a magnetic field $-(1/2)H_o$ in the vicinity of the core. These current pulses, are called half-select currents,

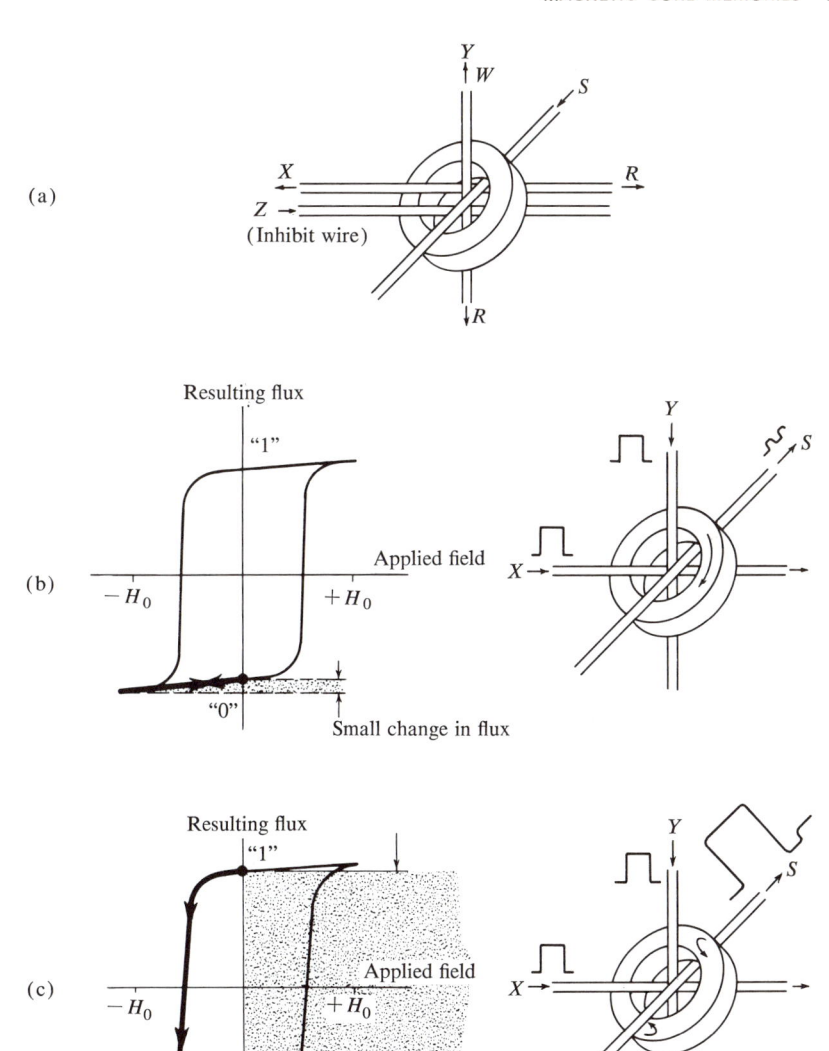

Fig. 9.27 (a) Three-dimensional threading of a magnetic core; (b) and (c) Sensing the information content by resetting to "0": (b) shows no change of state, (c) shows change from "1" to "0."

because adding their effects together in the vicinity of the core would produce a total full select field intensity of $-H_o$. If the core were in the logical zero-condition, the applied stimulus $-H_o$ would drive the core further in the negative direction of magnetization (counterclockwise in Fig. 9.23), and only slight change in magnetic flux would result [Fig. 9.27(b)]. If, however, the core were in the logical one-condition, the applied stimulus $-H_o$ would switch it to the logical zero-condition, and a large change in magnetic flux density, equal to $2B_r$, would induce a large voltage in the S ("sense") winding, as shown in Fig. 9.27(c). The presence of such large induced voltage in the S winding, and its polarity, indicates that the information content of the core was a one. Note that since the reading operation leaves the core in the zero-condition, the core must be set back into the one-condition each time a one is read. This is, again, characteristic of the "destructive read" memory. Because the operation of this memory requires that the X and Y lines be stimulated simultaneously, this particular memory organization has come to be known as "coincident current" memory.

For the "write-in" operations, the fourth Z line is used much the same way as the "inhibit" line in the previously described word-organized memory. Consider that the core has been "cleared" by reading its contents, so that it is in the zero-condition. In order to write-in a bit, the currents through the X and Y lines are reversed, so that each produces a magnetic field of magnitude $+(1/2)H_o$ in the vicinity of the core. The total field $+H_o$ would cause the core to switch from the logical zero- to the logical one- condition, so that a one is stored in the core. If, however, a zero is to be written into the core, then a current pulse producing a field $-(1/2)H_o$ is simultaneously applied in the Z (inhibit) line. This pulse reduces the net magnetic field in the vicinity of the core to $+(1/2)H_o$, so that the condition of the core is left unchanged in the logical zero-condition. Thus, in the writing operation switching is permitted or inhibited by the absence or presence, respectively, of the inhibit pulse.

Three-dimensional wiring of a section of a planar array is illustrated in Fig. 9.28. Notice that each core is placed in a 90° angle in relation to its closest neighbors. The Z winding runs parallel to the X winding; each X line runs through the cores of a single horizontal row; each Y line runs through the cores of a single vertical column; the S winding is woven through the array in a rather complicated manner; both the S and Z lines run through every core in the plane. A variety of practical problems which were overlooked in this rather gross description of the three-dimensional winding necessitate this peculiar fabrication of the array.

9.13
ALL-OR-NONE ORGANS IN LIVING ORGANISMS AND IN MACHINES

The functioning of living organisms is very complex and of mixed character—analog and digital. To disregard or to deny the importance of the analog aspects is certainly absurd. Consequently, on such a comparative basis the analogy between living organisms and digital machines is imperfect. Certain components of living organisms, however, may be more successfully compared with digital components of the digital machines.

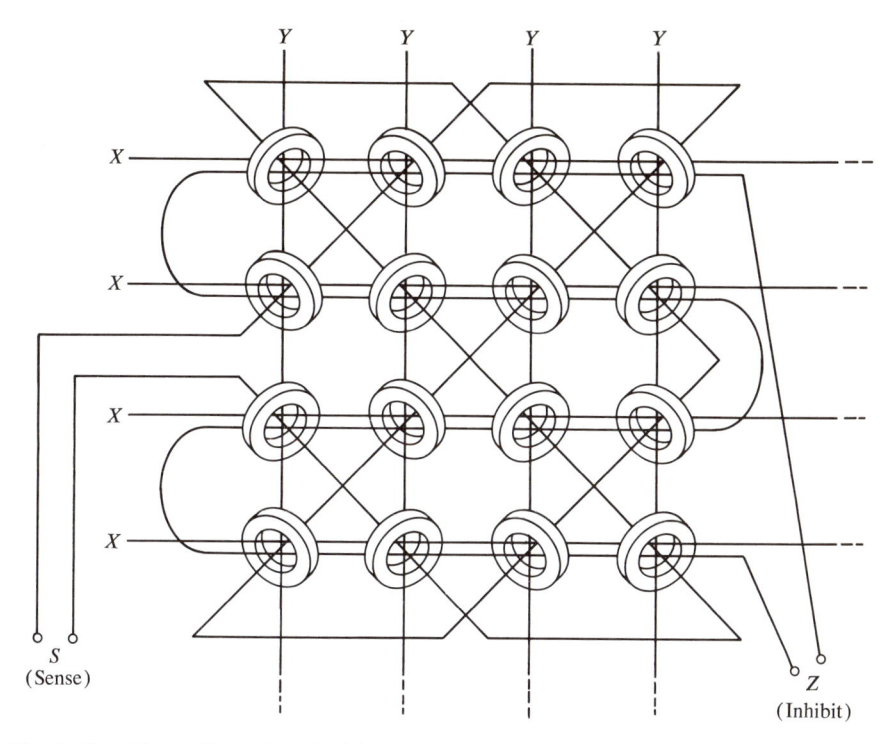

Fig. 9.28 Three-dimensional wiring of a planar array.

The "all-or-none" (binary) character of a fully developed electrical impulse emitted by neurons cannot be denied. It may be, however, that the neuron is not an absolute all-or-none organ, for if it is examined in detail, we see that the emission of an impulse is the result of complex electrochemical mechanisms within the neuron element. These mechanisms are definitely of an analog character. In addition, under a proper degree of stimulation,[19] it is possible to cause the neuron to respond in a *continuous* relation with the stimulus (proportionately or highly nonlinearly, depending on the strength of the stimulus). A critique of this sort, however, may be improperly rigid, since, strictly speaking, there is little in our technological or physiological experience which indicates that any absolute all-or-none organ exists. The diode, the e/m relay, the transistor, and the vacuum tube are components which may exhibit the all-or-none character only under very particular conditions of operation. Thus the concept of the all-or-none organ is attributed to an organ which exhibits this characteristic behavior under suitable operating conditions. With such proper qualification then (which is not intended to prejudge any open question on the subject) the neuron as well as a host of man-made devices may be characterized as all-or-none organs. The electrical, chemical, optical, or other internal mechanisms of

---

[19] "Subliminal stimulation."

the operation of such organs should be explained to understand the internal functioning of the organ, but they hardly seem necessary for a "black-box" description which makes the devices generically indentical.

A generic characteristic of such all-or-none organs is that the energy in their response is expected to be sufficient to stimulate several other identical organs. This implies that the energy in the response cannot be supplied by the stimulus alone, but originates in an independent source within the organ. The stimulus is a trigger-type signal which simply controls the flow of energy from an independent source within the organ. For example, the response energy is supplied by batteries in applications of the transistor as an all-or-none organ and by the chemical metabolism in a neuron.

With reference to size and efficiency, neurons are several thousand times smaller than the smallest switching microelement used today. They consume much less energy, but are close to a million times slower than today's fastest man-made switching devices. Great caution should be exercised when such comparisons are attempted, because the neurons and the components of digital machines have different modes of operation. An overall comparison between large computers and the central nervous system shows that the man-made machines may weigh several tons and consume many kilowatts of power, while the human central nervous system weighs about one pound, consumes less power, and is placed within the human skull. Thus, operationally, the human system possesses much greater versatility of function. This impressive discrepancy is traced to the weakness in our technology, primarily in the materials used, the system organization employed, and the shortage of new device concepts.

BIBLIOGRAPHY

Hurley, R. B., *Transistor Logic Circuits*. New York: John Wiley & Sons, Inc., 1961.

Huskey, H. D. and Korn, G. A., Eds., *Computer Handbook*. New York: McGraw-Hill, Inc., 1962.

Maley, G. A. and Earle, J., *The Logic Design of Transistor Digital Computers*. Englewood Cliffs, N.J.: Prentice-Hall, Inc., 1963.

McCluskey, E. J., *Introduction to the Theory of Switching Circuits*. New York: McGraw-Hill, Inc., 1965.

Problems

9.1 Draw the diode logic circuits for the following functions:
(a) $f = (x_1 \bar{x}_2 + \bar{x}_1 x_2) x_1$,

(b) $f = (x_1 x_2 + x_3)(x_2 x_3 + x_4)(x_3 x_4 + x_1)$.
Repeat after minimizing the functions.

9.2 In the idealized diode circuits given in Fig. 9.2.1, determine and draw the output voltage as a function of time, labeling the important values of voltage and time.

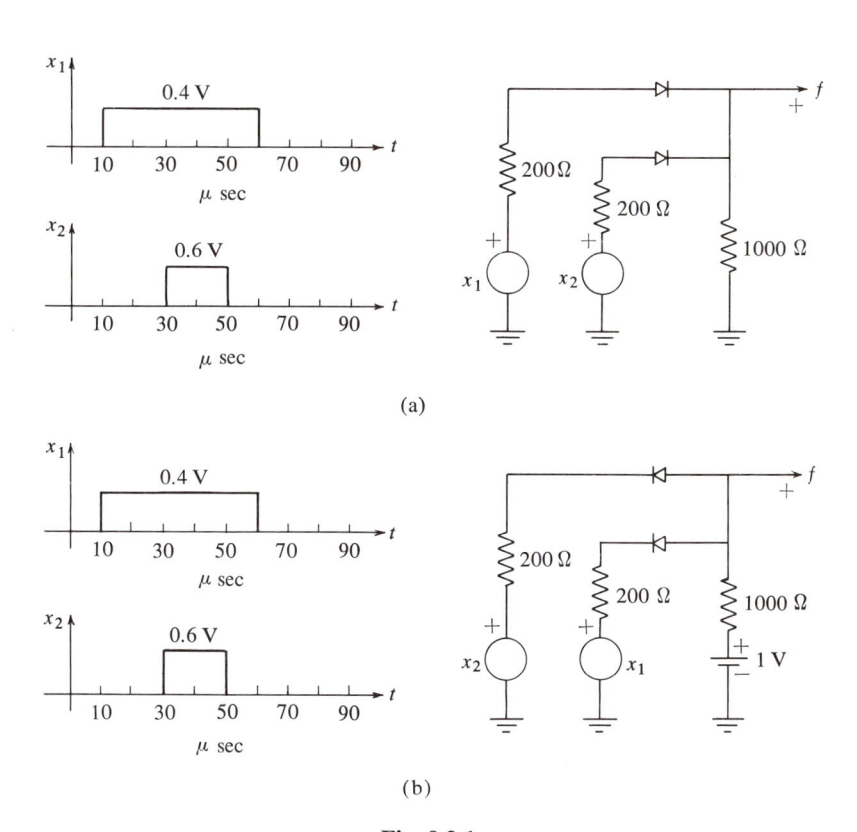

Fig. 9.2.1

9.3 In the OR diode circuit shown in Fig. 9.3.1, the internal resistances R_s of the signal sources are small compared to R, and the diodes are ideal.
(a) Find the value of the output voltage at $t = 0.4$ sec.
(b) If $R_s = 600 \, \Omega$, find the minimum value of the load resistor R, which will ensure only one conducting diode at $t = 0.4$ sec.
(c) If $R = 3000 \, \Omega$, at what time will D_3 begin to conduct? Use $R_s = 600 \, \Omega$.

9.4 Use a D flip-flop in conjunction with a combinational AND/NOT/OR circuit to realize
(a) a R-S flip-flop,

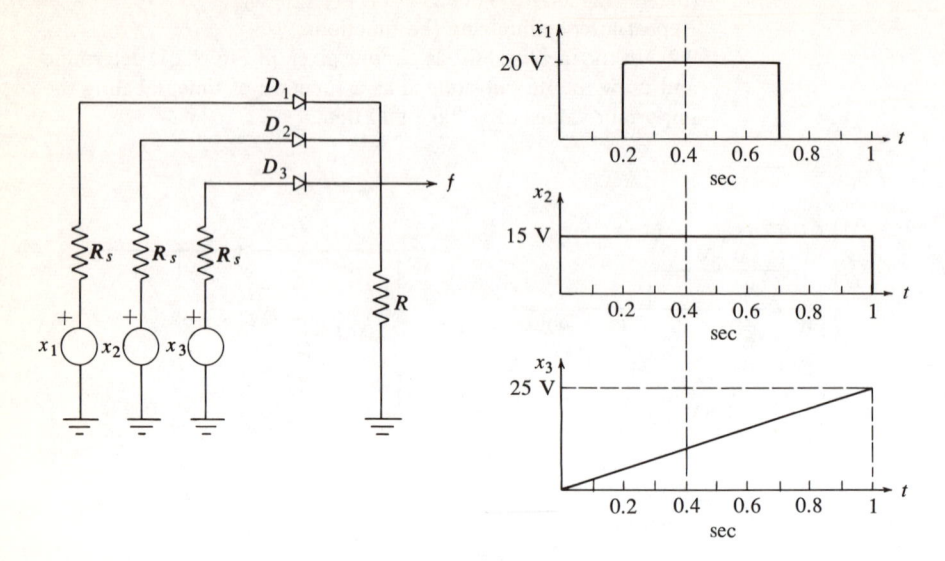

Fig. 9.3.1

(b) a *J-K* flip-flop,
(c) a *T* flip-flop,
(d) a *R-S-T* flip-flop.

9.5 Given the delay flip-flop as defined in Fig. 9.5.1 (notice that this D flip-flop provides two complementary outputs) derive, the excitation equations and the necessary gating AND/OR circuits in order to convert this D-flip-flop into (a) a R-S flip-flop, (b) a T flip-flop, (c) a R-S-T flip-flop.

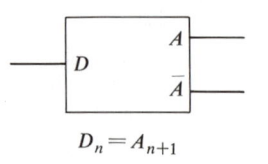

Fig. 9.5.1

$D_n = A_{n+1}$

9.6 Express the output of the sequential circuit shown in Fig. 9.6.1 algebraically in terms of x_1, x_2, A, and C (where C is the clock input). Then write the "transition table," that is, list the truth values of A_{n+1} and Z_n for all possible combinations $x_{1(n)} x_{2(n)} A_n$. (Notice that C is always one).

9.7 Determine the algebraic expression of the function f in Fig. 9.7.1 in terms of the input variables a, b, c.

9.8 Assume that the presence of $x = 1$ in the circuit of Fig. 9.8.1 is indicated by the presence of a positive voltage level on the x-input line, while $x = 0$ (that is $\bar{x} = 1$) is indicated by

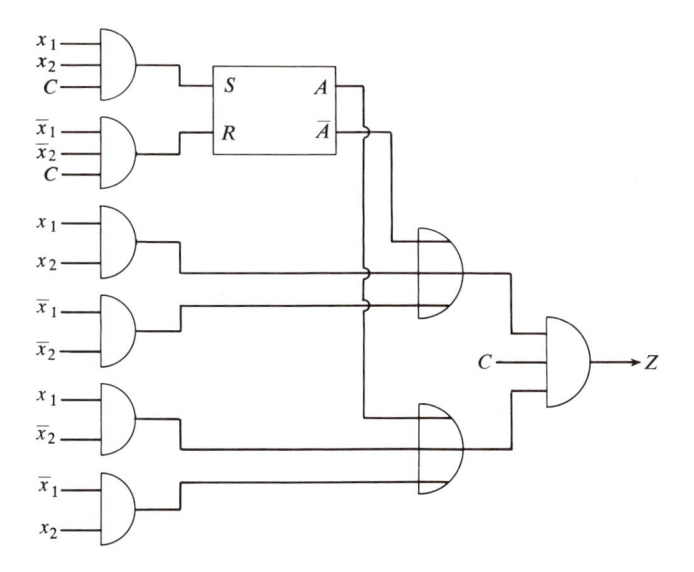

Fig. 9.6.1

the presence of a positive voltage level on the $\bar{x}$-input line. The delay flip flop shown provides an output on the line A if the last input was a 1, or on line $\bar{A}$ if the last input was a 0.
Describe the operation of the circuit.

9.9 Derive an algebraic expression for the output of the circuit shown in Fig. 9.9.1 and analyze the operation of the circuit. Compare it with the operation of a standard R-S flip-flop.

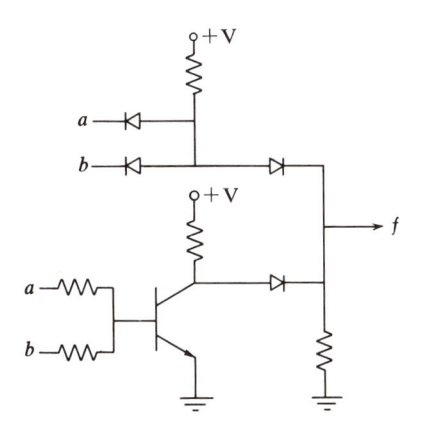

Fig. 9.7.1

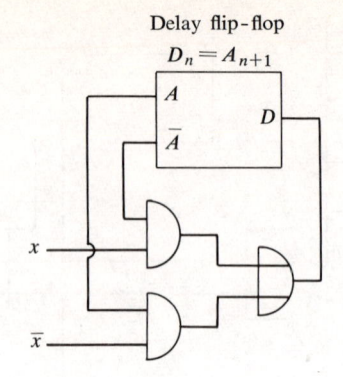

Fig. 9.8.1

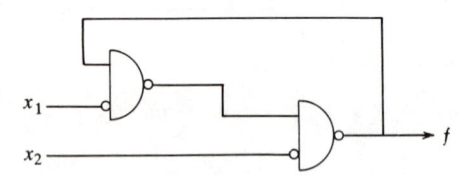

Fig. 9.9.1

10

Arithmetic Operations and Machine Organization

Since early childhood, our familiarity with how arithmetic is done and the subsequent obviousness of the matter has, for most of us, obscured the basic concepts in the workings of arithmetic. These concepts become all-important when we are about to design a machine and demand from it the ability to do arithmetic. For this reason, we examine the ways that arithmetic, more specifically digital computer arithmetic, is performed.

The organization and the essential functions of a digital data-processing machine are also discussed. Our overall objective is to examine the internal organization of a machine which performs arithmetic calculations, stores, edits, and transfers information.

10.1
OPERATING RULES IN ARITHMETIC

A number a may be transformed to a number c by operating on it by another number b in accordance to some transformation rules represented by the symbol $\star$ in the transformation equation $a \star b = c$. The arithmetic operations of addition, subtraction, multiplication, and division may be viewed as transformation operations $a \star b = c$, as shown

258 ARITHMETIC OPERATIONS AND MACHINE ORGANIZATION

in Fig. 10.1, where the operational symbol ★ is used to denote some procedural rules.

Look-up tables of addition, subtraction, multiplication, and division may be used to perform arithmetic, but they assume prohibitive dimensions if large numbers are involved or if more complex operations like taking a root or a differential are to be performed. Can you imagine, for example, using only look-up tables to calculate the national budget?

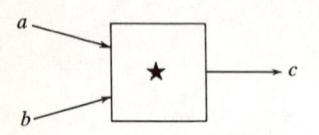

Fig. 10.1 Transformation $a * b = c$.

Operating rules and procedures in the decimal system have long been devised and established for doing arithmetic with pencil and paper quickly and efficiently. Thus, during a pencil and paper arithmetic operation, mental reference is made to simple tables of addition, subtraction, multiplication, and division, and to operational rules. The tables are limited only to the acceptable symbols, that is, in this case to the numbers from 0 to 9.

```
         3 7 5
        +6 6 4
         9 3 9
 Carry   1
         0 3 9
 Carry 1
       1 0 3 9
```

+	0	1	2	3	4	5	6	7	8	9
0	0	1	2	3	4	5	6	7	8	9
1	1	2	3	4	5	6	7	8	9	0
2	2	3	4	5	6	7	8	9	0	1
3	3	4	5	6	7	8	9	0	1	2
4	4	5	6	7	8	9	0	1	2	3
5	5	6	7	8	9	0	1	2	3	4
6	6	7	8	9	0	1	2	3	4	5
7	7	8	9	0	1	2	3	4	5	6
8	8	9	0	1	2	3	4	5	6	7
9	9	0	1	2	3	4	5	6	7	8

All digits below the dashed line transmit a carry of 1.

Addition Table

Fig. 10.2 Addition of decimal numbers by paper and pencil. The tables of addition are "closed" in the sense that no symbols other than the accepted ones appear. This is a common characteristic of every "radix" numerical system.

For example, the addition of two decimal numbers is commonly done in a "serial"-manner; pairs of digits of the augend and the addent, respectively, are "processed" successively, beginning with the least significant digits at the right end of the numbers and moving toward the left end, until all pairs of digits have been processed. The addition process for two decimal numbers, as done by pencil and paper, is illustrated in the example shown in Fig. 10.2. Each pair of digits is added "mentally" with reference being made to the simple addition table built for numbers from 0 to 9. Thus, the "decision" as to what is the sum of two digits is "automated" by the use of this "look-up" table. When the sum of two digits is greater than 9, a "carry" of 1 is created, which is later accounted for during the next addition, in connection with the sum of the next pair of digits.

A mechanical "black-box" which would operate in this fashion is shown in Fig. 10.3. Notice that the operation takes place in discrete time intervals. In

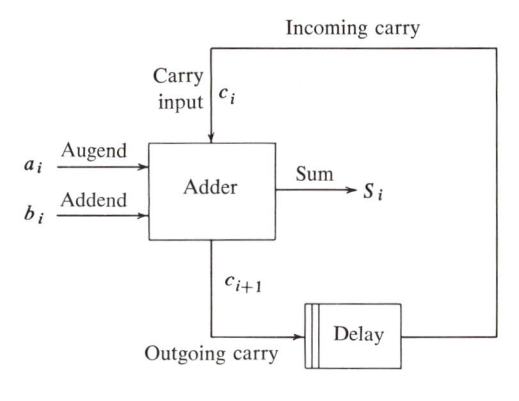

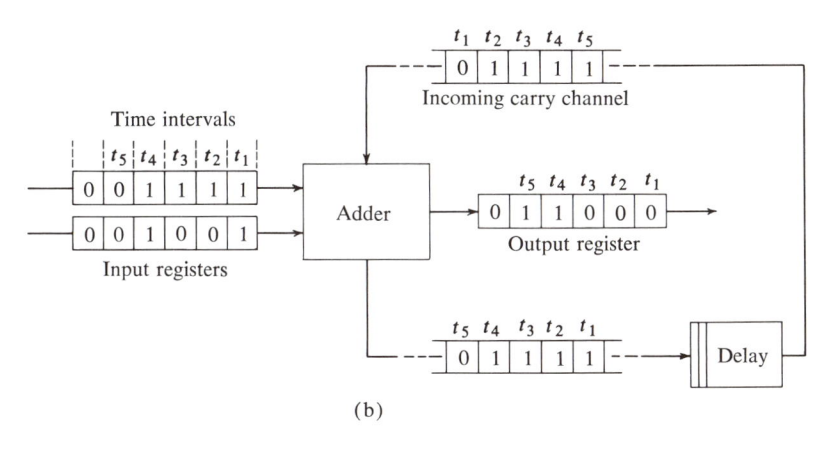

Fig. 10.3 Black-box representation of a serial adder.

each time interval two digits enter the black-box adder from the two input shift registers. The pair of digits is processed in accordance to established rules of addition in the decimal system, or by reference to the look-up table; a "sum-digit" and a "carry-digit" are produced. The sum-digit enters the output shift register, where it is stored and shifted to the right until the end of the addition. The carry digit enters a one time-interval delay device, which is a temporary storage device that delivers the carry digit to the "carry input" one time-interval later, in order that it may be used in the next operation.

The black-box adder appears as a three-input-two-output device, aided in its operation by a temporary storage device for the one time-interval delay of the carry digit. After the completion of the addition the resulting sum of the two numbers is read off the output shift register. The operation of the black-box adder will be discussed in more detail later, in relation to arithmetic operations in the binary number system.

We may recall at this point (Chapter 2, Section 2.4), that in practice the machine executes arithmetic operations using the A-register (accumulator) which serves as both input and output register. On command, the machine first transfers one of the operands into the A-register. Then, it recalls the other operand from the main memory, executes the arithmetic operation, and places the result back into the A-register.

Two important observations about the addition process emerge from the previous example: the need for some specific rules of addition or for a look-up addition table and the need for a temporary memory in order to perform the carry operation. Look-up tables and operational rules are established for any given number system. In Section 10.4 we examine such rules for the binary number system.

10.2
ARITHMETIC OPERATIONS

Essentially, arithmetic consists of the simple operations of addition, subtraction, multiplication, and division. Addition may be mechanized either by referring to built-in tables of addition or by implementing within the arithmetic unit of the computer the rules for finding the sum of two numbers. When arithmetic is done within a system of numeration, the rules of the arithmetic operation depend on the system. Numerical systems based on binary notation are very popular in computer design, because arithmetic operations in such systems are physically implemented economically and reliably by available electronic hardware.

The four operations of arithmetic, as performed by machine, are reduced to only *addition*, *complementation*, and *shifting*. Subtraction, multiplication, and division, and other more complicated mathematical operations such as taking roots, integrating, differentiating, or solving transcendental equations are possible by instructing the computer to perform sequences of such elementary operations. The variety of options available in sequencing elementary operations to perform more complex ones greatly determines the versatility and the speed of the arithmetic portion of a digital computer.

Complementation is the operation of taking the "complement" of a number written in a numerical system of radix r. To determine the

"$(r-1)$-complement" of a number N written in a system with radix r (symbolically represented as $\bar{N}_{r-1}$), we subtract each digit of the given number from $r-1$. This operation in the binary system consists of simply changing all ones to zeros and all zeros to ones. In the decimal system we subtract all the digits of the number from nine.

The "r-complement" (radix complement) of an integral number N (symbolically represented as $\bar{N}_r$) is found by adding 1 to the $(r-1)$-complement and performing the carry operations, if any. If the number is not an integer, a unit is added to the least significant nonzero digit of the $(r-1)$ complement. For example, in the binary system, if $N = 101101$, then $\bar{N}_{r-1} = 010010$ and $\bar{N}_r = 010011$. Similarly, in the decimal system, if $N = 937$, then $\bar{N}_{r-1} = 062$ and $\bar{N}_r = 063$. Also, if $N = 12.34$, then $\bar{N}_{r-1} = 87.65$ and $\bar{N}_r = 87.66$.[1]

Shifting. Because of the place-value notion associated with the representation of numbers in a numerical system with radix, "shifting" the position of the radix point toward the right or left corresponds to multiplying or dividing the number by some power of the radix. Thus, if $N = 937.$, then $93700. = N \times 10^2$ and $9.37 = N \times 10^{-2}$. Similarly, in the binary system, if $N = 1011.$ (eleven), then $101100. = N \times 2^2$ (forty-four) and $10.11 = N \times 2^{-2}$ (two and three-quarters).

From the above discussion it is suggested that the "shifting" operation has to do with multiplying, dividing, or taking powers of numbers and that, together with the operations of "addition" and "complementation," they comprise the three elementary operations which the arithmetic section of the computer must perform.

Numerical data or program instructions appear within the computer in the form of "machine words," which are coded bit-sequences commonly of a fixed length. The machine will interpret a "word" as numerical data or as a program instruction, depending on the place of its origin. For example, if the word is copied from the portion of the main memory reserved for data, it will be interpreted as data.

Between arithmetic operations numerical data may be required to be stored temporarily while awaiting entry to the arithmetic section of the machine. For that, devices of temporary storage, like the shift registers discussed in Chapter 9, Section 9.9, are used. Binary coded numerical data may be entered at the one end of a shift register, shifted along step-by-step in successive time intervals regulated by shift-triggering electric pulses, and exit from the other end of the register, as illustrated in Fig. 9.20.

In the following sections we shall examine the basic arithmetic operations, and the internal organization of the arithmetic unit of an information-processing machine.

---

[1] Taking the "radix complement" corresponds to subtracting N from r_n, where n is the number of significant digits to the left of the radix point. The "$(r-1)$-complement" is also referred to as the "diminished radix complement."

10.3 PROCESSOR

Information processing consists of elementary operations performed on machine words representing numerical or other kinds of data. This data manipulation takes place in the "processor," or as it is otherwise referred to, the "arithmetic unit" of the machine.

A processor would normally have three "registers." Each register is a temporary memory unit for the storage of one piece of information, such as a computer "word." The processor receives the "operands" in two of these registers. The third register is called the "accumulator" since it is used to accept the result of the arithmetic or the logical operation which is performed by the processor. In modern machines, especially in the larger ones, this basic scheme is used in many variations and often in conjunction with the operation of the control and main memory units.

The processor of an information-processing machine is normally capable of performing arithmetic and mathematical operations, such as addition, multiplication, and integration. In the course of its normal operation, the processor also performs operations which are prima facie nonmathematical, such as shifting, complementing, truncating, rounding-off, and others, such as comparison, cataloging, tabulating election results, and making library search, to name but a few.

The machine will execute all these operations by performing sequences of elementary logical transformations, like the ones discussed in earlier chapters. For example, the machine may be required to perform "selection," which means that the processor must be able to select between two given numbers a and b, depending on the "value" of a binary "indicator" variable q. In symbolic logic notation this operation may be expressed by the equation $c = a \cdot q + b \cdot \bar{q}$. Thus, if q is "one," the number a is selected, ($c = a$), while, if q is zero, then the number b is selected, ($c = b$). In this manner, "selection" is accomplished by performing one NOT, two AND, and one OR logical transformations. In mathematical language, the operation of "selection" is expressed by the algebraic equation $c = aq + b(1-q)$.

This underlying truth about the methodology of information processing by machines is further illustrated in the next section with several examples. Now that we have an adequate grounding in symbolic logic, in binary arithmetic and in the principles of Boolean algebra, we are in a position to demonstrate the internal relation of information processing by machine to logical transformations and to Boolean algebra, by probing the ways by which the machine executes arithmetic operations.

10.4 BINARY ARITHMETIC

All four basic arithmetic operations, and even more complex mathematical ones, are reducible to the three elementary operations of *addition*, *complementation*, and *shifting*. Binary addition is reduced to performing logical transformations, like those discussed in previous chapters. Subtraction is done by complementation and addition. Complementation itself in the binary system of notation is reduced to performing the logical transformation NOT for each binary digit. Shifting the digits of a binary number by stepping to the left or to the right in value-positions is essentially equivalent to multiplying or dividing the number by some integral and positive power of 2 (that is, of the radix), because of the position-notation which is used in our systems of numeration (Chapter 3). Combined with addition and subtraction, it makes for a simple and efficient way of performing multiplication and division, respectively.

In specifying the computing circuits, we will normally use the black-box description. This in no way reduces the value of the discussion, and it is in keeping with the basic objective of this text, which is to focus the attention on the fundamental principles that underline the operation and the design of information-processing machines. Any attempt to discuss hardware detail here would be severely curtailed if we are to maintain any measure of balanced text proportions. Besides, a nonpenetrating discussion of computer electronic hardware would be soon outdated by the fast progressing microcircuit technology, which has made black-box-like functional integrated circuitry a reality.

a. The addition of two binary numbers

Adding two binary numbers by hand is done in a fashion exactly identical to that used in the addition of two decimal numbers: we begin at the extreme right of the numbers, adding digits in pairs and carrying out the operation to the left until the addition is completed. The machine operates in exactly the same fashion.

In the binary numerical system, the guide rule of the operation is the following addition table:

+	0	1
0	0	1
1	1	10
		↑
		carry

The carry appears here, as it does in the addition table of any other numerical system, when the sum of two digits becomes greater than the value of the highest acceptable symbol in the system. With binary numbers, this happens when we add "one" and "one." The result is "two," which is greater than the value of the highest acceptable symbol, namely "one." If a carry is obtained from the

264 ARITHMETIC OPERATIONS AND MACHINE ORGANIZATION

addition of a pair of digits, it must be "preserved" and used in the addition of the next pair of digits down the line as we progress from right to left.

The "adder," the component of the processor which will perform addition, is a three-input and two-output device, as illustrated in Fig. 10.3(a). The inputs a_i and b_i are the binary digits in the ith position of the augend and the addend, counting from right to left. The input carry c_i has been preserved from the addition of the previous pair of digits, [the $(i-1)$th]. The output digit S_i is the resulting digit in the sum, which corresponds to the addition of the ith pair of digits in the augend and the addend, while c_{i+1} is the new carry produced in the ith addition. This new carry must be preserved, or "delayed" by one addition interval time, so that it may be used in the next addition down the line to the left.

Let us now examine closer the contents of the black box marked as "adder" in the Fig. 10.3(a). Most often it is easier to begin with an example and then move to a more general discussion. Let us then consider the addition of 1001 (decimal 9) to 1111 (decimal 15). The addition routine, in accordance with the addition table, is illustrated in Fig. 10.3(b), and goes as follows:

Binary		Decimal
1111	*Augend*	15
1001	*Addend*	+ 9
11000	*Sum*	24
1111	*Carry*	1

Routine: Begin from column 1 (far right) toward the left.
1 plus 1 is 10; write down 0 and carry 1;
1 plus 0 is 1, and 1 to carry is 10; write down 0 and carry 1;
1 plus 0 is 1, and 1 to carry is 10; write down 0 and carry 1;
1 plus 1 is 10, and 1 to carry is 11; write down 1 and carry 1;
write down the 1 carried. END.

The process of binary addition, illustrated above with a specific example, will now be more generally expressed in the form of Boolean algebraic equations. The Boolean equations will then lead us in specifying the hardware implementation of the adder.

The two operands may be represented generally as follows:

Augend: $a_k a_{k-1} a_{k-2} \ldots a_i \ldots a_1$,
Addend: $b_m b_{m-1} b_{m-2} \ldots b_i \ldots b_1$.

Before the operation begins, the two operands are stored in the two receiving registers of the processor. The augend would require a register with at least k

bit-storing locations, while the addend would occupy m bit-storing locations in the other register. Each register is a form of a temporary memory, which may be exemplified by an array of electromagnetic relays. Energized relays (closed contact) would signify "1's," while unenergized relays (open contact) would signify "0's." A string of magnetic cores, or semiconductor diodes and transistors may also be used in building such registers. Information may be "shifted" along these input registers to feed the adder with pairs of digits of the two operands. (See Fig. 9.19, Chapter 9.)

Let S_i be the ith digit in the sum S of the two operands, counting from the extreme right to left. The carry digit, which has resulted from the addition of the $(i-1)$th pair of digits, and which is to be added in the ith column, is c_i, while c_{i+1} is the carry digit produced in the addition of the ith column (that is, from the addition of a_i, b_i, and c_i). Accordingly, the previous numerical example of binary addition may be illustrated in the following functional table:

c_i	a_i	b_i	c_{i+1}	S_i
	1	1	1	0
1←	1	0	1	0
1←	1	0	1	0
1←	1	1	1	1
1←			0	1

Generalized truth tables for the addition of two binary numbers are shown in Table 10.1.

Table 10.1. Truth tables of binary addition.

	c_i	a_i	b_i	c_{i+1}	S_i
(Assume[a] $c_1 = 0$) →0	0	0	0	0	0
	0	0	1	0	1
	0	1	0	0	1
	0	1	1	1	0
	1	0	0	0	1
	1	0	1	1	0
	1	1	0	1	0
	1	1	1	1	1

[a] In the beginning of any addition, (1st pair of digits), a "clear" condition, $c_1 = 0$, is assumed, since there is no carry generated as yet.

Notice that c_{i+1} is "one" if two-or-more, that is, the majority of the inputs are in the "one" state. Also notice that S_i is "one" if an odd number of "one" states are present at the inputs. By mapping the above truth tables as shown in Fig. 10.4, one may put them in the form of the following two Boolean algebraic equations:

$$c_{i+1} = a_i b_i + c_i(a_i \bar{b}_i + \bar{a}_i b_i), \tag{10.1}$$

$$S_i = \bar{c}_i(a_i \bar{b}_i + \bar{a}_i b_i) + c_i(a_i b_i + \bar{a}_i \bar{b}_i). \tag{10.2}$$

In the above equations, c_i can be substituted in terms of a_{i-1} and b_{i-1}, c_{i-1} can be substituted in terms of a_{i-2} and b_{i-2}, and so forth, so that both the sum S_i and the carry c_{i+1} of the ith addition may be expressed in terms $a_1, a_2 \ldots, a_i$, and $b_1, b_2 \ldots, b_i$.

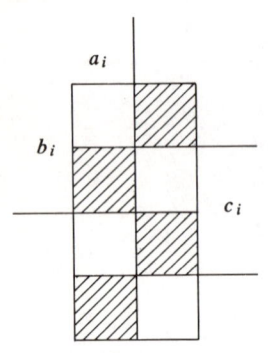

$S_i = \bar{a}_i b_i \bar{c}_i + a_i b_i c_i + \bar{a}_i \bar{b}_i c + a_i \bar{b}_i \bar{c}_i$
(No adjacent cells—Canonical sum)
$= \bar{c}_i(a_i \bar{b}_i + \bar{a}_i b_i) + c_i(a_i b_i + \bar{a}_i \bar{b}_i)$

Karnaugh map for the sum output S_i

$C_{i+1} = a_i b_i + c_i d_i + c_i b_i$
$= a_i b_i + c_i a_i(b_i + \bar{b}_i) + c_i b_i(a_i + \bar{a}_i)$
$= a_i b_i + c_i(a_i \bar{b}_i + \bar{a}_i b_i) + a_i b_i c_i$
$= a_i b_i + c_i(a_i \bar{b}_i + \bar{a}_i b_i)$

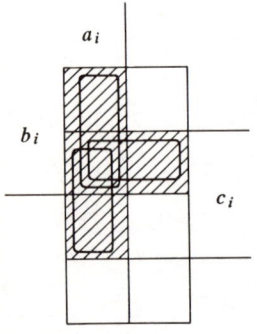

Karnaugh map for the carry output c_{i+1}

Fig. 10.4 Karnaugh maps and Boolean expressions for the sum and carry outputs of a full adder.

BINARY ARITHMETIC 267

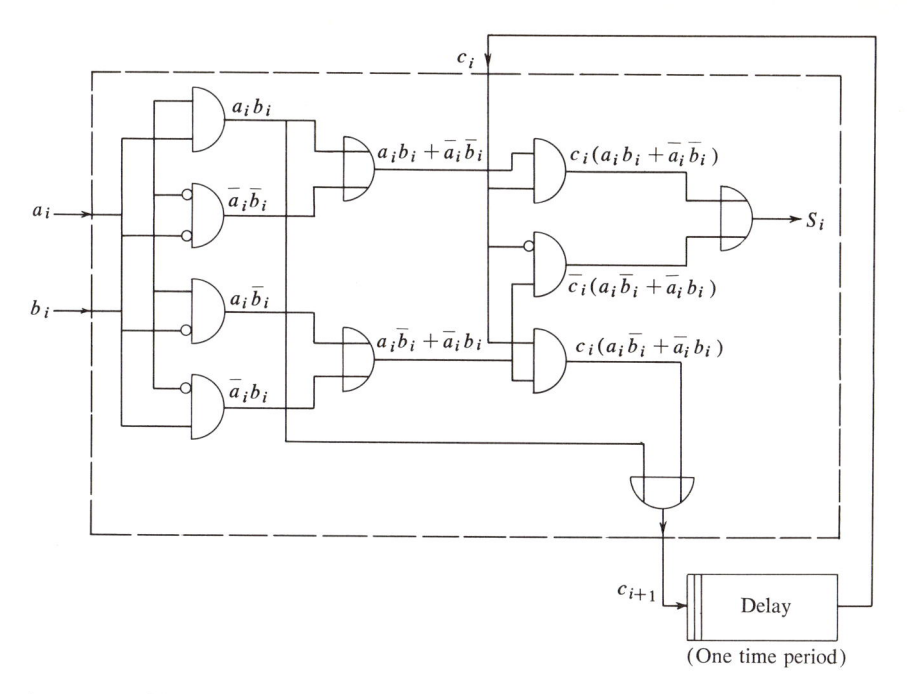

Fig. 10.5 Adder of two binary digits.

It should be apparent now that the above two functional equations (10.1) and (10.2) can be implemented in hardware to provide an adder, in a straightforward manner, using only simple circuits which will implement the elementary logical transformations AND, OR, NOT. Such circuits, which would deliver a signal designating a "one" to the output, only if certain combinations of signals exist at the inputs, were discussed in Chapters 8 and 9. They constitute the elementary building blocks of information-processing circuits,[2] and are implemented with mechanical, electronic, magnetic, optical, and various other means.

A straightforward implementation of Equations (10.1) and (10.2) into an adder, not the most economical one however, is shown in Fig. 10.5. The equations, however, may be manipulated and realized in more economical ways.

Notice that the circuit shown in the Fig. 10.6 realizes only portions of Equations (10.1) and (10.2). It provides the following sub-carry and sub-sum outputs:

$$c_i' = a_i b_i, \tag{10.3}$$

$$\bar{S}_i' = \overline{a_i b_i}(a_i + b_i)$$
$$= a_i \bar{b}_i + \bar{a}_i b_i. \tag{10.4}$$

The circuit of Fig. 10.6 is commonly known as the "half-adder," because the circuit combination shown in the Fig. 10.7 utilizes two such half-adders to make

---
[2] They also find use in a wide variety of other applications.

up the logical portion of a full adder. Notice that in the circuit of Fig. 10.7, and in accordance with Equations (10.3) and (10.4), we obtain

$$c_i'' = c_i S_i' = c_i(a_i \bar{b}_i + \bar{a}_i b_i) \tag{10.5}$$

and thus,

$$c_{i+1} = c_i' + c_i'' = a_i b_i + c_i(a_i \bar{b}_i + \bar{a}_i b_i), \tag{10.6}$$

which is the carry of the binary addition [Equation (10.1)]. From the circuit of Fig. 10.7 we also obtain

$$S_i = S_i' \bar{c}_i + \bar{S}_i' c_i = \bar{c}_i(a \bar{b}_i + \bar{a}_i b_i) + c_i[(\bar{a}_i + b_i)(a_i + \bar{b}_i)]$$
$$= \bar{c}_i(a_i \bar{b}_i + \bar{a}_i b_i) + c_i(a_i b_i + \bar{a}_i \bar{b}_i), \tag{10.7}$$

which is the sum of the binary addition [Equation (10.2)].[3]

The circuit of Fig. 10.6 has been discussed in Chapter 8, Section 8.7, in connection with the implementation of binary addition. The operation of both circuits is now justified in terms of Table 10.1 and the above equations. The circuits illustrate most convincingly the relation which exists between symbolic logic, Boolean algebra, and the hardware implementation of arithmetic binary addition. In the following we illustrate briefly the implementation of binary subtraction, multiplication, and division. The illustration is largely based on the above-discussed implementation of the binary addition.

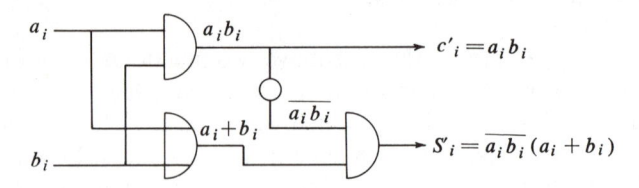

Fig. 10.6 Half-adder.

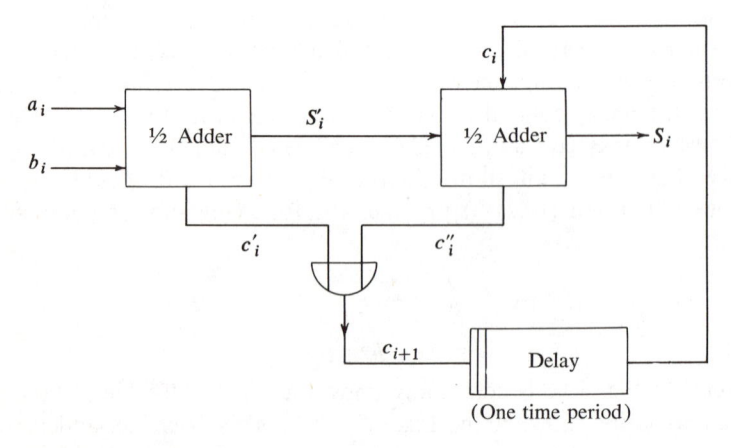

Fig. 10.7 Full-adder.

---

[3] Notice that $S_i' = a_i \bar{b}_i + \bar{a}_i b_i = (\bar{a}_i + b_i)(a_i + \bar{b}_i)$ because of DeMorgan's theorem.

b. The subtraction of two binary numbers

We now consider the subtraction of the number 101 (decimal 5) from the number 1011 (decimal 11). The operation is completed if we take the *complement* of the subtractor, add it to the subtrahend, and then perform the *end-around-carry* operation.

Taking the complement has been explained in Section 10.2. It involves changing each digit 0 to the digit 1, and each digit 1 to the digit 0. Thus, the complement of 101 is the number 010. The only logical operation involved in taking the complement of a binary number is the transformation NOT. It may be implemented in hardware by a signal inverter, which will transform each 1 and each 0 signal at its input to 0 and 1 signals, respectively, at its output.

The operation of addition has already been discussed. Thus the number 010 may be added to the number 1011 using the adder discussed previously.

The end-around-carry operation which is needed in order that the subtraction cycle described above be completed consists of taking the digit 1 at the extreme left of the sum above, and adding it to the remainder of the sum. The operation again involves a simple addition.

These three steps, which are required for completing the subtraction of two binary numbers, are illustrated in the following example:

$$\text{Complement (101)} = 010$$

$$\text{Addition:} \quad \begin{array}{r} 1011 \\ 0010 \\ \hline 1101 \end{array}$$

$$\text{End-around-carry:} \quad \underset{\hookrightarrow 1}{1101}$$

$$\overline{110}$$

c. The multiplication of two binary numbers

Multiplication in the binary system follows the same general rules as those of decimal multiplication. Of course, here we work only with two digits, 0 and 1. Looking at the multiplication table

×	0	1
0	0	0
1	0	1

or at the corresponding truth table

a_i	b_i	$c_i = a_i b_i$
0	0	0
0	1	0
1	0	0
1	1	1

we should recognize the logical transformation AND. Since only "1 times 1 makes 1," we conclude that, when a binary number is multiplied by the digit 1, the resulting product is the same number. A number which is multiplied by the digit 0 results in a string of "zeros."

To multiply a binary number by another, we multiply the multiplicand by each digit of the multiplier beginning from the right successively toward the left. Each time, we record the partial product shifted one position to the left in relation to the previous one. At the end we add the so-arranged partial products. Thus, the multiplicand is repeated and shifted one place to the left each time the corresponding digit in the multiplier is a one, while the product becomes a string of "zeros," again shifted one place to the left each time the corresponding digit of the multiplier is a zero. We again use a specific example to illustrate the process.

Let us assume that we want to multiply the binary number 1001 (decimal 9) by the binary number 1011 (decimal 11). The multiplication is illustrated as follows:

$$
\begin{array}{r}
1001 \longrightarrow \text{decimal } 9 \\
\times\ 1011 \longrightarrow \text{decimal } 11 \\
\hline
1001 \\
1001 \\
0000 \\
1001 \\
\hline
1100011 \longrightarrow \text{decimal } 99
\end{array}
$$

Having in our possession the adder described earlier, mechanizing binary multiplication becomes a fairly simple engineering task. In addition to the adder, we should need a set of "shift registers" to record and shift the multiplicand or a string of zeros each time we encounter a one or a zero, as we "read" the multiplier from right to left. Electromagnetic relays, a properly constructed electronic circuit around a string of ferrite cores, and a great variety of other electronic devices are used as shift-registers.

d. The division of a binary number by another

Again, the general rules of decimal division are applicable also in the division of a binary number by another. Let us illustrate the binary division, by dividing the binary number 10000101 (decimal 133), by the binary number 1101 (decimal 13).

```
dividend ⟶  10000101 │ 1101   ⟵── divisor
             0000      01010  ⟵── quotient
             ─────
             10000
              1101
             ─────
              0111
              0000
             ─────
              1110
              1101
             ─────
              00011
              00000
             ─────
              00011  ⟵── remainder
```

We observe that only two multiples of the divisor are used: 1 × (divisor) = (divisor), and 0 × (divisor) = 00000000 . . . . Each time, we compare the divisor with the partial remainder, and we place one or zero in the quotient if the divisor is smaller or larger than the partial remainder, respectively. Essentially, binary division is reduced to subtracting and shifting.

In the above brief discussion about the means and the components required for mechanizing arithmetic operations, some of the finer points, such as fractions, positive and negative numbers, and others, have been ignored. Only the essential underlying processes which are involved in applying symbolic logic and Boolean algebra reasoning to the mechanization of arithmetic operations were illustrated, and the corresponding hardware components were discussed.

Such features as decimal points, powers, fractions, roots, positive and negative numbers, and so on, may be handled by proper coding. For example, negative numbers may be accounted for if it is agreed that the extreme left-hand digit of each number will tell the sign of the number. If that digit is zero, then the number is positive. If it is a one, then the number is a negative one.

The circuits for performing arithmetic operations must include means for recognizing the sign of each number, the location of the decimal point, watching that the size of the number does not exceed the capacity of the machine and many other similar features. (See example in Chapter 12.)

10.5 SERIAL AND PARALLEL ADDITION

When two positive binary numbers are added in the familiar column method, as discussed in the previous section, the operation begins with the least significant digits at the extreme right of the two numbers and proceeds to the left. This is the serial mode of operation, in which each addition step requires knowing the carry of the previous step. A single full-adder network with three inputs a_i, b_i, c_i and two outputs c_{i+1}, S_i, like the one shown in Fig. 10.7, is used repetitively to perform a serial addition. This mode of operation is economical because it requires comparatively little hardware, but it is rather slow. A much faster parallel mode of operation, one in which the addition of two numbers is done by simultaneous processing of all corresponding pairs of digits of the two numbers, is discussed later in this section.

The role of the accumulator or A-register is an important one, especially when cumulative computations are performed in the arithmetic unit of the machine. Any time the next computation to be performed requires the result of a previous computation, it is very convenient to have that result available in a special register (the A-register) for immediate use in the next computation. (See Section 10.6.) The A-register operates differently in serial and in parallel adders. A serial accumulator register makes available the stored bits one at a time for arithmetic or logic operations, while a parallel accumulator register makes available all stored bits simultaneously for parallel operations to be performed. These properties are elaborated further in the following discussion, where the two modes of operation are viewed separately.

a. Serial Addition

Consider two given positive binary numbers to be added. In machines with fixed n-bit word length, each number must be extended, with 0's added at the left end, to an n-bit word. Thus, the two operands $a_k a_{k-1} \ldots a_1$ and $b_m b_{m-1} \ldots b_1$ are represented within the machine as n-bit words $a_n \ldots a_k a_{k-1} \ldots a_1$ and $b_n \ldots b_m b_{m-1} \ldots b_1$, where the added higher order bits are all 0's. Binary addition of two such given numbers within the machine involves the processing of all n pairs of bits.

Serial addition proceeds by processing pairs of bits and carry at a time in a column fashion, beginning at the extreme right of the two operands and moving to the left. The logical operations suitable for forming sum and carry digits were discussed previously in Sections 10.4 and 8.7. It may be recalled that the

SERIAL AND PARALLEL ADDITION 273

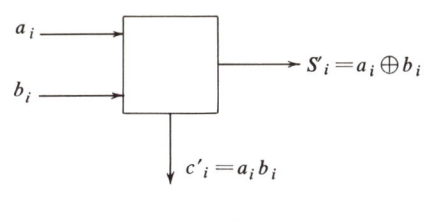

(a) Half-Adder

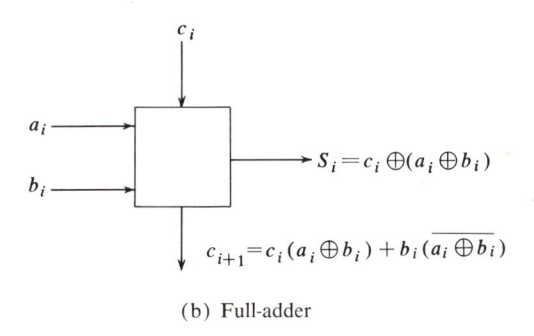

(b) Full-adder

Fig. 10.8 Schematics and relations of the logical half-adder and full-adder blocks.

half-adder is a two-input and two-output combinational network, as shown in Fig. 10.8(a), which performs the following logical operations:

$$c_i' = a_i b_i = \overline{b_i \bar{S}_i'}, \tag{10.8}$$

$$S_i' = a_i \bar{b}_i + \bar{a}_i b_i$$
$$= a_i \oplus b_i. \quad \text{Exclusive-OR, also referred to as the} \tag{10.9}$$
"ring sum" of a_i and b_i.

Equations (10.8) relating the output c_i' to the other output S_i' is of particular importance in some sequential modes of implementation (see Section 8.7). If S_i' is formed first, then a_i is not needed to form c_i'.

The logical portion of a full-adder is a three-input and two-output combinational network, as shown in Fig. 10.8(b), which performs the following logical operations:

$$c_{i+1} = a_i b_i + c_i(a_i \bar{b}_i + \bar{a}_i b_i)$$
$$= a_i b_i + c_i(a_i \oplus b_i)$$
$$= c_i(a_i \oplus b_i) + b_i\overline{(a_i \oplus b_i)}, \tag{10.10}$$

$$S_i = \bar{c}_i(a_i \bar{b}_i + \bar{a}_i b_i) + c_i(a_i b_i + \bar{a}_i \bar{b}_i)$$
$$= \bar{c}_i(a_i \oplus b_i) + c_i\overline{(a_i \oplus b_i)}$$
$$= c_i \oplus (a_i \oplus b_i). \tag{10.11}$$

The proof of Equations (10.10) and (10.11) is left as an exercise for the student (Problem 10.1). The last of Equations (10.10) allows the determination of c_{i+1} from b_i, c_i, and the ring sum $S_i' = a_i \oplus b_i$. Also notice that c_{i+1} is realizable directly by a majority element. (Chapter 8, Section 8.2.)

A single full-adder network suffices for the entire arithmetic operation. The adder is used n times repetitively, until all n pairs of digits of the two operands are processed. During the operation, the operands are stored in two n-bit shift registers, each register feeding the adder with one bit at a time, as illustrated in Fig. 10.3. At the same time as those bits are shifted out of the register, all other bits in the register also shift one place, thus leaving the left end cells of the register vacant. Consequently, the first sum digit of the result can be returned and placed in the cell formerly occupied by a_n, and subsequently shifted to the right. Similarly the b_1 digit can be returned and placed in the cell formerly occupied by b_n, and subsequently shifted to the right. When the shifting and adding operation is completed by having processed all n pairs of digits of the operands, the sum of the two given numbers will appear in the register formerly held by $a_n \ldots a_k a_{k-1} \ldots a_1$. The other register will still hold the same number, that is, $b_n \ldots b_m b_{m-1} \ldots b_1$. This scheme, as illustrated schematically in Fig. 10.9, is commonly used in modern computers. The register which holds the result of the arithmetic operation is called a "serial accumulator register."

The machine performs an arithmetic operation by following the program instructions (see Chapter 2, Section 2.4). On command, the operands are first copied from the main memory into the two registers shown in Fig. 10.9. The digits of the two operands are shifted by the timing pulses into the adder and are

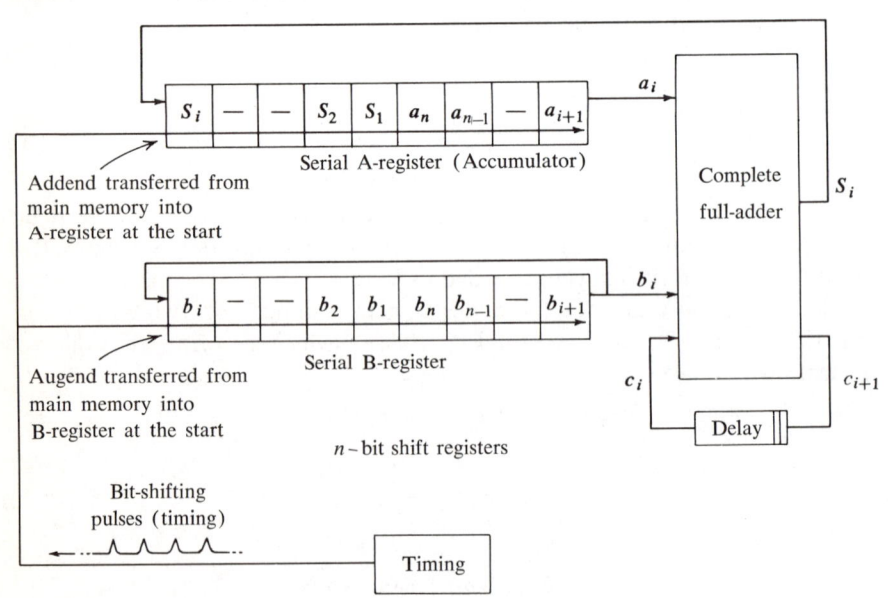

Fig. 10.9 Schematic of a serial full-adder operation.

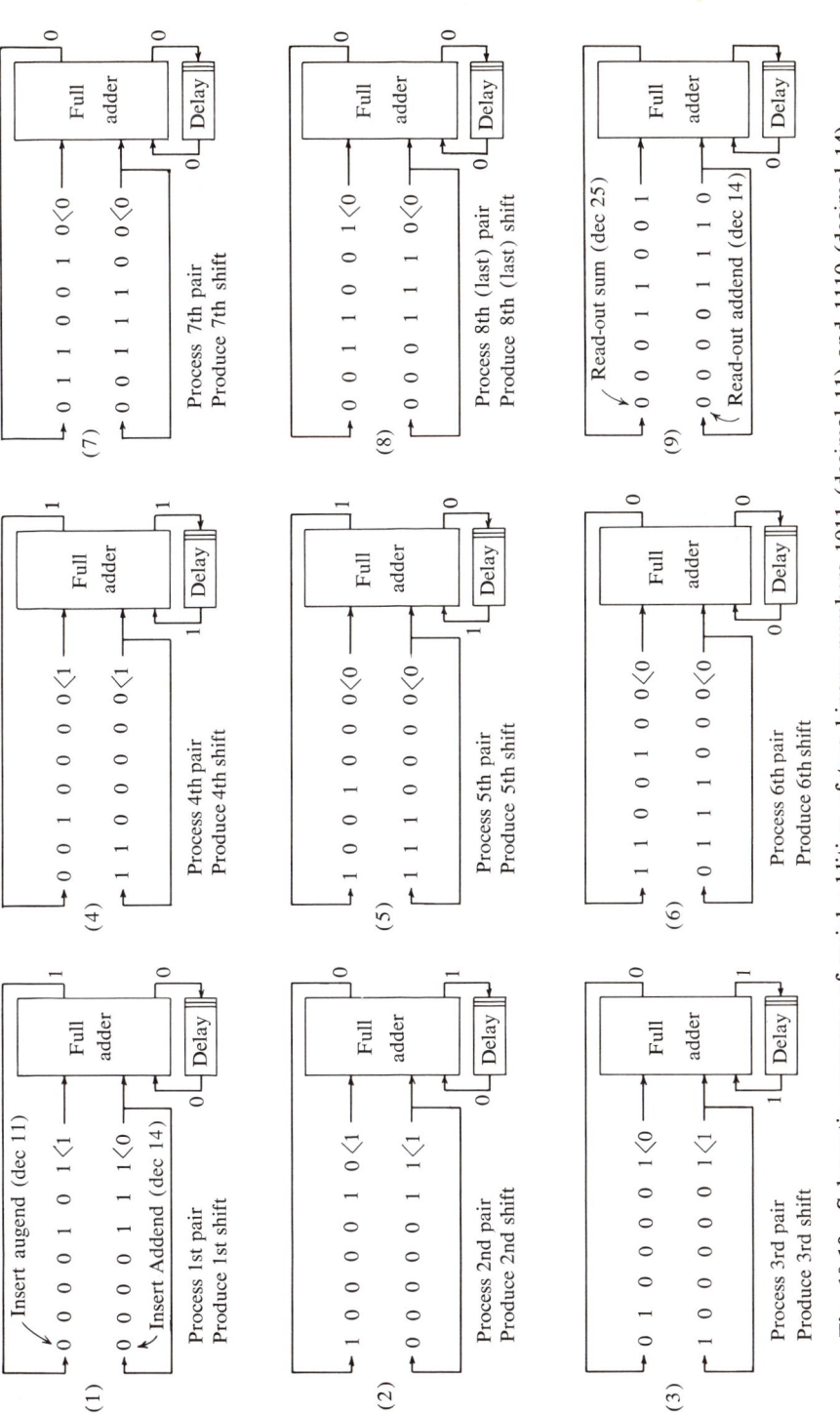

Fig. 10.10 Schematic sequence of serial addition of two binary numbers 1011 (decimal 11) and 1110 (decimal 14).

processed. At the completion of the operation the A-register will contain the result of the arithmetic operation. In the case of a cumulative computation, the result is kept in the A-register, while a new operand is brought into the other register, and a new arithmetic operation takes place (see example in Section 10.6).

Figure 10.10 illustrates the addition of two binary numbers, 1011 (decimal 11) and 1110 (decimal 14). Two eight-bit shift registers are used to hold the augend and addend. The complete addition is illustrated in a sequence of nine frames each corresponding to the processing of a pair of bits and to the shifting of one position to the right. The two right-end bits (one and zero) are processed first to produce a sum bit of one and a carry bit of zero, as shown in frame (1). The sum bit is returned in the top register (the accumulator) while the second of the bits processed is returned to the bottom register (B-register). The bits in the two registers are then shifted one position to the right and the procedure is repeated. The serial addition is completed when all eight pairs initially stored in the two registers have been processed.

It should be noted here that if the processing of the nth pair of digits produces a nonzero carry into the $(n+1)$th position, this is considered as an "overflow condition." Since the machine is not made to handle words longer than n-bits, the computation stops and special provisions are taken. One such case of overflow is dealt in conjunction with the computer design example discussed in Chapter 12.

b. Parallel Addition

Let us consider once again that the accumulator register of the arithmetic unit contains the number $a_n \ldots a_i a_{i-1} \ldots a_1$, and that another register (B-register) holds the number $b_n \ldots b_i b_{i-1} \ldots b_1$. The two numbers are to be added in the arithmetic unit which now operates in a parallel addition mode. As may be noticed from Equations (10.10) and (10.11),

$$c_{i+1} = c_i(a_i \oplus b_i) + b_i\overline{(a_i \oplus b_i)}, \tag{10.12}$$

$$S_i = c_i \oplus (a_i \oplus b_i). \tag{10.13}$$

The predominance of the exclusive-OR operation strongly suggests that the addition process may be started by first forming all $a_i \oplus b_i$ terms *simultaneously*. Since a_i would not be needed again after the ring sums have been formed, the $a_i \oplus b_i$ terms could be stored in the A-register, in the cells formerly occupied by a_i. Thus, in one step (one time interval), the ring-sum operations may be carried out simultaneously and the quantity $r_n \ldots r_i r_{i-1} \ldots r_1$, where $r_i = a_i \oplus b_i$, may be stored in the A-register.

Equations (10.12) and (10.13) may now be rewritten as follows:

$$c_{i+1} = c_i r_i + b_i \bar{r}_i, \tag{10.14}$$

$$S_i = c_i \oplus r_i. \tag{10.15}$$

The carries can be formed next. One may choose to compute the carries beginning with $c_1 = 0$, and applying Equation (10.14) repetitively for $i = 1, 2, \ldots, n$. Notice, however, that the second term of Equation (10.14) does not involve a knowledge of any previous carry, and therefore it can be computed in one step as the logical product of the contents of the B-register and of the complement of the contents of the A-register. Moreover, notice that when the second term, $b_i \bar{r}_i$, is one, the first term must be zero. (And c_{i+1} is one.) This means that a carry c_i developed in the ith stage can propagate into the $(i + 1)$th stage only if the second term, $b_i \bar{r}_i$, is zero. As long as zero values are encountered for $b_i \bar{r}_i$, the carry may propagate to the next stage; if a one value is encountered for $b_i \bar{r}_i$, the propagating carry would be halted, but a new carry propagation would be initiated because c_{i+1} would also become one.

Based on these observations the carries may be computed by a sequence of logical operations, initiated by a clock pulse. Once the carries become available, the sum digits are computed in one step by performing simultaneously the modulo 2 additions (that is, exclusive-OR operations), denoted by Equation (10.15). In summary, the two given positive binary numbers can be added by parallel means with two timing pulses. The originally given quantities are stored in the A- and B-registers. At the conclusion of the operation, the A-register will contain the sum, while the B-register may contain the same quantity as originally stored. The "parallel accumulator register" differs from the serial A-register, in that it makes available all of the stored bits *simultaneously*, instead of one at a time. The operation of a parallel A-register may be extended to the addition or subtraction of two binary numbers with arbitrary signs. One such case is discussed in the computer design example of Chapter 12.

10.6
ORGANIZATION AND OPERATION OF A SMALL COMPUTER

In order to extend our insight of the internal organization and operation of information-processing machines (Chapter 1, Section 1.9) we shall examine here the execution of the cumulative addition of given positive numbers by a small computer. We shall assume that 200 positive numbers have been punched into cards (see Chapter 1, Section 1.10) and have been transferred and stored in the main memory of a small stored program computer. We wish to have the computer add these numbers cumulatively, so that all subtotals and the grand total will be available at the end of the operation. One could imagine that the requested operation is part of a larger inventory problem executed by the computer. We may add here, that this simple example of cumulative addition was selected because we wished to show the essential features of operation and organization of information-processing machines with the minimum discussion of hardware complexity.

Let the computer have a 1000-word storage capacity. Each memory word-cell is identified by a three-digit address, *xxx*, from 000 to 999. Let also the portion of the memory from 000 to 099 be allocated for the storage of program instructions, and the portion from 100 to 999 be allotted for storing data. The given numbers to be added occupy the memory locations from 100 to 299.

a. Operation

We shall use the parenthesis to signify the "contents of" a memory cell, as, for example, (101) should signify "the contents of memory location 101". The sequence of operations that should be executed by the machine are as follows:

$$(100) + (101) \rightarrow 101$$
$$(101) + (102) \rightarrow 102$$
$$(298) + (299) \rightarrow 299$$
$$\text{STOP}$$

Each operation

$$(xxx) + (yyy) \rightarrow yyy$$

signifies "Add contents of memory location *xxx* to contents of memory location *yyy* and place result in memory location *yyy*."

At the end of the sequence, the grand total is available in memory location 299, while the memory cells 101 to 298 contain all the cumulative subtotals.

The above sequence is executed by the machine in smaller steps. However, before we break the operations down to small steps, we should say something about the *timing* of the sequence.

Synchronous and asynchronous operations: If during computations the time is divided in equal time intervals, so that each one of the required operations takes place in one unit of time, the operation is referred to as being "synchronous." Time is divided in equal intervals by a "master clock," which is a well-regulated pulse generator. The clock generates a pulse train, whose period is equal to one unit of time. Clock pulses are then used to trigger the operations, and thus to time all the events taking place within the machine.

The operation described previously, namely,

$$(xxx) + (yyy) \rightarrow yyy$$

may be considered to take place in time during what may be called one "machine cycle." For simplicity assume that the duration of a machine cycle is one unit of time. A machine cycle includes a complete "fetch and execute" cycle and may be divided into two parts, with one part devoted to "fetching" the instruction from the memory and interpreting its contents, and the other part devoted to the execution of the task as prescribed by the instruction.

The time unit must be long enough, so that no matter what the task is, there is always enough time to permit the completion of a machine cycle. Since elementary tasks require different time durations for their completion, the time unit must be tailored after the longer of the elementary tasks performed by the machine. For example, the time that it takes to read information off a rotating drum memory varies depending on the relative position of the magnetic reading head and of the requested information on the drum at the start of the reading cycle. Because, then, of the restriction of performing elementary operations in fixed duration of time intervals, which may be longer than necessary for most machine cycles, synchronous operation does not permit efficient use of the machine from a time standpoint.

Another approach is to design the machine so that as soon as a machine cycle is completed, an appropriate signal is generated to initiate the next cycle. In this fashion the time allotted for a machine cycle is variable, being held for each task to a practical minimum, thus permitting more efficient use of the machine's circuits from a time standpoint. This mode of operation is the "asynchronous" one and permits much higher speeds of information processing than the synchronous operation although at the cost of more complex circuitry.

Since the majority of modern machines are synchronous, we will assume that our small computer is also a synchronous one. Each one of the previously listed operations $(xxx) + (yyy) \rightarrow yyy$ is a machine cycle which is completed in one time interval. In the first part of the cycle, that is, during the "fetch" portion, the instruction is fetched and interpreted by the control unit. During the "execute" portion of the cycle, the task is completed. The task of adding two numbers involves several subtasks. For our simple computer then the time sequence of the operations may be written as follows:

Time interval	Operations	Descriptions
1	$(100) \rightarrow A$	Copy (100) into A-register
	$(101) + (A) \rightarrow A$	Add (101) to (A) and place sum into A-register
2	$(A) \rightarrow 101$	Copy (A) into mem. loc. 101
	$(102) + (A) \rightarrow A$	Add (102) to (A) and place sum into A-register
3	$(A) \rightarrow 102$	Copy (A) into mem. loc. 102
	$(103) + (A) \rightarrow A$	Add (103) to (A) and place sum into A-register
...	...	...
199	$(A) \rightarrow 298$	Copy (A) into mem. loc. 298
	$(299) + (A) \rightarrow A$	Add (299) to (A) and place sum into A-register
200	$(A) \rightarrow 299$	Copy (A) into mem. loc. 299
	STOP	STOP

The above time-sequence of operations may be represented in a simpler manner, by using superscripts to denote the time intervals, as follows:

$$\text{START:} \qquad (100)^1 = (A)^1$$
$$(101)^1 + (A)^1 = (A)^2 = (101)^2$$
$$(102)^2 + (A)^2 = (A)^3 = (102)^3$$
$$(103)^3 + (A)^3 = (A)^4 = (103)^4$$
$$\cdots \qquad \cdots$$
$$(299)^{199} + (A)^{199} = (A)^{200} = (299)^{200} \qquad (10.16)$$
$$: \text{STOP}$$

More generally, one may write

$$(xxx)^k = (A)^k,$$
$$(A)^k + (yyy)^k = (yyy)^{k+1} = (A)^{k+1}. \qquad (10.17)$$

It should be recalled at this point, that the machine must be instructed to "stop" when the information-processing task has been completed. Computers stop either because a built-in detector observes some adverse condition during operation, such as an attempt to divide by zero, in which case the operation is stopped automatically, or because it is instructed to do so. Normally, the machine is instructed to stop when some condition is satisfied, as for example by counting 200 time intervals for the computations shown in Equations (10.16). In any case, the machine is incapable of stopping because it "reasons" that the job is done.

The use of index registers. Modern computers have large memories, often containing tens of thousands of words of storage, and perhaps hundreds of instructions available. However, the additional features added by more available instructions do not enable the machine to do anything essentially new, only they are convenient and make the programmer's job a lot easier. The use of an "index register" is a good example of such a very convenient feature.

Much of the usefulness of computers stems from their ability to modify their own instructions. They can do so in various ways, but the *modification of addresses* is the only important one. A simple case of address modification through the use of an index register is illustrated in Fig. 10.11 with the execution of the cumulative addition shown in Equations (10.16). The small computer may perform the cumulative addition of the 200 numbers by repetitive "looping." As the machine loops it may require on each successive circling to obtain data from a new source. This may be accomplished by successively modifying certain instructions along the way. Such address modification may be accomplished by arithmetic manipulations. However, it is simply accomplished by "indexing." "Indexing" refers to adding a modifier to the address part of an instruction before execution. Special equipment is required in the control unit for this purpose. The resulting "effective address" is used to locate the new source of data for the execution of the instruction.

ORGANIZATION AND OPERATION OF A SMALL COMPUTER

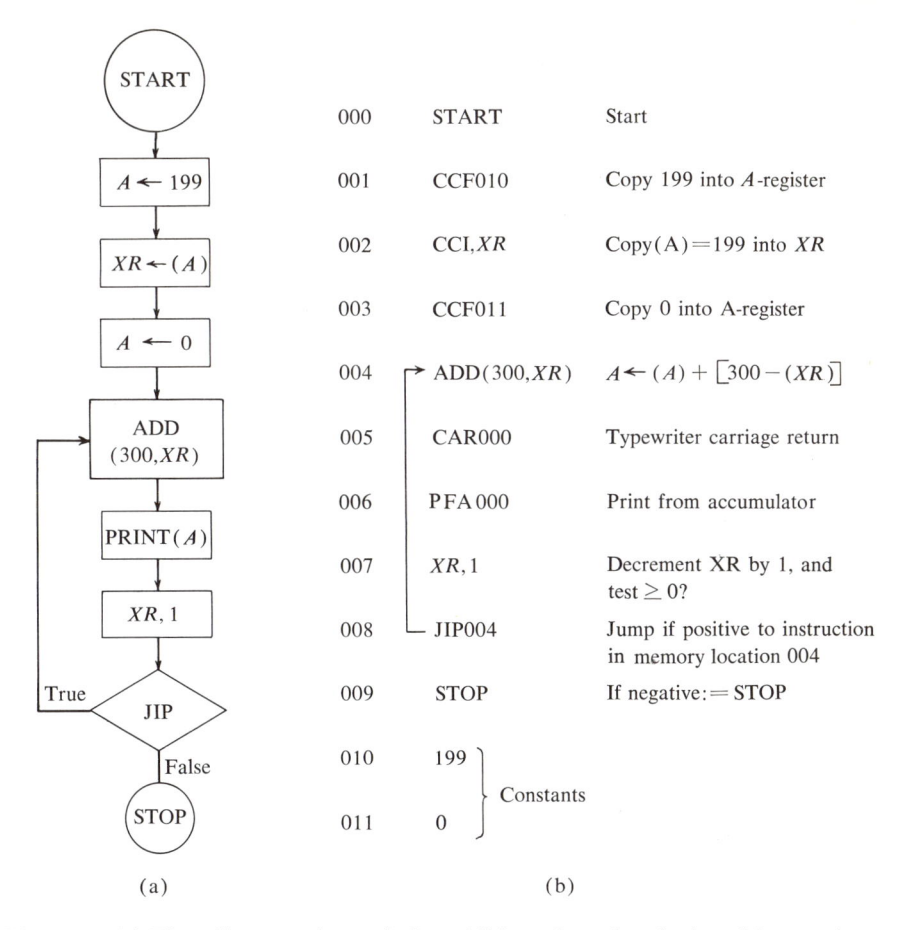

	000	START	Start
	001	CCF010	Copy 199 into A-register
	002	CCI,XR	Copy$(A) = 199$ into XR
	003	CCF011	Copy 0 into A-register
	004	ADD(300,XR)	$A \leftarrow (A) + [300-(XR)]$
	005	CAR000	Typewriter carriage return
	006	PFA000	Print from accumulator
	007	XR,1	Decrement XR by 1, and test ≥ 0?
	008	JIP004	Jump if positive to instruction in memory location 004
	009	STOP	If negative: $=$ STOP
	010	199	Constants
	011	0	

(a)   (b)

Flow diagram (a) shows: START → $A \leftarrow 199$ → $XR \leftarrow (A)$ → $A \leftarrow 0$ → ADD(300,XR) → PRINT(A) → XR,1 → JIP (True loops back to ADD; False → STOP).

Fig. 10.11 (a) Flow diagram of cumulative addition of two hundred positive numbers stored in memory locations 100 to 299; (b) corresponding sample program using mnemonics of RPC 4000–PINT (Chapter 2).

An "index register" is another special word storage in the machine's memory, only large enough to hold an address. With the hypothetical small computer executing the above cumulative addition, the index register would, therefore, hold three decimal digits (xxx).

Machines that have index registers also have instructions available for: (a) copying numbers into and out of the index register, (b) incrementing or decrementing the numbers in these registers, and (c) adding or subtracting the contents of the index register to the address of a given instruction. For example, denoting the index register with XR, we may say that

$$XR \leftarrow (A)$$

means: "Copy the contents of the accumulator into the index register." Also

$$XR, 1$$

282 ARITHMETIC OPERATIONS AND MACHINE ORGANIZATION

may mean: "Decrement the contents of the index register by one." As we may recall (Chapter 2), the command ADD 300 means: "Add the contents of memory location 300 to the contents of the A-register and place the sum into the A-register," that is, $A \leftarrow (A) + (300)$. With the use of the index register, the command ADD(300, XR) may mean: "Add the contents of memory location 300-(XR) to the A-register and place the sum into the A-register.

If only one indexing quantity is available, each instruction contains a special "flag" digit indicating whether or not the instruction is to be indexed, for example, one means "index" and zero means "no index." With more than one indexing quantity, each instruction must devote a certain short portion in its address to indicating which index register is to be used.

Figure 10.11 illustrates the flow diagram for the cumulative addition of 200 numbers making use of an index register.

Memory Cycles. Before we discuss the organization of this small computer we shall briefly describe two basic machine cycles referred to as "memory cycles." These refer to the sequence of actions taken by the machine during *storing* or *copying* information to and from the main memory. These cycles are of fundamental importance to the operation of any automatic information-processing machine.

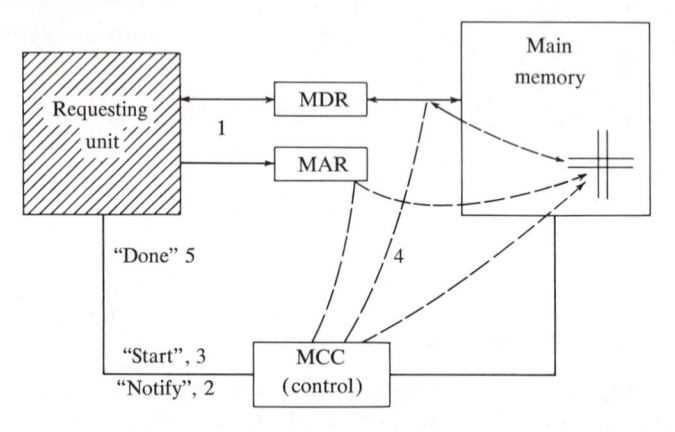

Fig. 10.12 Memory cycles schematic. Dashed lines are control lines.

The two memory cycles are described below and in conjunction with Fig. 10.12. The numbers shown along with the connecting channels in Fig. 10.12 correspond to the numbers and the actions described below.

Store Cycle. Actions taken in order to store a word into the main memory.
1. Requesting unit places the word to be stored in MDR (Memory Data Register), and the address in MAR (Memory Address Register). The MAR holds the address of the cell currently under consideration.

2. It notifies MCC (Memory Control Component), that this is going to be a "store" cycle. The MCC is a small control unit, which is used to supervise the execution of operations within the memory, initiate memory cycles, locate cells, take requested action, keep track of operations, and issue the "done" signal to the main control unit after the task is completed.
3. It issues the signal "start" to the MCC, so that the execution of the request may begin immediately after.
4. The MCC consults the MAR about the address, locates the cell, and executes the order by transferring the contents of the MDR into the cell.[4]
5. The MCC issues the "done" signal to the requesting unit and to the main control unit, thus notifying "mission accomplished and ready for the next request."

Copy Cycle. Action taken in order to copy a word from the main memory.
1. Requesting unit places address in the MAR.
2. It notifies the MCC that this is going to be a "copy" cycle.
3. It issues the "start" signal to the MCC, so that the execution of the request may begin immediately after.
4. The MCC consults the MAR about the address, locates the cell, and executes the order by copying the contents of the cell into the MDR.
5. The MCC issues the "done" signal to the requesting unit, thus notifying "mission accomplished and ready for the next request." It is then up to the requesting unit to retrieve the datum from the MDR.

It may be noted from the above that a memory cycle does not begin until the requesting unit has supplied full information to the registers, checked it, notified the MCC about the type of the action requested, and issued the "start" signal. Then, while the MCC is working on the specific request, the memory unit and the two registers are isolated from all other units, so that no interference may occur from overlapping demands. In the case of the copy cycle it is when the requesting unit has recovered the datum from the MDR that the registers become available to accept new information.

b. Organization

In the course of transferring information to and from the memory unit, a serious *traffic* problem may be generated. During information processing by the machine, the facilities of the arithmetic unit are used over and over, and data are being transferred repetitively between the memory and the processor. Traffic problems of information transfer between principal units of the machine may be eased by the use of *selector* units.

A selector unit between the memory and the processor, for example, operating under the supervision of the control unit, would make sure that, given an

---

[4] This completely obliterates the previous contents of the cell.

284 ARITHMETIC OPERATIONS AND MACHINE ORGANIZATION

address, only the contents of the corresponding memory location are transferred into the processor. Also, the selector would make sure that the results of computations are returned from the accumulator to the appropriate memory locations.

A single selector unit, operating between the processor and the memory, would direct the information traffic in both directions. Another selector unit may operate between the memory and the control unit for the selection of stored instructions, when the instruction address is provided to the control unit. This selector unit supplies an *instruction register* (I-register), which holds the instruction until the prescribed task is completed.[5]

The two selector units are shown in the organization diagram, illustrated in Fig. 10.13. A "timer" and a "cycle counter," which provide the timing pulses in a synchronous operation, are also shown.

In the organization shown in Fig. 10.13, information coming from the memory is interpreted as "data" or as "instructions" depending on the location

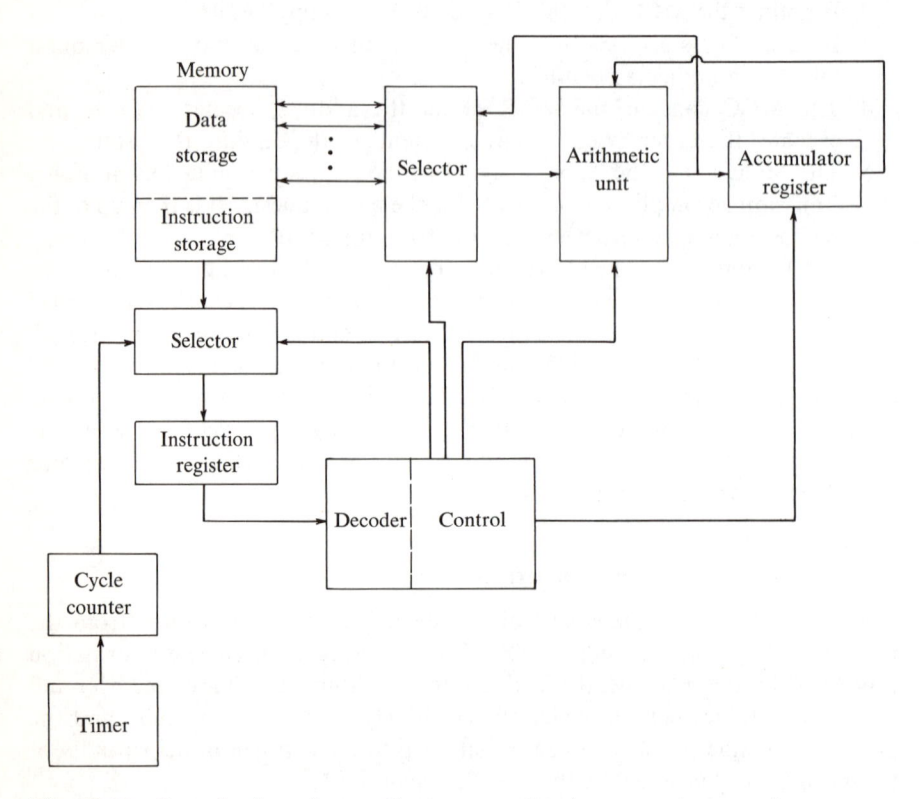

Fig. 10.13 Organization of a small computer showing two selector units.

---

[5] The machine is often capable of modifying a stored instruction as it goes along. In this case a modified instruction may be returned into the memory for future use.

of its origin within the memory, and on the path that it follows, that is, on the selector unit that it uses. However, if the operations are timed in alternating "fetch" and "execute" half-cycles, then information coming from the memory during the first half-cycle ("fetch"-cycle) is automatically regarded as *instructions*; during the following half-cycle ("execute"-cycle) the information is automatically regarded as *data*. Under this scheme of operation, more efficient use of the memory and the selector facilities is accomplished. There exists no more need to partition the memory by allotting separate portions for the storage of instructions and data. Instead, instructions and data may be mixed and stored anywhere within the memory. Also, a single selector unit may be used to divert information coming from the memory either to the I-register or to the processor unit. The revised organization is shown in Fig. 10.14.

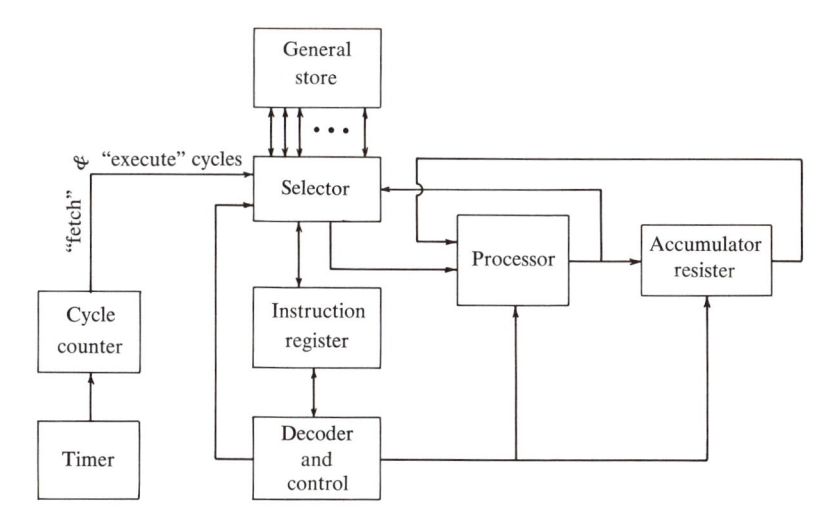

Fig. 10.14 Organization of a small computer using a single selector unit.

BIBLIOGRAPHY

Caldwell, S. J., *Switching Circuits and Logical Design*. New York: John Wiley & Sons, Inc., 1958.

Humphrey, W. S., *Switching Circuits*. New York: McGraw-Hill, Inc., 1958.

Huskey, H. D. and Korn, G. A., Eds., *Computer Handbook*. New York: McGraw-Hill, Inc., 1962.

McCluskey, E. J., *Introduction to the Theory of Switching Circuits*. New York: McGraw-Hill, Inc., 1965.

Richards, R. L., *Arithmetic Operations in Digital Computers*. Princeton, N.J.: D. Van Nostrand, Inc., 1955.

Problems

10.1 Using algebraic manipulations, verify Equations (10.10) and (10.11). Repeat by using Karnaugh maps.

10.2 Two 2-bit registers with bits A_1, A_0, and B_1, B_0, respectively, and a three-bit sum register with bits C_2, C_1, C_0, are given. Design a two-level AND/OR/NOT circuit which will place the sum of A and B registers into the C register in a parallel operation.

10.3 Perform the following operations in pure binary code using r-complements for the representation of negative numbers:

(a) $+19-7=+12$

(b) $-19-7=-26$

(c) $-19+7=-12$

(d) $-14+22=+8$

Repeat the above operations in 8421 and XS3 codes using $(r-1)$-complements for the representation of negative numbers.

10.4 An integer a is said to be congruent to an integer b (modulo m), if their difference is evenly divisible by m. The notation often used is: $a = b$ (mod m). On the basis of this definition, find the minimum positive x if:

(a) $16253 = x$ (mod 9)

(b) $16253 = x$ (mod 8)

(c) $16253 = x$ (mod 7)

(d) $16253 = x$ (mod 4)

(e) $16253 = x$ (mod 2)

10.5 Prepare the multiplication tables in the trinary and in the octonary numerical systems. Then perform the following operations:

(a) $121 \times 11 =$
(b) $121 : 11 =$ } In the trinary system

(c) $143 \times 41 =$
(d) $143 : 41 =$ } In the octonary system

10.6 You are asked to design an arithmetic unit which will perform the addition of two eight-bit positive binary numbers. For a serial and a parallel adder respectively:
(a) describe the addition process and form an algorithm,
(b) construct a flowchart of the process,
(c) construct a block-diagram indicating the operation of the shift registers, delays, and of the other necessary components, down to the greatest detail you can possibly design.
Illustrate your design using the positive binary numbers 10111001 (decimal 185) and 11101 (decimal 29).

11
Sequential Machines

Before we examine the formal presentation of sequential machines, we shall follow a rather intuitive approach by examining the specific design requirements of a simple sequential circuit. In more complex design situations, an optimum solution may not be possible without the sophistication of the formal method of design. However, our purpose will have been well served with this approach. In the simple, specific design example discussed first in Sections 11.2 through 11.7, we should be able to trace in a direct and practical manner the essential points and concepts in the intricate techniques employed in the design of sequential machines. The more abstract descriptions which will follow then should allow us to generalize the concept of a sequential machine and acquire an overview of its properties and its limitations.

11.1
DEFINITIONS

In Chapter 8 we examined information-processing circuits, for which the values of the outputs at any given time t_1 are exclusively and uniquely determined from the values of the inputs at that same time t_1. Boolean algebra (Chapter 7) derived from the premises of Symbolic Logic (Chapter 6) was used to analyze these circuits which are referred to as "combinational," or "logical," or sometimes, "contact" circuits.

In the operation of "sequential" circuits[1] time becomes of the essence. Generally speaking, the output values of these circuits at any given time t_1, depend on the input values at all previous times t, such that $t \leq t_1$, that is, on the "sequence" of events that preceded time t_1. We may also say that, at any given time t_1, a sequential circuit will be characterized by an internal "condition" or "state," which depends on the past history of the values of its inputs. Therefore, then, at any given time, the values of the outputs are determined from the values of the inputs and from the state of the circuits.

It becomes apparent in the above definition, that sequential circuits must possess the capabilities of *memory* in order that they may remember past events. "Memory" is the ability to retain information over some period of time. "Static memory" will retain information for an indefinite period of time, until a proper stimulus causes it to deliver the stored information. It, then, may do so either "destructively" or "nondestructively," that is, either by wiping out the stored information in the process or by simply copying it for delivery. A single magnetic core is an example of a static, destructive, memory element for the storage of one bit of information.[2]

"Dynamic memory" will retain information only for some definite characteristic time, and then it will deliver the stored information automatically. The information may be fed back into the dynamic memory and thus circulated for a number of cycles or new information may be entered. "Delay" devices, such as transmission lines, are typical examples of dynamic memories. A delay device was used for example as a dynamic memory in conjunction with the full-adder operation described in Chapter 10 (Section 10.5, Fig. 10.9) to take care of the carry operation.

Every information-processing machine which stores information (data or instructions) for various periods of time is a sequential machine. Such machines, composed of an assortment of logical circuits and memory elements in various levels of complexity, are capable of performing a great variety of useful information-processsing tasks. The full-adder described in Chapter 10 is an example of a small sequential machine. It is composed of a combinational circuit which performs the logical transformations expressed by Equations (10.10) and (10.11), and of a delay (memory) element for the performance of the carry operation. In following, we shall discuss the design of a small sequential machine in order to familiarize ourselves with the procedures involved.

---

[1] The terms sequential "circuits" or "machines" or "systems" may be used interchangeably. At times, though, the characterization "machines" or "systems" is reserved for large scale designs, which include a great number of combinational and sequential circuits with added special memory, control, and input–output capabilities.

[2] Since information is represented and processed in digital form and serially, that is, bit at a time, such single-bit memory devices are of basic importance.

11.2
A SMALL "SPECIAL-EVENT" SEQUENTIAL MACHINE

Consider the following design problem: "Design a one input-one output machine to accept at its input a sequence of 'ones' and 'zeros.' The machine issues an output signal of 'one' whenever a 'zero' appears at the input following a string of an odd number of 'ones.' Otherwise the machine's output is a 'zero'." This is a special-event sequential machine which responds to a sequence of special events at its input, namely to a sequence of an odd number of ones followed by a zero. We shall assume that the presence of a pulse at the input or output line will represent a one, while no-pulse will represent a zero.[3]

First we must think of all the possible circumstances with which the machine may be met. This kind of an intuitive approach should lead us to the identification of the possible "states" of the machine.

11.3
STATES

Let the "state" of a machine refer to its condition at a given time as described by its past history. The description of the sequence of the inputs up to an instant t_1 would lead to the description of the state of the machine at that instant. Of course, this assumes that there is also an adequate description of some "starting" state at the beginning of counting the time.[4] How far back the machine can "remember" of its input history depends on its available memory. It turns out that our machine has a very *short* memory, which allows it to remember either one of only two states, as will be shown later in Section 11.5.

In order to begin somewhere with the description of the operation of our machine, let us assume that we define the "starting state" of the machine as being the condition of the machine just after it has issued an output pulse; we shall call it the "CLEAR" state.[5] We choose this to be the "starting" state because we can easily place the machine in this state by simply supplying it with the input sequence 010. No matter what the state of the machine at the time, the machine should issue an output pulse at the end of this short sequence and be cleared.

---

[3] See also Section 11.7.

[4] This is characteristic of "deterministic" machines, that is, those whose future behavior is absolutely predictable if given their present state and inputs.

[5] We agree to always specify a "starting" state when we specify a machine. We may choose that to be any *known* state. The above CLEAR state was chosen because it is an easy one to establish and to define.

It seems at this time that at any instant of its operation the machine should be found in a different state, depending on the different input sequence preceding. However, it will become obvious in a short while that several superficially different states are actually *equivalent* and may be grouped together in one essential state or "superstate." Two states are "equivalent" if, started in either one, the machine would display identical behavior for *all* possible sequences of input combinations; this really means that the machine will make identical transitions to the same or equivalent states and issue the same outputs in both cases.

In describing the operation of the machine one may use only the essential states or superstates. These superstates are fewer for machines which possess smaller memory capacities. We shall see later that this machine has a very short memory and possesses only two superstates.

11.4
STATE FLOW DIAGRAM

By constructing a "state flow diagram" we may trace pictorially the operation of the machine through all possible states. We may begin at the START, with the machine being at the CLEAR state, as shown by the block marked with "A" in Fig. 11.1. Each time the machine receives an input it undergoes a *transition*, that is, a change of state, and issues an output.

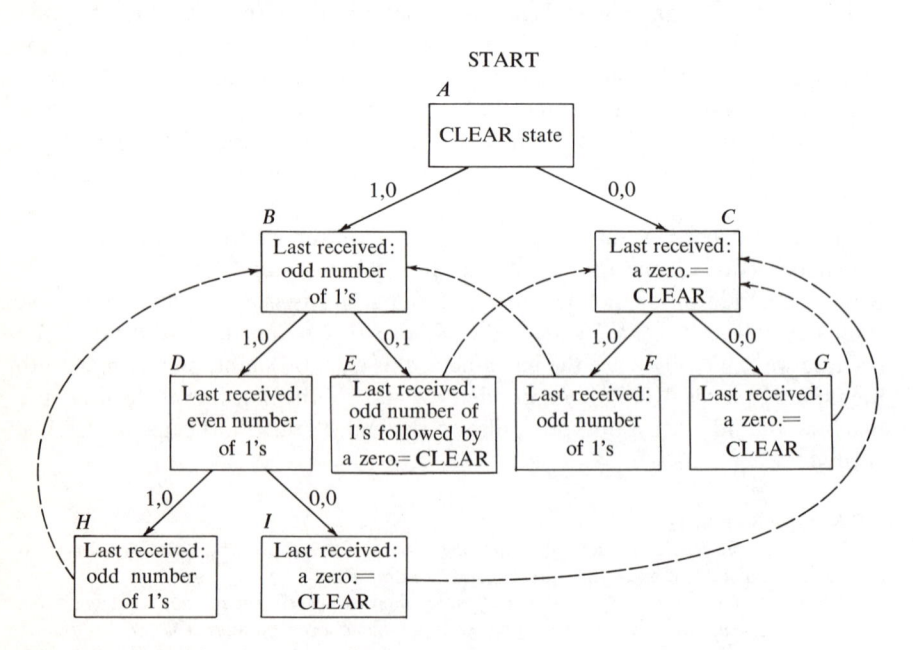

Fig. 11.1 Primitive state flow diagram.

Since there are only two alternatives of input values, that is, the machine may receive either a zero or a one, then, given the state of the machine at any time, there should be only two possible state transitions.[6] These are designated in the diagram by two arrows leaving each block. We use two binary digits with each transition arrow to designate input and output. Thus, beginning with the CLEAR state at the START we mark two possible transitions: At left, a 1,0 transition, (input = one, output = zero), leading to the state characterized as "Last received odd number of ones." At right, a 0,0 transition leading to the state characterized as "Last received a 0."

Continuing on Fig. 11.1, we mark all possible transitions as shown in the corresponding "transition-table" below. We use the capital letters *A, B, C, D, E, F, G, H,* and *I,* to mark the corresponding state-blocks as shown in Fig. 11.1. The table is followed by several important observations [Notes (1) through (5)] concerning several transitions, as indicated in the "remarks" column of the table.

The state flow diagram shown in Fig. 11.1 is in closed form, that is, all possible transitions and all possible states have been accounted for. However, as indicated clearly by the observations below, a simplification is possible by eliminating the equivalent "redundant" states.

From	*Input,output*	*To*	*Remarks*
A	1, 0	B	
A	0, 0	C	
B	1, 0	D	
B	0, 1	E	See Note (1) below
C	1, 0	F	See Note (2) below
C	0, 0	G	See Note (3) below
D	1, 0	H	See Note (4) below
D	0, 0	I	See Note (5) below.

Note (1): State *E* is equivalent with state *C*. Therefore, a broken line arrow is used to indicate that the state-block *E* may be eliminated from the diagram, and that the 0,1 transition from state *B* leads directly to state *C*.

Note (2): State *F* is equivalent with state *B*. Therefore, the 1,0 transition from state *C* leads directly to state *B*.

Note (3): State *G* is equivalent with state *C*. Therefore, the 0,0 transition arrow "loops" around and returns to state *C*.

Note (4): State *H* is equivalent with state *B*. Therefore, the 1,0 transition from state *D* leads directly to state *B*.

Note (5): State *I* is equivalent with state *C*. Therefore, the 0,0 transition from state *D* leads directly to state *C*.

---

[6] That is, two different directions of change of state.

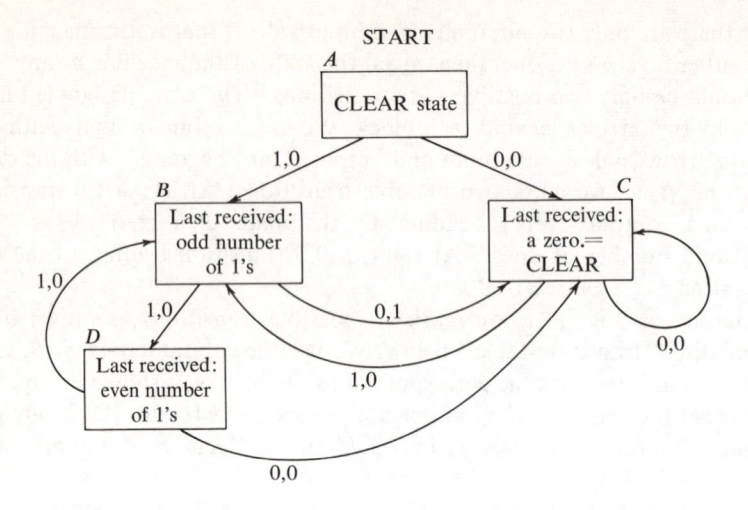

Fig. 11.2 Simplified state flow diagram.

The diagram shown in Fig. 11.1 is called a "primitive" state flow diagram because it was constructed from the original statement of the problem by tracing all possible transitions and states, without regard to simplification. After eliminating the apparent redundancy noted above, there results the simplified state flow diagram, shown in Fig. 11.2.

Observing the diagram in Fig. 11.2, we may realize that one can start the machine at any state, as long as that state has been identified. If the starting state is unknown, we may have to wait until the first 1 is issued from the output, signifying that the machine has reached a CLEAR state. It may be suggested, for simplicity, that the machine user be asked to CLEAR the machine at the start by supplying a 010 sequence at the input. Otherwise, any unfinished sequence of inputs from a previous user may lead to an erroneous reading.

We may also observe in the diagram of Fig. 11.2 that the initial CLEAR state (block A) is equivalent to the "Last input was a 0" state (block C). Notice that the machine never returns to block A. This indicates that further simplification is possible. Before we proceed with any further simplification, however, we should discuss the "state transition table," which offers a formal method of simplification in place of the intuitive way followed previously.

11.5
STATE TRANSITION TABLE

The "state transition table" contains the same information as the state flow diagram but in a tabular form, which makes it easy to identify and eliminate redundancy. As illustrated in Fig. 11.3, the state transition table

STATE TRANSITION TABLE

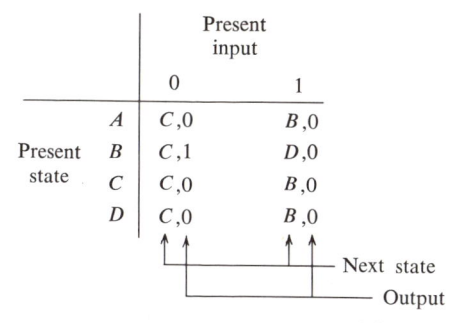

		Present input 0	1
	A	C,0	B,0
Present	B	C,1	D,0
state	C	C,0	B,0
	D	C,0	B,0

Fig. 11.3 Transition table.

is formed in a matrix configuration with "present input" values[7] and "present state" symbol designations as its coordinates. At the intersection of each "present input-state" pair we place the "next-state" designation and the "next-output" value, the way it is illustrated in Fig. 11.3. For example, we may observe that if the machine is at the state A and accepts an input 0, it will change to the state C and issue an output 0, as may be easily verified in the state flow diagram of Fig. 11.2. Also, if the machine is at the state A and accepts an input 1, it should move to the state B and issue an output 0.

By tracing the transitions marked in the flow diagram of Fig. 11.2, we construct the transition table shown in Fig. 11.3. By examining the various rows of the transition table, we may notice that the states A and C respond identically to either input, and therefore, are equivalent. This fact may be easily verified on the flow diagram as well as by simple reasoning: "Both states are in the CLEAR condition because the last input was a 0." Also notice that the states C and D are equivalent. This is so because any time the circuit has counted an *even* number of 1's, it is automatically CLEARED.

11.6
MINIMIZED STATE FLOW DIAGRAM AND TRANSITION TABLE

After the redundancies noted above are eliminated, there remain only two states, namely B and C. We shall call the state C, the RESET-state, because when in this state, the machine is CLEAR and ready to begin anew. We shall also call the state B, the SET-state, because the machine moves from it to the RESET-state and issues a 0 or a 1, depending on whether the input is a 0 or a 1, respectively. The minimized state transition table and flow diagram are shown, respectively, in Fig. 11.4 and Fig. 11.5.

---

[7] In a generalized sequential machine the input values would represent the set of all possible combinations of input values. In a one-input machine as this one, this simply corresponds to the values 0 and 1.

It should be noted that the facilities of both the state table and state diagram techniques must be used interchangeably and with an understanding of their advantages and limitations. State tables make it easier to identify equivalent states and eliminate redundancy, while a state diagram sometimes may show patterns that the state table will not.

	Input	
	0	1
SET – B	C,1	C,0
RESET – C	C,0	B,0

Fig. 11.4 Minimized transition table.

It may now be observed by comparing the diagrams of Fig. 11.1 and Fig. 11.5, that the initially defined states *A, C, D, E, G* and *I*, were found to be equivalent and were eventually all grouped into the single state "RESET." This state serves as

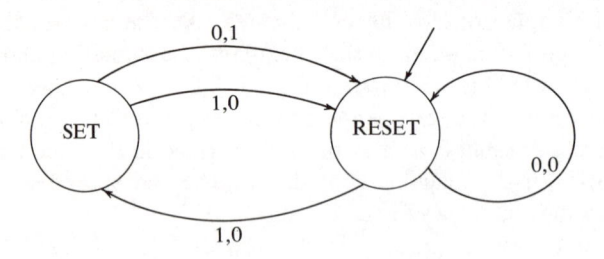

Fig. 11.5 Minimized state flow diagram.

the "starting" state, and it may be established at any time by supplying the machine with a 010 input sequence, as mentioned previously in Section 11.3, or even more simply by supplying the machine with a single 0. The states *B, F,* and *H* were also found to be equivalent and were substituted by the SET-state. The "starting" state is marked by a short arrow in the diagram of Fig. 11.5.

11.7
CIRCUIT IMPLEMENTATION

By inspection of the minimized flow diagram shown in Fig. 11.5, we observe that our machine possesses essentially only two stable states. Therefore, the machine's circuit implementation would require the memory capacity of a single binary storage element, that is, of a single flip-flop, which can "remember" either one of two stable conditions. It seems that the *R-S* flip-flop, discussed in Chapter 9, Section 9.7, may be well suited for this application.[8]

---

[8] Other types of flip-flops, as for example the T-flip-flop, may also be used. (See Problem 11.1.)

A sequential machine must perform in accordance to its transition table and flow diagram, like those shown in Figs. 11.4 and 11.5. It must provide a "next state" A_{n+1}, and an output Z_n, which are functions of the "present state" A_n and the "present input" X_n. The minimized transition table which describes the operation of the machine is presented again in Fig. 11.6, using the "set" and

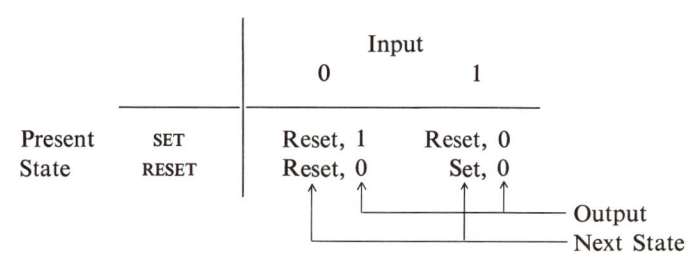

		Input	
		0	1
Present State	SET RESET	Reset, 1 Reset, 0	Reset, 0 Set, 0

Output
Next State

Fig. 11.6 Minimized Transition table using "set"-"reset" state designations.

"reset" designations for the two stable states. The table may be rearranged in the familiar truth table form by assigning: 0 = Reset, 1 = Set, as shown in Fig. 11.7.

The memory portion of the machine is driven to its "next state" through logical circuits which drive the memory input lines. For our machine we should need to design appropriate logical circuits in order to drive the S and R input lines. The design of such logical circuits is carried out using the familiar techniques of Karnaugh mapping, minimization and circuit implementation with any available complete set of elementary circuit blocks. (See Chapters 7 and 8.)

S-line drive. In order to design the necessary logical circuit to drive the S input line for the R-S flip-flop, we observe on the truth table of Fig. 11.7 that the S line must be pulsed if, and only if, $A_n = 0$ and $X_n = 1$.

A_n ↓	X_n ↓	A_{n+1} ↓	Z_n ↓
Present State	Input	Next State	Output
0	0	0	0
0	1	1	0
1	0	0	1
1	1	0	0

State Assignment:
0 = RESET
1 = SET

Fig. 11.7 Transition table in the form of a truth table.

296 SEQUENTIAL MACHINES

Therefore:

$$S = X_n \tilde{A}_n,$$

or simply

$$S = X\tilde{A}, \tag{11.1}$$

since in the operation of logical circuits time is not a factor. The logical function for the S-drive is best derived with the help of a Karnaugh map. In more complex designs this may be essential. The Karnaugh map for the Set-line drive is shown in Fig. 11.8(a), as derived by mapping the 1's of the "next state" column of the truth table (Fig. 11.7).

R-line drive. As we may observe from the truth table of Fig. 11.7, the R-line must be pulsed if, and only if, either $A_n = 1$, $X_n = 0$, or $A_n = 1$, $X_n = 1$. Notice that the case $A_n = 0$, $X_n = 0$, is a "don't care" condition, because the state of the flip-flop should remain the same, that is a 0, whether or not the R-line is pulsed. The Karnaugh map for the Reset-line drive is derived by mapping the 0's of the "next state" column of the truth table, as shown in Fig. 11.8(b). While mapping for the S or R drive we must keep an eye

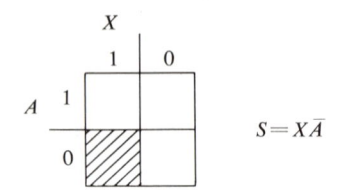

(a) Logical equation to SET the flip-flop. $S = X\bar{A}$

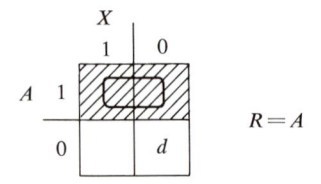

(b) Logical equation to RESET the flip-flop. $R = A$

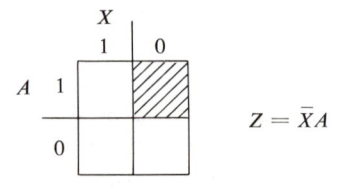

(c) Logical equation for output. $Z = \bar{X}A$

Fig. 11.8 Karnaugh mapping and logical equations.

for possible "don't care" conditions, like the (0,0) case above. The R-map in Fig. 11.8(b) suggests that

$$R = A. \tag{11.2}$$

Output. From the "output" column of the truth table (Fig. 11.7), we observe that the output Z is 1 if, and only if, $A_n = 1$, $X_n = 0$. The Karnaugh map for the output, as shown in Fig. 11.8(c), suggests that

$$Z = \bar{X}A = \overline{X + \bar{A}} \quad \text{(DEMORGAN THEOREM)} \tag{11.3}$$

Circuits. The machine, as suggested by Equations (11.1) to (11.3), is implemented in the network shown in Fig. 11.9(a) using only AND/NOT elementary logical blocks, and also in Fig. 11.9(b) using only AND/NOR elementary logical blocks. One may easily verify that either implementation performs the task, as prescribed by the specifications of the special-event machine in Section 11.2.

(a)

(b)

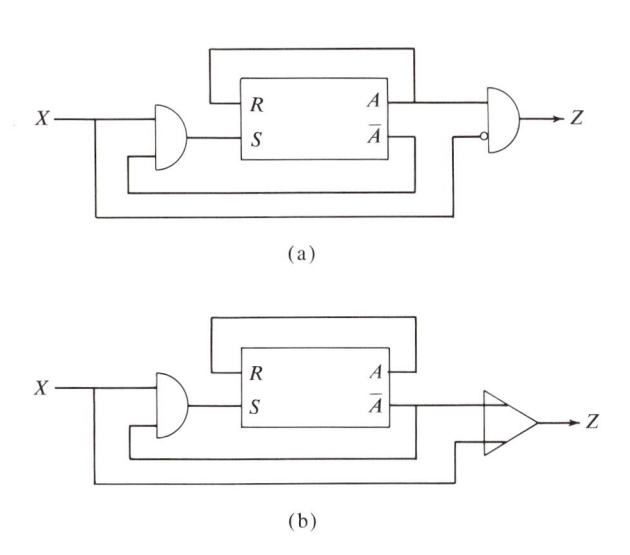

Fig. 11.9 A special event sequential machine; implementation using (a) AND/NOT elementary logical blocks and (b) AND/NOR elementary logical blocks.

Pulse and level signals. Before we proceed any further with the discussion of sequential machines, we must describe the nature of input and output signals with which we deal. Input and output signals in logical circuits are commonly coded in accordance with positive logic (Chapter 9) and therefore the occurrence of a positive pulse represents a 1 while a no-pulse represents a 0.

298 SEQUENTIAL MACHINES

Flip-flops may also be driven by pulses. The occurrence of an input on the S-line, for example ($S = 1$), is represented by the appearance of a pulse on that line. Thus there are three possible alternative driving conditions for the R-S flip-flop, namely: "pulse the SET-line," "pulse the RESET-line," "do not pulse either input line," since the simultaneous occurrence of a pulse in both input lines of the R-S flip-flop is not allowed.

In electronic flip-flops, such as the electronic R-S flip-flop, the state of the flip-flop is represented by voltage levels on the output lines. "High" voltage level on the A-line corresponds to the 1-state ($A = 1$), and "high" voltage level on the $\bar{A}$-line corresponds to the 0-state ($\bar{A} = 1$). The two outputs are the logical complements of one another. The level signals are often used to gate pulse signals through logical circuits. Such is the case, for example in Figs. 11.9(a) and 11.9(b), with the $\bar{A}$ logical level signal gating the input signal X through the input AND gate to the S-drive line.

On several occasions, however, level signals may have to be converted to pulse signals. Notice, for example, that in the sequential networks of Figs. 11.9(a) and 11.9(b), the A-output is shown "directly" connected to the R-input of the flip-flop. This connection would prevent the flip-flop from remaining in the $A = 1$ state for any length of time; a "high" level on the A-output would force the flip-flop automatically into the $\bar{A} = 1$ state. This situation may be remedied with the aid of synchronization clocking pulses, in a manner discussed previously in Chapter 9, Section 9.7, that is, by AND-ing the output voltage level with a clock pulse.

Actually the operation of the flip-flop is clocked through the input and output AND gates, as shown in Fig. 11.10. The clocking connections are identical to those shown in Fig. 9.15(f). Notice that when $A = 1$, the flip-flop maintains its state until a clocking pulse causes the R-line to be activated and the flip-flop driven into the $\bar{A} = 1$ state.

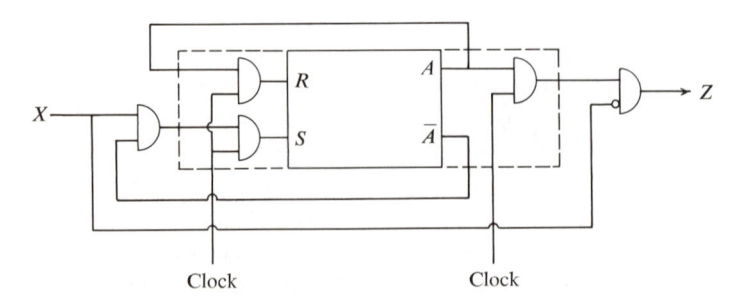

Fig. 11.10 A special event sequential machine; implementation showing clocking arrangements. [Compare clocking connections with those shown in Fig. 9.15 (f).]

The above designed special-event machine is a simple sequential machine with a very short memory. It possesses only two recognizable, distinguishable states. When the machine is in the "RESET" state, it cannot tell if it got there by a last input of zero or by a last input of an even number of ones. When in the "SET" state, the machine cannot tell which input sequence has caused it to be in that state; it can only tell that the last input was a sequence of an odd number of 1's.

11.8 GENERALIZATIONS

A fair number of important concepts and techniques were brought forth with the above example of designing a simple sequential machine. The need for "memory" in sequential machines, and the concept of "state" were first demonstrated. The important techniques of state flow diagrams, transition tables and their minimization by means of eliminating redundant states, were illustrated. The essential techniques for designing the logical (combinational) portion of the sequential machine were also briefly described. The design involves the derivation of the input logical equations for the memory driving lines (often referred to as "secondary excitations") and of the output logical equations.

The synthesis method for sequential machines may be stated in three general steps.
1. Construct a state flow diagram and a transition table to describe the machine requirements.
2. Use techniques of modifying and simplifying the flow diagram and transition table by checking and eliminating redundancies of equivalent states.
3. Transform the minimized transition table into maps, which are then manipulated and read in the familiar ways of the Karnaugh maps (Chapters 7 and 8) to provide us with logical expressions for the memory inputs ("secondary excitations") and for the outputs.

Step 2 involves the "merging" of equivalent states into superstates, and employs the use of "merger" and "secondary state" diagrams not defined here. Step 3 should lead to logical expressions for the memory inputs and for the outputs. Y-maps, like the Set- and Reset-maps in the previous design example, are essential in this step.

The general organization of a sequential machine is shown in Fig. 11.11. Information is supplied to the machine through the input lines X. Information concerning the states of the machine is supplied to the Logic from the Memory by the state lines Y. During operation the logic portion of the machine accepts the input signals X plus information from the memory section about the "present state" (Y lines), to produce the outputs Z and set up the "next state" of the memory section (and thus of the machine) through signals on the y-lines.

Notice the striking resemblance of Fig. 11.11 to the general organization of an information-processing machine (Chapter 1, Fig. 1.4). All general information-processing machines with memory capabilities are sequential machines and operate in the fashion described above and illustrated in the general organization scheme shown in Fig. 11.11. They operate by receiving information from the

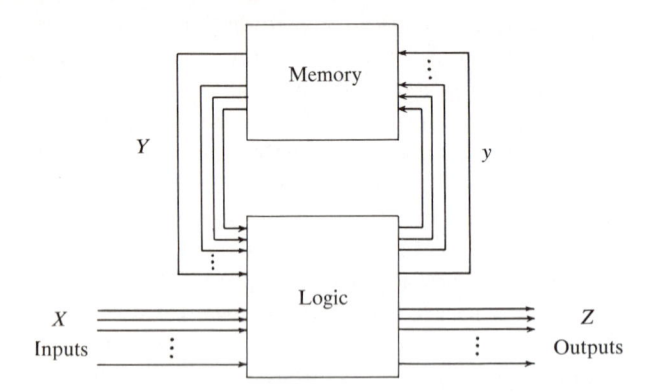

Fig. 11.11 General organization of a sequential machine.

"environment,"[9] setting up their memory states, and then by undergoing cycles of processing information about "present states" and "inputs" to produce information outputs to the "environment" and set up the "next states." The logic section shown in Fig. 11.11 corresponds to both the "Control" and "Arithmetic" units of the general information-processing machine (Chapter 1, Fig. 1.4).

A great variety of modern-life machines, from the elevator and the coffee-vending machine to the most sophisticated and complex digital information-processing machines available commercially belong to the general class of "finite state" sequential machines. They all possess a finite memory capability, and thus, at any given time, they may be in any one of a finite number of non-redundant, *distinguishable*, internal states. Such machines perform in discrete procedural steps from one state to another, all states belonging to a well-defined finite set of internal states. Finite State Machines will be defined more completely and discussed further in a rather general and abstract manner in the following section.

---

[9] The "environment" may be the user of the machine or the physical environment of it.

In contrast, machines with *infinite*, that is, unlimited, storage capacity can perform information-processing tasks beyond the capability of finite state machines. A special class of machines possessing an external memory of infinite storage capacity, better known as "Turing Machines," is of particular interest if the question of "computability" is posed. The question of computability refers to the information-processing limitations of finite state machines and to the definition of problems that cannot be solved by machines. Turing machines will be discussed further in Section 11.10 and the question of computability in Section 11.11.

All logical and sequential circuits which we have studied in this textbook are *deterministic*; their functional relations are nonprobabilistic, in the sense that the same "cause" always produces the same "response," which thus is exactly *predictable*. Relations among states, inputs and outputs were expressed by algebraic, graphical, or tabular means. Nonprobabilistic Boolean algebraic equations, truth and transition tables, graphs and maps of various kinds, were used to express the anambiguous and completely predictable ways in which the present states and outputs were related to the inputs and to the previous states of these circuits.

There exist a class of sequential machines whose inputs are subject to probabilistic constraints, and so are their outputs. The relations between inputs, states, and outputs under these circumstances are *nondeterministic*. Changes of state and output values are characterized by "transition probabilities" instead of complete determinancy as was the case with the machines and circuits with which we have dealt. It is not our intention to deal with nondeterministic sequential machines in this book.

11.9
FINITE STATE MACHINES

In this section we provide a brief characterization of sequential, deterministic machines, which possess a finite number of possible distinguishable states. After having examined the essential features of these machines by studying a special-event small machine, we now keep our presentation general and abstract so as not to inadvertently tie it to the properties of any particular example.

At any given time,[10] the machine may be found in any one of a finite number of possible internal "configurations" or states (see Section 11.3). We may represent the set of the available distinguishable[11] states by $S = \{S_1, S_2, \ldots S_q\}$.

---

[10] Time is always measured in discrete intervals.
[11] Nonredundant, nonequivalent.

302 SEQUENTIAL MACHINES

The machine also has a finite number of input and output lines, each of which can take only one of two available logical values (0 or 1) at any given time. Thus, the combination of values which appears at the input or the output at any given time is one of a finite set of input or output combinations, represented, respectively, by $X = \{X_1, X_2, \ldots X_n\}$ for the inputs and $Z = \{Z_1, Z_2, \ldots Z_m\}$ for the outputs.

A finite state machine is completely defined by the above three sets S, X, and Z, plus two functional relationships f and g defined as follows:

f is the "next state" function, such that

$$\text{next state } S = f(S, X), \tag{11.4}$$

g is the "output" function, such that

$$\text{output } Z = g(S, X). \tag{11.5}$$

The f and g relations, which describe the operation of the machine, are often stated tabularly or graphically. The "next states" f and the "output" values g are listed or mapped for all possible combinations of states and input values.

For reasons of clarity in representing the general properties of finite state machines, we use an arbitrary example of a finite state sequential machine defined tabularly in Fig. 11.12. The tables in Fig. 11.12 should contain all essential information about the operation of the machine, although not necessarily in the most concise form.

A state *transition table* (Section 11.5) may be derived as shown in Fig. 11.13. Information about input X and state S and about the resulting state and output is transferred from the tables of Fig. 11.12. The "starting" state is placed on the top of the "present state" column. Since the behavior of the machine depends

$S = \{A, B, C\}$ Set of available distinguishable states
$X = \{0, 1\}$ Set of input combinations (single 0 or 1 in this case)
$Z = \{0, 1\}$ Set of output combinations (single 0 or 1 in this case)

Input X	Present state S	Next state f	Output g
0	A	B	0
0	B	C	1
0	C	A	1
1	A	C	1
1	B	B	0
1	C	A	1

Fig. 11.12 Description of a one input-one output finite state binary machine.

```
                          X  ←──── Input
                        0     1
Starting state ──→  A  B,0  C,1
                    B  C,1  B,0
                    C  A,1  A,1
Present state ──────↑
                              └──── Next state
                              └──── Output
```

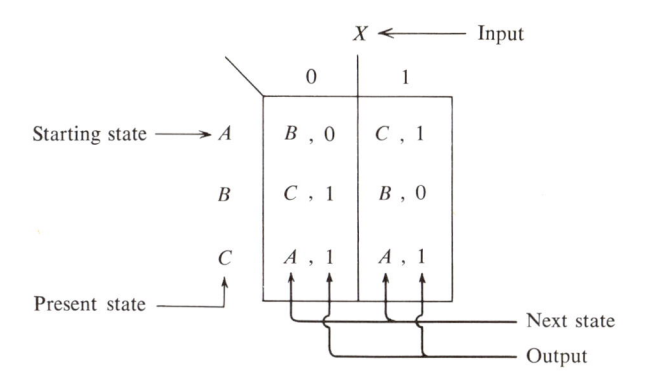

Fig. 11.13 Transition table.

in general on the starting state, that is, the state of the machine just prior to the application of any inputs, we agree to always specify a starting state when we specify a machine and place it at the top of the transition table.

A *state flow diagram* (Section 11.4) may be drawn next from the information supplied by the transition table. The state diagram is shown in Fig. 11.14. Notice that the starting state is marked with a short unlabeled arrow. The state diagram facilitates the observation of patterns of transitions between states.

After the transition table and the flow diagram have been formed, the designer of the machine must search for redundant equivalent states and must eliminate them. The minimized table and diagram are then tested for consistency.

The machine is supplied with an arbitrary sequence of inputs and its behavior is calculated using the minimized transition table and state diagram. The machine's behavior must conform with the specifications of the problem. An example of a test sequence using the transition table and diagrams in Figs. 11.13 and 11.14, is shown in Fig. 11.15.

The machine is implemented from the minimized state table using techniques described previously in this chapter.

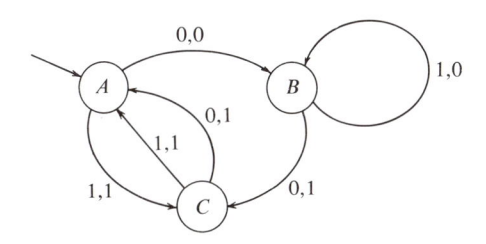

Fig. 11.14. State flow diagram.

Input sequence → | 0 | 1 | 0 | 1 | 1 | 1 | 0 | 0 | 1 | 0 | 1 | 1 |
State sequence → A B B C A C A B C A B B B
Output sequence → | 0 | 0 | 1 | 1 | 1 | 1 | 0 | 1 | 1 | 0 | 0 | 0 |

Fig. 11.15 Test sequence.

11.10
TURING MACHINES

Since the pioneering work of A. M. Turing[12] in the early 1930's, many investigators such as Shannon, Moore, and Davis, have studied a class of mathematical machines, called the Turing machines, with the intended purpose of answering the question as to what can and what cannot be computed. For that purpose, Turing devised a slow and simple imaginary computer that he proved to be theoretically capable of performing all the operations of any computer. Any problem for which we can set up a "cookbook way of solution," that is, an algorithm, can be solved by a Turing machine and vice versa. Although these machines have been largely defined in an abstract mathematical manner, a few attempts have been made to physically realize a few special examples.

Turing devised his simple computer to emulate a human calculator, and he used it to demonstrate the close kinship of computer theory and mathematical logic. Take, for example, the way that a man carries out the solution of a computational mathematical problem. He may use pencil and paper on which he writes and erases numbers and other symbols. While operating, he makes reference to mathematical facts and tables which he has in store either in his mind or in reference books. He also follows a set of instructions about the nature and sequence of his computations.

The Turing machine is a finite-state machine with an external infinite memory that operates in an analogous fashion. As illustrated in Fig. 11.16, the machine has a marking and erasing device, a scanning device, a logical control unit, and it is associated with an infinitely long tape which serves as the memory unit. The tape is divided lengthwise in square cells and can be moved in either direction along its length. It is because of this infinite memory that the Turing machine is an idealized one.

---

[12] Turing, A. M., "On computable numbers, with an application to the Entscheidungsproblem," *Proc. London Math. Soc.*, Ser. 2–42, 230–265 (1936).

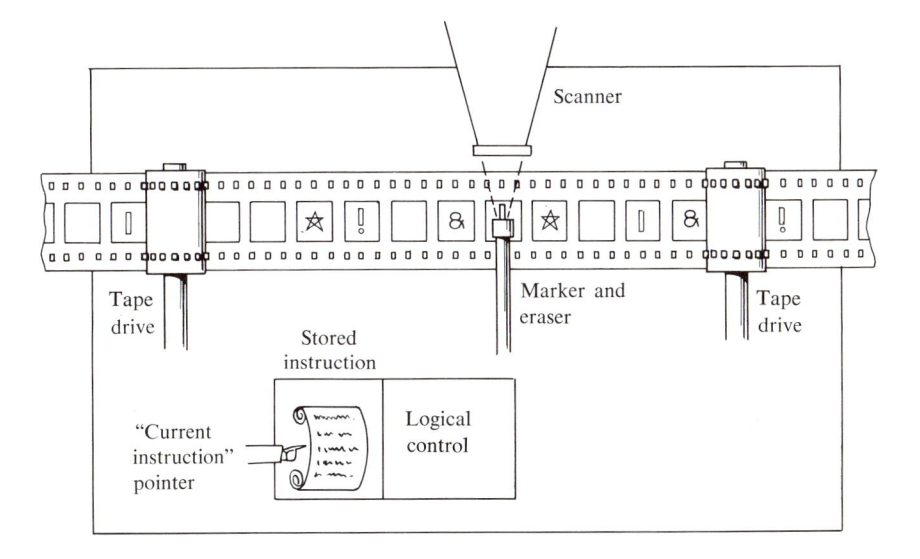

Fig. 11.16 Turing machine.

The machine operates in "cycles." During each cycle it scans one square on the tape and takes a specific action in the following manner: The scanner relays the contents[13] of the square to the control unit (that is, to the finite-state portion of the machine). The control unit then consults the list of stored instructions by reading the instruction pointed by the "current instruction" pointer. Depending on the contents of the scanned square the instruction specifies one of four actions to be taken:

1. print a symbol on the square under scan, erasing a previous symbol if necessary,
2. move the tape one square to the right (or more),
3. move the tape one square to the left (or more),
4. halt.

Each action is followed by a specified change in instruction configurations, that is, by a relocation of the "current instruction" pointer to a new instruction. A Turing machine endowed with adequate instructions and an infinitely long tape is capable of solving any specified mathematical problem for which an algorithm has been set up.

---

[13] The machine uses a finite set of symbols, that is, a finite alphabet of symbols.

11.11
UNSOLVABILITY AND LIMITATIONS OF COMPUTING MACHINERY

It can be said that no conceivable machine is truly "general purpose" in the sense that all theoretically solvable information-processing problems can be solved by that machine. In any conceivable practical computing machine the number of available memory cells is *finite* and the "machine words" are of finite length, so that only finite amounts of information may be stored by the machine at any given time. Also it is a fact of life that there exists an upper ceiling on information-processing rates imposed by the inherent "uncertainty" which is present in all observations and by excessive power demands at high processing rates.[14] Therefore some information-processing tasks which may require the facilities of extremely large memories and very high processing rates become uncomputable.

Several "games" of strategy, like checkers, chess, bridge, and the like, impose information-processing tasks which require memory capacity which is beyond any practical reality. For example, the decision-tree (Chapter 5) for chess requires 10^{120} decisions, a number which is of the order of the number of all the molecules in the universe! In cases as these, a rigorous solution (or "sure" victory) is unthinkable even theoretically. A different approach, known as a "heuristic" approach, is followed with such problems. In that, each step leading to possible solution (or victory) is decided on the basis of the "probability of success," the probability derived from previous "experience" and from calculations on limited data.

The apparent limitations of practical finite state machines are derived from their *finite* memory capacity and the limited length of "machine words" (Chapter 4). It can be proven, for example, that no finite-state machine can be designed which can multiply any two given numbers arbitrarily large. In fact, depending on the size of the numbers involved, it may become necessary to carry arbitrarily large amounts of information during the formation and processing of the "accumulated partial sums."

Multiplication is one of a number of limitations of finite-state machines, all due to finite memory capacity. However, what happens if the restriction of finite memory is lifted? Are there any problems which cannot be solved by *any* computing machine, even one with unlimited memory? Several variations of equivalent Turing machines have been devised and used by various researchers attempting to study the ultimate limitations of practical finite-state computing

---

[14] Ligomenides, P. A., "Wave-mechanical uncertainty and speed limitations," *IEEE-Spectrum*, vol. 4, no. 2, pp. 65–68, February 1967.
———"Wave-mechanical limitations on information retrieval rate," *Proceedings I.R.E.E. Australia*, vol. 29, no. 3, pp. 65–79, March 1968.

machines with finite or infinite memory capacity and to gain further insight into possible modes of operation.

Finite state machines with finite or external infinite memory all require an algorithm, that is, a set of mechanical rules, or an "effective procedure," that is, a method of solution where each step may lead to another, which will instruct the machine during the solution of a problem. However, just as there are certain games for which there is no optimum strategy which will guarantee victory, so there are certain classes of problems for which there is no algorithm. Such problems are also unsolvable on computing machines. In addition to problems for which no algorithm exists, there are also many other, less obvious, cases of problems uncomputable by machines. (See *Computability and Solvability*, by M. Davis, in Bibliography.)

BIBLIOGRAPHY

Arbib, M. A., *Brains, Machines and Mathematics*. New York: McGraw-Hill, Inc., 1964.

Davis, M., *Computability and Solvability*. New York: McGraw-Hill, Inc., 1958 (bibliography included).

Gillespie, R. D. and Aufenkamp, D. D., "On the Analysis of Sequential Machines." *IRE Trans*. EC-7, vol. 2, June, 1958, pp. 119–122.

Maley, G. A. and Earle, J., *The Logic Design of Transistor Digital Computers*. Englewood Cliffs, N.J.: Prentice-Hall, Inc., 1963.

Marcus, M. P., *Switching Circuits for Beginners*. Englewood Cliffs, N.J.: Prentice-Hall, Inc., 1962.

McCluskey, E. J., *Introduction to the Theory of Switching Circuits*. New York: McGraw-Hill, Inc., 1965.

Shannon, C. E. and McCarthy, J., *Eds. Automata Studies*. Princeton, N.J.: Princeton Univ. Press, 1956.

Wang, Hao, *A Survey of Mathematical Logic*. Peking: Science Press, 1963.

Problems

11.1 Starting with the minimized state flow diagram shown in Fig. 11.5, implement the sequential machine represented by this diagram using
(a) a T flip-flop,
(b) a J-K flip-flop.

11.2 Design a sequential circuit to perform as a J-K flip-flop using a R-S flip-flop as the memory element.

11.3 Design a single input-single output sequential circuit whose output is "ON" if three *consecutive* inputs are "ON."

Draw the necessary state flow diagrams and transition tables and justify any minimization performed.

11.4 The sequential black-box shown in Fig. 11.4.1 receives at its input a series of three-bit characters, and it is required to detect those characters with exactly two ones. Design the sequential circuit which will detect such characters and produce an output of one when this occurs; otherwise it produces an output of zero. Provide the transition tables, the minimized state flow diagram, and the block diagram of the circuit, using (a) AND/OR/NOT logic and type T flip-flops, and (b) NOR logic and type R-S flip-flops.

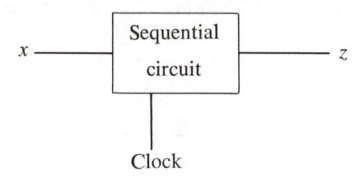

Fig. 11.4.1

11.5 In the two-input sequential circuit shown in Fig. 11.5.1 only one of the inputs changes at a time. Its output is one only when x_2 becomes one and x_1 is already one; otherwise its output is zero. Design the circuit and show the transition tables and the minimized state flow diagram.

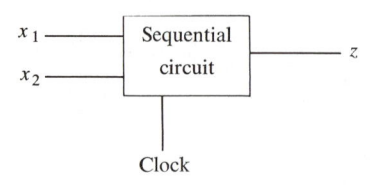

Fig. 11.5.1

11.6 Analyze the single-input sequential circuit shown in Fig. 11.6.1, assuming that X is a pulse input.

(a) Assign symbols to the sixteen possible states, such as s_1, s_2, s_3, ... , s_{16}, and derive the transition table. In it, for every combination of "present states" $A_n B_n C_n D_n$ derive the combination of "next states" $A_{n+1} B_{n+1} C_{n+1} D_{n+1}$ and the output Z_n, corresponding to $X = 0$ and $X = 1$ in each case.

(b) List the sequence of states and outputs which will occur if a series of sixteen ones (pulses) is applied to the input X. The initial state is (0001).

(c) Repeat part (b), if the initial state is (0000).

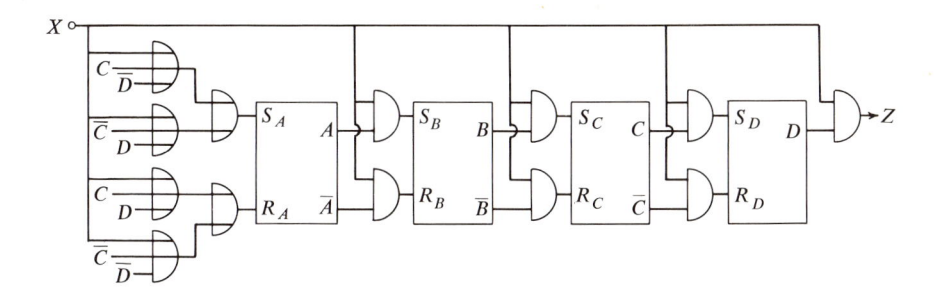

Fig. 11.6.1

12

The Design of a Small Computer

The first eleven chapters of this book have treated various aspects of organization, programming, coding, and the logic design of combinational and sequential circuits of general purpose information-processing machines. In this chapter we apply the methods of logic design to the design of a small computer. Our prime consideration will be the articulation of the steps which lead to the engineering design of a small machine that can add and subtract two signed binary numbers.

It is intended that this design problem should serve to the student as a point of departure for further study on the methodology of designing information-processing machines. The design steps are stated distinctly but concisely so that the student may, on his own time, elaborate on the details.

12.1
STATEMENT OF THE DESIGN PROBLEM

Consider a computing machine, shown schematically in Fig. 12.1, which operates serially and synchronously. Two N-bit binary numbers X and Y are to be read into the machine in serial fashion, bit by bit and synchronously, least significant bit first and sign bit last. Negative numbers are converted and presented to the computer in "two's

STATEMENT OF THE DESIGN PROBLEM

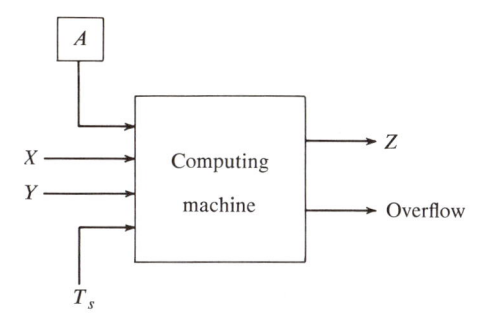

Fig. 12.1 Schematic of a small computer.

complement" form (Chapter 10, Section 10.2). The two signed numbers X and Y are to be added or subtracted by the machine.

The flip-flop A indicates whether *addition* or *subtraction* is to take place. For $A = 1$ the output is to be $X + Y$, while for $A = 0$ the output is to be $X - Y$. A timing signal T_s is used to mark the end of the numbers being computed; when $T_s = 1$, the sign-bits of X and Y are being read-in, as illustrated in the timing chart of Fig. 12.2.[1] The clocking of the whole operation is controlled by a central "timer" unit not shown in Fig. 12.1. The timing pulse T_s is derived from the central timer (clock) by frequency division.

We are asked to design and draw the sequential circuit for the small computer of Fig. 12.1, which will provide the sum $X + Y$ or difference $X - Y$ in serial fashion, that is, least significant bit first and sign bit last. We should also make provision for an "overflow" indication. We may assume that X, Y, and Z all have the same number of bits.

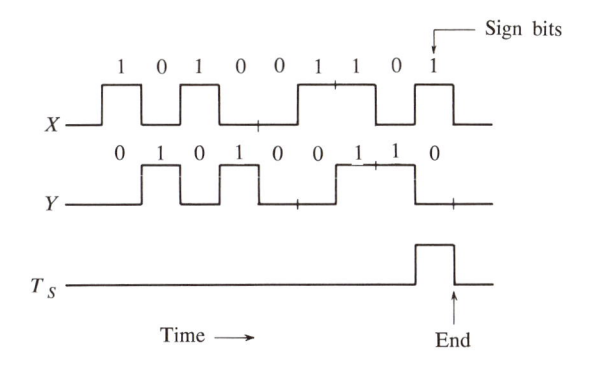

Fig. 12.2 Timing chart.

---

[1] "Positive logic" or "positive true" may be assumed, whereby a positive pulse signifies a 1, and no-pulse signifies a 0.

312 THE DESIGN OF A SMALL COMPUTER

12.2
TRUTH TABLES

As a first step in the design of the small computer, we should draw the truth tables. In accordance with the specifications given we should draw truth tables for the following cases: (a) the addition of X and Y (when $A = 1$, $T_s = 0$); (b) the subtraction of X and Y (when $A = 0$, $T_s = 0$); and (c) the processing of the signs for addition and subtraction (when $T_s = 1$ and A is either 1 or 0). Keep in mind that all negative numbers are presented to the computer in their two's complement form.

(a) The conditions for binary addition, discussed previously in Chapter 10, Section 10.4a, Equations (10.1) and (10.2), are such that the sum-output Z_i is one only if an odd number of ones are present among the input bits X_i, Y_i, and C_i (carry). Also, the next carry C_{i+1} is one only if two-or-more, that is, the majority of the input bits X_i, Y_i, and C_i are one. These conditions are tabulated in the truth tables of Fig. 12.3(a).

(b) The conditions for binary subtraction were also discussed in Chapter 10, Section 10.4b. It was stated then that the operation is completed if we take the ones complement of the subtractor (Y), we *add* it to the subtrahend (X), and then we perform the end-around-carry operation. However, one may verify easily that the above procedure is equivalent to taking instead the *two's complement* of the subtractor (Y) and simply add it to the subtrahend (X). Then by simply dropping the highest order bit, as needed, we obtain the difference $X - Y$ without the need for the end-around-carry operation.

The conditions for the binary subtraction using the two's complement for the subtrahend, are shown in the truth tables of Fig. 12.3(b). Notice that the difference output Z_i is identical to the sum output Z_i of Fig. 12.3(a). This is so because by having the two's complement of the subtrahend at the input, the computer performs addition instead of subtraction to produce the difference $Z = X - Y$.

(c) The conditions for the processing of the signs are shown in Figs. 12.3(c) and 12.3(d), for addition and subtraction, respectively. A 1 bit at the sign position of the number designates a positive number, while a 0 bit designates a negative number.

12.3
MAPS AND BOOLEAN ALGEBRAIC EXPRESSIONS

The next step consists of obtaining the combinational switching functions by mapping the truth tables. The methodology of mapping and minimization of the Boolean expressions was discussed in some detail in Chapter 7. Here we have five binary input variables, namely A_i, X_i, Y_i, C_i,

MAPS AND BOOLEAN ALGEBRAIC EXPRESSIONS 313

Digit processing $T_s = 0$

Equivalent decimal number	A_i	X_i	Y_i	C_i	Z_i	C_{i+1}
8	1	0	0	0	0	0
10	1	0	1	0	1	0
12	1	1	0	0	1	0
14	1	1	1	0	0	1
9	1	0	0	1	1	0
11	1	0	1	1	0	1
13	1	1	0	1	0	1
15	1	1	1	1	1	1

Addition $A = 1$

(a)

Equivalent decimal number	A_i	X_i	Y_i	C_i	Z_i	C_{i+1}
0	0	0	0	0	0	0
2	0	0	1	0	1	1
4	0	1	0	0	1	0
6	0	1	1	0	0	0
1	0	0	0	1	1	1
3	0	0	1	1	0	1
5	0	1	0	1	0	0
7	0	1	1	1	1	1

Subtraction $A = 0$

(b)

Sign processing $T_s = 1$

Sign designation: $(X_i, Y_i, Z_i) = 1$: plus
$(X_i, Y_i, Z_i) = 0$: minus

Equivalent decimal number	A_i	X_s	Y_s	C_i	Z_i	C_{i+1}
8	1	0	0	0	0	1
10	1	0	1	0	0	0
12	1	1	0	0	0	0
14	1	1	1	0	1	0
9	1	0	0	1	0	0
11	1	0	1	1	1	0
13	1	1	0	1	1	0
15	1	1	1	1	1	1

Addition $A = 1$

↑ Addition
↑ Sign of X & Y
↑ Carry from last operation
↑ Sign of the sum
↑ Next carry

Explains the "sign processing"

(c)

Equivalent decimal number	A_i	X_s	Y_s	C_i	Z_i	C_{i+1}
0	0	0	0	0	1	0
2	0	0	1	0	0	0
4	0	1	0	0	1	1
6	0	1	1	0	1	0
1	0	0	0	1	0	0
3	0	0	1	1	0	1
5	0	1	0	1	1	0
7	0	1	1	1	0	0

Subtraction $A = 0$

↑ Subtraction
↑ Sign of X & Y
↑ Carry from last operation
↑ Sign of the difference
↑ Next carry

(d)

Fig. 12.3 Truth (Transition) tables.

314 THE DESIGN OF A SMALL COMPUTER

and T_s. We also have two binary output variables Z_i and C_{i+1}. We shall map each output variable using two 4-variable submaps by setting $T_s = 0$ and $T_s = 1$, respectively. (Chapter 7, Section 7.9).

To facilitate the mapping operation, each binary input combination $A_i X_i Y_i C_i$ of the truth tables is designated by its equivalent decimal number, as shown in the left-end columns of the truth tables (Fig. 12.3). The decimal numbers are

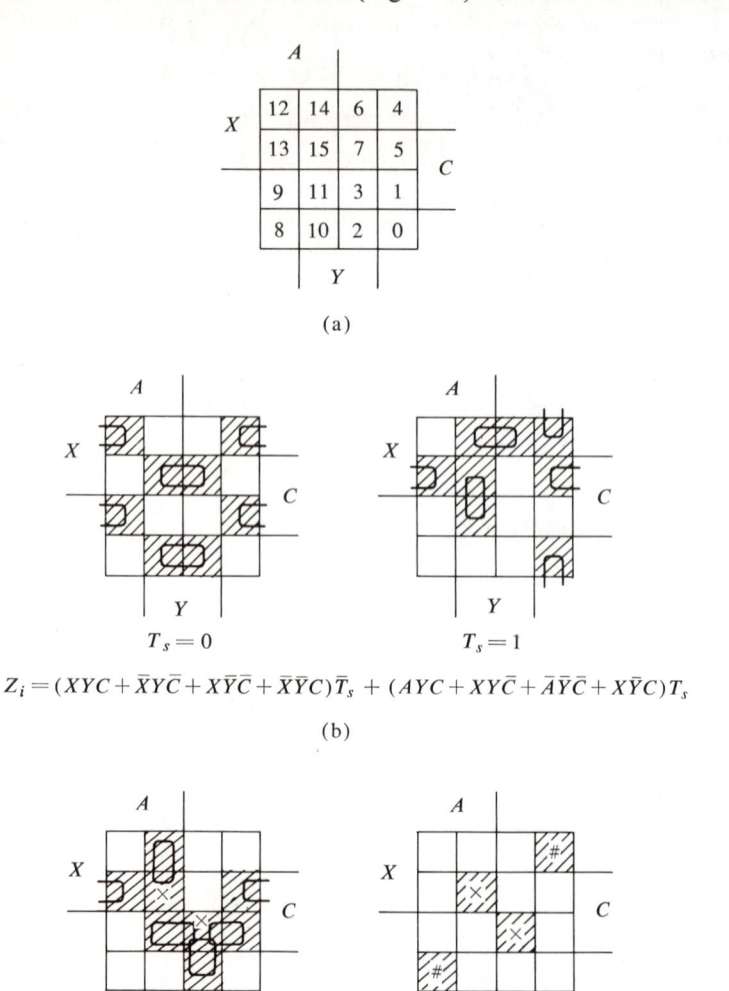

(a)

$$Z_i = (XYC + \bar{X}Y\bar{C} + X\bar{Y}\bar{C} + \bar{X}\bar{Y}C)\bar{T}_s + (AYC + XY\bar{C} + \bar{A}\bar{Y}\bar{C} + X\bar{Y}C)T_s$$

(b)

$$C_{i+1} = (AXY + X\bar{Y}C + \bar{X}YC + \bar{A}\bar{X}Y + \bar{A}\bar{X}C)\bar{T}_s + AXYC + \bar{A}\bar{X}YC + (\bar{A}X\bar{Y}\bar{C} + A\bar{X}\bar{Y}\bar{C})T_s$$

(c)

Fig. 12.4 Mapping and demapping of output fuctions.

used to mark the corresponding cells on a pilot map as shown in Fig. 12.4(a). Using, then, the marked pilot map one can locate immediately the appropriate cells during mapping operations.

Mapping and demapping of the output variables Z_i and C_{i+1}, in accordance to the truth tables of Fig. 12.3, is shown in Figs. 12.4(b) (for Z_i) and 12.4(c) (for C_{i+1}). The minimized expressions which resulted from demapping are

$$Z_i = (XYC + \bar{X}Y\bar{C} + XY\bar{C} + \bar{X}\bar{Y}C)\bar{T}_s \\ + (AYC + XY\bar{C} + \bar{A}\bar{Y}\bar{C} + X\bar{Y}C)T_s, \qquad (12.1)$$

$$C_{i+1} = (AXY + X\bar{Y}C + \bar{X}YC + \bar{A}\bar{X}Y + \bar{A}\bar{X}C)\bar{T}_s \\ + AXYC + \bar{A}\bar{X}YC + (\bar{A}X\bar{Y}\bar{C} + A\bar{X}\bar{Y}\bar{C})T_s. \qquad (12.2)$$

In order to perform the carry operation we may produce C_{i+1} by designing a combinational circuit, which will implement Equation (12.2), and using a delay element in conjunction with its output (Chapter 10). An alternative, and maybe more economical way, is to perform the carry operation by using an *R-S* flip-flop. In this case we should not need to implement the logical circuit of Equation (12.2). Instead we design the logical circuits for driving the input SET and RESET lines of the flip-flop, as described in the next section.

12.4
SET-RESET EQUATIONS FOR THE "CARRY FLIP-FLOP"

We shall use a *R-S* flip-flop to perform the carry operation. The input variables of the *C*-flip-flop[2] are C_{SET} for the SET drive line and C_{RESET} for the RESET drive line. The state of the *C*-flip-flop is determined by the value of its output, where $C = 1$ corresponds to the "SET" state, and $C = 0$ corresponds to the "RESET" state.

The value of the input variable C_i corresponds to the "present state" of the flip-flop, while the value of the output variable C_{i+1} corresponds to the "next state" of the flip-flop. Therefore the C_{i+1} columns of the truth tables of Fig. 12.3 provide the "next state" of the *C*-flip-flop as a function of the inputs X_i, Y_i, and of the "present state" C_i. Therefore the functions for the SET-line drive variable, C_{SET}, and for the RESET-line drive variable, C_{RESET}, are mapped by mapping the ones and the zeros of the C_{i+1} columns correspondingly. One should observe the "don't care" conditions, that is, the transitions during which there is no change of state ($C_i = C_{i+1}$).

Figure 12.5 shows the mapping and demapping of the input drive variables

---

[2] Meaning "Carry" flip-flop. (Not to be confused with the CFF, that is the complementary flip-flop, Chapter 9, Section 9.10.).

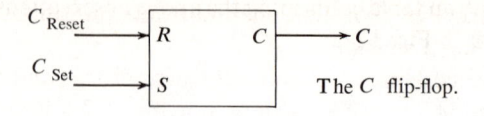

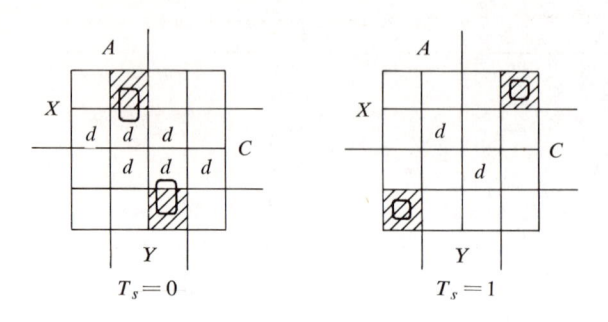

316   THE DESIGN OF A SMALL COMPUTER

[Diagram: C flip-flop with inputs $C_{\text{Reset}} \to R$ and $C_{\text{Set}} \to S$, output C. The C flip-flop.]

[Karnaugh maps for $T_s = 0$ and $T_s = 1$]

$$C_{\text{Set}} = AXY\bar{T}_s + \bar{A}\bar{X}Y\bar{T}_s + A\bar{X}\bar{Y}CT_s + \bar{A}X\bar{Y}CT_s$$

[Karnaugh maps for $T_s = 0$ and $T_s = 1$]

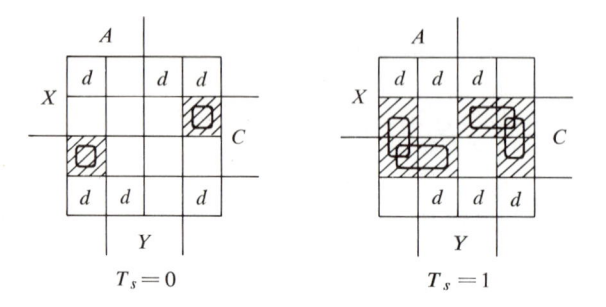

$$C_{\text{Reset}} = A\bar{X}\bar{Y}C\bar{T}_s + \bar{A}X\bar{Y}C\bar{T}_s + A\bar{Y}CT_s + A\bar{X}CT_s + \bar{A}\bar{Y}CT_s + \bar{A}XCT_s$$

Fig. 12.5   SET-RESET equations for C flip-flop.

C_{SET} and C_{RESET}. The functions, which must be implemented by logical circuits, are:

$$C_{\text{SET}} = AXY\bar{T}_s + \bar{A}\bar{X}Y\bar{T}_s + A\bar{X}\bar{Y}CT_s + \bar{A}X\bar{Y}CT_s, \qquad (12.3)$$

$$C_{\text{RESET}} = A\bar{X}\bar{Y}C\bar{T}_s + \bar{A}X\bar{Y}C\bar{T}_s + A\bar{Y}CT_s + A\bar{X}CT_s + \bar{A}\bar{Y}CT_s + \bar{A}XCT_s. \qquad (12.4)$$

$C_{\text{SET}} = 1$ implies that a carry $C = 1$ is to be generated for the next operation. Similarly, $C_{\text{RESET}} = 1$ implies that a carry $C = 0$ is to be generated for the next operation. If $C_{\text{SET}} = 0$ and $C_{\text{RESET}} = 0$, the state of the flip-flop remains unchanged, and the carry of the next operation is the same as the carry of the last operation. The condition $C_{SET} = 1$ and $C_{RESET} = 1$ is not allowed (that is, is undefined). The transition table confirming the above observations for the SET-RESET C-flip-flop is shown in Fig. 12.6.

Present state C_i	SET input C_{SET}	RESET input C_{RESET}	Next state C_{i+1}	
0	0	0	0	
0	0	1	0	don't care
0	1	0	1	
0	1	1	Not allowed (undefined)	
1	0	0	1	
1	0	1	0	
1	1	0	1	don't care
1	1	1	Not allowed (undefined)	

Fig. 12.6 Transition table—C-flip-flop

12.5 BLOCK DIAGRAM

In accordance to the Z-Equation (12.1) we can easily form a minimal combinational circuit, which will provide the sum output $(Z = X + Y)$ or the difference output $(Z = X - Y)$, depending on whether A is 1 or 0. The techniques of implementing the circuit configuration using AND/OR/NOT or NOR or NAND gates, have been discussed in detail in Chapter 8. Because the circuit implementation of Equation (12.1) is straightforward and the particular configuration chosen may be dictated by several practical considerations of economy, speed, and so forth, we will not show here any specific circuit implementation. Also, in order to illustrate the essential connections in the block diagram as clearly as possible, we will represent the logical equations by logical blocks. Thus, the Z-logic is performed by the logical block shown in the block diagram of Fig. 12.7.

Two more logical circuits are required to produce the drive SET and RESET pulses for the C-flip-flop. These circuits are designated by the C_{SET}-logic and C_{RESET}-logic logical blocks in the diagram of Fig. 12.7.

All three logical blocks have five inputs (A, X, Y, T_s, and C_i) and one output (Z, C_{RESET} and C_{SET}, respectively). The connection of the logical blocks and of the R-S carry-flip-flop are shown in Fig. 12.7.

To complete the discussion on the block diagram, a word should be said about the *overflow* condition. If a carry is to be generated during the processing of the sign-bits, that is, when $T_s = 1$, this carry will cause "overflow." This is so because

318 THE DESIGN OF A SMALL COMPUTER

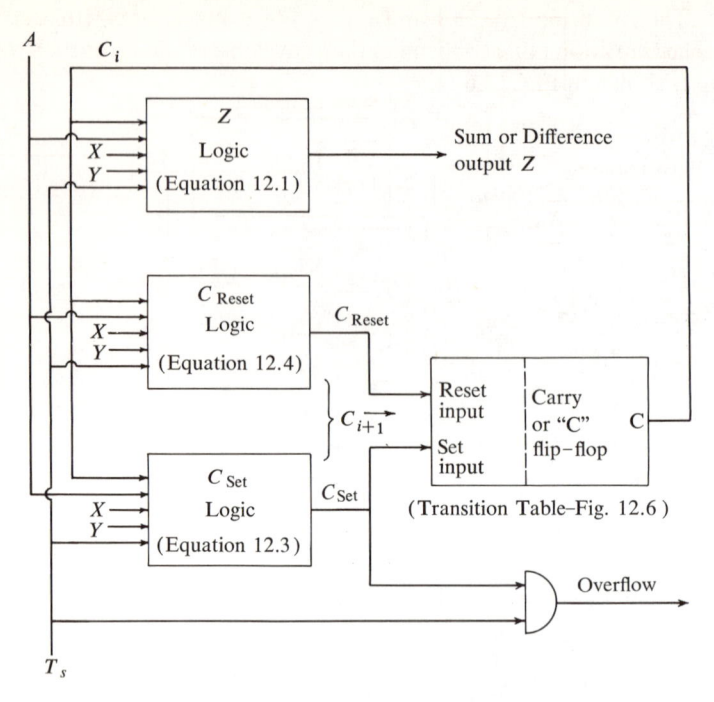

Fig. 12.7 Block diagram. Notice that the X and Y input lines may carry the X and Y bits, or their complements $\bar{X}$ and $\bar{Y}$, or the sign bits.

we have assumed that X, Y, and Z all have the same number of bits, that is, that the Z-register has the same size as the X and Y registers. Therefore the simultaneous occurrence of $T_s = 1$ and $C_{SET} = 1$ should signal an "overflow" condition. To take care of this requirement, the C_{SET} and the T_s lines are ANDed, as shown in Fig. 12.7. From Equation (12.3) we may observe that an "overflow" signal is issued if any one of the last two terms is a one. The "overflow" signal may be used to stop the machine or to initiate a corrective procedure outside the small computer. To correct an "overflow" condition is another problem by itself, which would require the design of additional combinational and sequential circuits.

12.6 MACHINE ORGANIZATION

The organization of the small computer that will perform addition or subtraction of two numbers X and Y is shown in Fig. 12.8. The diagram includes the input and output shift registers, timers, and the A-flip-flop whose state dictates which operation is to be performed.

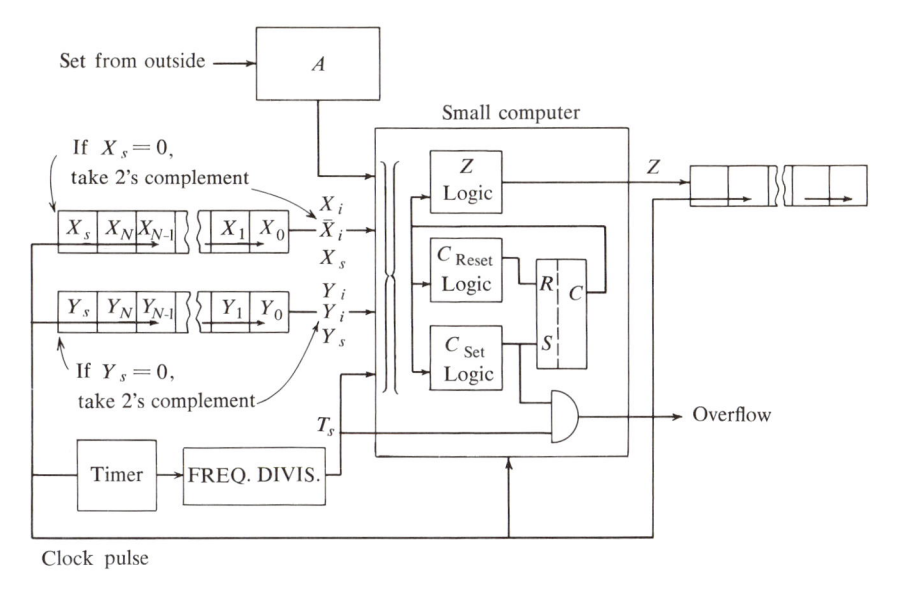

Fig. 12.8 Organization of the small computer.

The numbers X, Y, and Z are all $(N + 1)$-bit long including the sign-bit. The input numbers X and Y are stored initially in the input registers as shown in Fig. 12.8, with the sign-bit stored at the "highest" position (that is, the left-end position). It should be recalled, though, that if the sign-bit of an input number is zero, then the two's complement of the number is stored in the register instead. Thus, the X and Y input lines carry the bits X_i, Y_i, or their complements $\bar{X}_i$, $\bar{Y}_i$, or the sign-bits X_s, Y_s.

The signal T_s is produced by "frequency division." The "FREQ. DIVIS." block receives the synchronization pulse train from the "TIMER" (that is, the central clock), and produces one T_s-pulse for every $N + 1$ clock pulses it receives. The T_s-pulse is used to signal the arrival at the small computer's inputs of the sign-bits of X and Y. The synchronization pulses from the TIMER are used to clock all the activities taking place during the cycles of operation, such as for example bit shifting in the X, Y, and Z registers, as illustrated in Fig. 12.8.

12.7
MACHINE OPERATION—STATE DIAGRAMS

The operation of the small computer may be alternatively represented by state flow diagrams (Chapter 11, Section 11.4). The patterns of transitions between states are seen more clearly on state diagrams than in transition tables.

The state flow diagram during *addition* ($A=1$) is shown in Fig. 12.9. It is constructed by reading the "output" Z_i and the "next state" C_{i+1} for all possible combinations of "present inputs" T_s, X_i, Y_i, and "present states" C_i, from the transition tables of Figs. 12.3(a) and 12.3(c). Observe that there are two

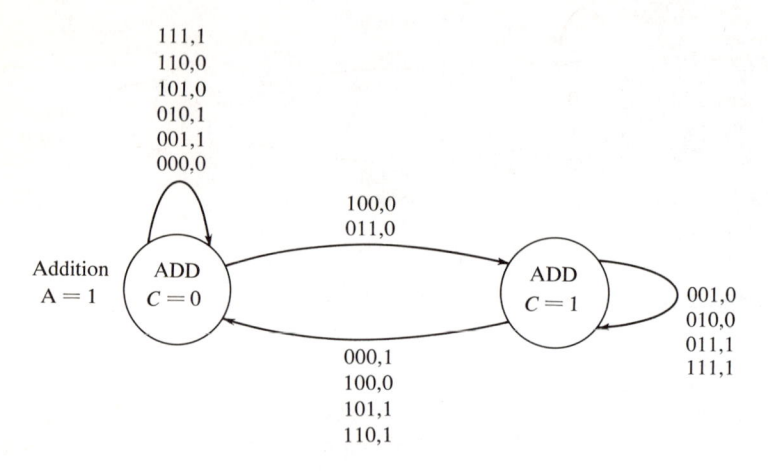

Fig. 12.9 State flow diagram during addition ($A=1$).

stable states corresponding to the two states of the C-flip-flop, that is, $C=0$ and $C=1$. Each transition is represented by an arrow, and characterized by a "present state" which is at the beginning of the arrow, by a "next state" which is at the end of the arrow, and by an input-output relation. The input-output relation is represented on the state diagram by two groups of bits recorded in the middle of the transition arrow and separated by a comma (XXX, X). The first group of bits consists of the three input bits T_s, X_i, Y_i, while the second group is, in this case, composed of the single output bit Z_i.

The state flow diagram for *subtraction* ($A=0$) is similarly constructed and shown in Fig. 12.10. Information is read off the transition tables for SUBTRACTION, in Figs. 12.3(b) and 12.3(d).

12.8
COMMENTS

The design of a sequential machine begins with a task statement explaining what the machine is to do in relating output to input. It is characteristic of sequential machine design problems that there is always a variety of ways of describing a problem. A good formulation of the task statement may be of paramount importance in facilitating the solution of the problem. A primitive transition table or state flow diagram should be formed initially from

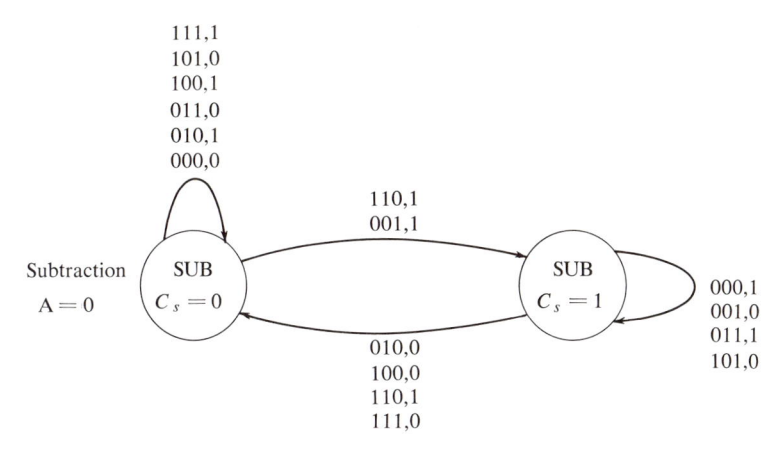

Fig. 12.10 State flow diagram during subtraction ($A = 0$).

the wording of the task statement. The table and the diagram are then to be used interchangeably during the procedure of state minimization and circuit implementation. The steps in the design of sequential, finite-state, machines were illustrated with simple examples in Chapters 11 and 12. It should be realized, however, that designing larger sequential machines requires a certain amount of ingenuity in formulating the problem properly and selecting the appropriate graphical or tabular means for solution. The methodology involves numerous techniques and graphical or tabular means, of which only the essentials were exemplified in Chapters 11 and 12.

In a synchronous design it is assumed that the input, output, and internal state changes coincide with the occurrence of clock pulses. It is also assumed that all internal state changes are completed in the interval between clock pulses. If a single clock is used throughout the machine, clock pulses will not ordinarily appear in the tables, graphs and design equations. Exceptions may be made if a system involves more than one clock pulse channel and it becomes necessary to distinguish between different clock pulses.

Asynchronous sequential designs pose special design problems. Transition tables must be made in a way which does not permit incorrect outputs during secondary transitions. Pulses signaling the end of elementary operations must appear in transition tables and state diagrams. It should be mentioned here that the possibility exists of designing sequential machines containing synchronous and asynchronous portions working together. In very large information-processing systems, designs of limited synchronization with asynchronous operation at the microlevel[3] and synchronous operation at the macrolevel, would take

---

[3] "*Microlevel*" refers to the level of basic circuits which perform the elementary arithmetic operations. "*Macrolevel*" refers to larger order operations and transfers of information which involve whole units of the machine.

excellent advantage of the speed capabilities of the associated electronic devices and circuits and should lead to an optimized design containing both types of operation. Challenging systems design problems arise in such cases.

As it has already been shown, the only function provided by sequential machines which is not provided by combinational machines is the capacity to delay, and thus "remember" information. Both examples of sequential machines of Chapters 11 and 12 display this ability with the use of flip-flops, which themselves are sequential machines of a very elementary form.

The small computer of this chapter is only a portion of an information-processing system. It must operate in conjunction with other information-processing and storage units. Any particular information-processing system will have the capacity of performing a variety of basic operations, such as add, subtract, multiply, divide, read-in, read-out, store, and stop. The machine determines, from the instruction provided to it, which of the alternative basic operations is to be performed in any particular machine cycle, and delegates (through its control unit) the task for execution by the appropriate unit. From an organizational point of view the capability of an information-processing machine to solve a great variety of problems rapidly and flexibly depends to a considerable degree on the number of different operations which it can perform without undue complication on programming, and hence, on the variety of instructions available to the programmer.

In summarizing, it was demonstrated that beginning with a fairly simple design problem in arithmetic, it has been possible to sketch the general approach to block-diagramming a machine to solve the problem. It has been our intent to provide the reader with a perspective concerning the general nature of requirements in logic and sequential machine design. Additional reading will place this information in better perspective with regard to the problem of machine organization and sequential machine design.

BIBLIOGRAPHY

Marcus, M. P., *Switching Circuits for Beginners*. Englewood Cliffs, N.J.: Prentice-Hall, Inc., 1962.

Ware, W. H., *Digital Computer Technology and Design*. vol. 1, New York: John Wiley & Sons, Inc., 1963.

Problems

12.1 Design a binary counter for: 001, 010, 011, 100, 101, REPEAT.

The three input counter will receive a 001 and produce a 010, will receive a 010 and produce a 011, etc. The outputs of the counter may be connected respectively to the inputs and the counter triggered by a clock pulse will count from one to five in binary repeatedly.

12.2 Formulate the problem and design a small sequential machine which will operate as an electronic combination lock.

Postscript

Man now faces the prospect of having unlimited energy to put to work for him, of possessing unlimited variety of "made to order" materials, and of being aided by revolutionary means for the generation, storage, processing, and communication of information. Among the areas of technological innovation, the "information revolution" is more important than the others because of its direct social impact. The psychological and sociological impact of automation is already being felt in relations among employers, employees, computer manufacturers, and the public. Telephone industries, steel industries, chemical, petroleum and refining industries, and many others have already been automated or are being automated. More and more production or performance is entrusted to automatic machines, taking their orders from giant computers and reporting back to them. Military strategy has been revolutionized by the introduction of computer-aided decision-making. Digital computers can choose the correct course of action from thousands of alternatives with speed, objectivity, and accuracy unmatched by any human brain. Coupled to the fertile field of cybernetics, the tremendous capabilities of digital computers should produce astounding changes in our way of life in the years ahead.

While digital computers may be viewed simply as being tools for collection, organization, manipulation and programmed processing of data, cybernetic machines may be considered more generally as products of the science of optimal directional control of complex dynamic systems and processes. In addition to reducing the need for human labor, as other machines do, digital computers and cybernetic machines also assist man work out the optimal solutions to the complex problems of production, economics, and science. Mechanical, chemical, nuclear, economic, manufacturing, and a multitude of other processes from an ever-expanding spectrum of industrial, economic, sociological, and scientific activities can now be analyzed mathematically and expressed in terms of quantitative and topological relations. Digital computers and cybernetic machines analyze, optimize and control such processes to good advantage.

* * *

Calculating machines have come a long way since the early seventeenth century. Present-day computers achieve startling performances (such as a quarter-million calculations per second), have tremendous memory capabilities (very promising for use in illness-diagnosing machines, where the empirical knowledge of medical science may be loaded in a computer memory for this purpose), and are capable of language translation into English at rates of 1800 words per minute, more than fifteen times faster than human translators. Today's machines may be programmed to play checkers or may "read" handwritten instructions, but possess a relatively low status of "intelligence" compared to cybernetic machines. The actions of a computer are programmed in advance and are thus rigidly mathematically predictable. Choice may be incorporated but it will always follow some definitely tractable mathematical law or rule. On the other hand, cybernetic machines possess the possibility of initiative and liberty in actions.

The same technological breakthroughs that brought about such fantastic feats as interplanetary missions, submarine cruises under the North Pole, or automatic computers and automated factories, are now promising new feats in the area of cybernetics, general-purpose robots, and man-made "thinking" machines that will challenge the human imagination.

The comparison of computer functions to human behavior has been inspired by the computers' ability to retain a memory from which information may be drawn for processing when needed, in a pattern similar to the human pattern. To a certain extent, digital computers may be thought of as duplicating the human pattern of thinking—but only to a very limited extent—and to that extent they may also be thought of as simulating human behavior. In the area of "thinking" machines, however, it is the cybernetic machines that may provide the answers. However, even in the area of cybernetic machines, the term "thinking" machines should be used with great caution, if at all. Opinions appear from time to time in the popular press which unqualifiedly maintain that computers can "think." It seems unwise to apply a term with such a broad spectrum of meanings to today's greatly restricted computer processes.

* * *

The digital computer as a tool is the most complex, yet the most efficient means for making information available in an organized and comprehensible manner. This ability is used often today to reduce or replace human labor. Information processing machines replace man today in the same way that the automobile has replaced the horse. Thus, the direct impact of information processing machines and other technological achievements on our life has already become the subject of sociological and philosophical discussions of the utmost importance. Similar concerns have come to the minds of many engineers and scientists, as witnessed by their publications on the matter and by the formation of various professional societies and institutions for the study of the problem.

The idea of man's "replacement" by cybernetic machines and computers has been debated in newspapers, in magazines, and in radio and television, not to mention the vast number of scientific publications which have argued this subject. Highly controversial discussions by the world's leading mathematicians, engineers, philosophers, physicists, biologists, medical researchers, sociologists and economists are representative of the people in the various disciplines that have argued the present and the future of cybernetic machines. Opposing views have been presented regarding the subject, resulting generally in the emergence and strengthening of the conviction that the possibilities of cybernetics are almost unlimited as a means of amplifying man's natural resources and capabilities. A common path to research has also been opened for many different disciplines, mainly for biology, physiology, and engineering.

*   *   *

Technological developments may be applied for good or for evil. Degrees of "goodness" can be identified in terms of the character of the utilization. The manner of utilization of new technological developments, such as computers and information machines which have a direct relation to automation, are decisions that can only be made after serious consideration of the many facets of their social impact. Through study and understanding, automation can be made a force for improving man's life by offering him the opportunity to use the full measure of his capabilities and by allowing him more free time for creative work. The predicted second industrial revolution, brought about by computers and cybernetic machines, may thus be made a gradual and acceptable industrial evolution which we are already witnessing today. The introduction of automation in our life can be such a gradual evolution, provided that we act wisely, basing all decisions upon the accumulated knowledge and past experiences. In this task we shall use computers as tools. But the decisions will always be ours.

Appendix A

TABLE A.1.
CHRONOLOGICAL TABLE OF COMPUTING DEVELOPMENTS[a]

Year	Country	Developer	Development
?	Near East	?	Notched sticks, knotted strings
?	Far East	?	Abacus
960	France	Gerbert	Moorish calculating ideas
1580?	Scotland	Napier	Napier's bones (logarithms)
1632?	England	Oughtred	Slide rule
1642	France	Pascal	Number wheel and ratchet
1666	England?	Moreland	Multiplication by repeated addition
1671	Germany	Leibnitz	Stepped cylinder (calculus)
1770	Germany	Hahn	First dependable four-process calculator (using Leibnitz cylinder)
1786	Germany	Muller	Idea for a difference engine
1801	France	Jacquard	Loom controlled by punched cards
1812	England	Babbage	Models of automatic "analytical" engines
1814	Germany	Hermann	Planimeter
1820	France	Thomas	Crank machine using Leibnitz cylinder
1850	Germany	Mannheim	Improved slide rule

[a] With kind permission. Ned Chapin: *An Introduction to Automatic Computers*, pp. 232–235. D. Van Nostrand, 1957.

TABLE A.1.—*Continued*

Year	Country	Developer	Development
1851	Switzerland	Schilt	Keyboard machine with springs and ratchet
1853	Sweden	Scheutz	Difference engine
1854	England	Boole	"Laws of Thought" (Boolean Algebra)
1857	U.S.A.	Hill	First key-driven machine
1863	Sweden	Wiberg	Difference engine
1869	U.S.A.	Sinclair	Tabular freight computer
1872	U.S.A.	Baldwin	First reversible four-process calculator (no keyboard)
1872	U.S.A.	Barbour	Model of a direct multiplying machine
1878	U.S.A.	Verea	First model of widely used direct multiplying method
1878	England	Kelvin	Mechanical integrator
1878	Germany	Ohdner	Rectractable teeth system like Baldwin's
1884	U.S.A.	Felt	Simple key-driven reciprocating machine ("comptometer")
1885	U.S.A.	Grant	Crank-operated reciprocating adding machine (also difference engine)
1886	Germany	Selling	Direct multiplying machine similar to Verea's
1886	U.S.A.	Burroughs	Listing calculator with keyboard
1889	France	Bollee	Difference engine with direct multiplication
1897	Germany	Stoigor	"Millionaire Machine" like Bollee's
1901	U.S.A.	Hopkins	Ten-key adding machines ("Standard")
1902	U.S.A.	Rechnitzer	Multiplication-division machine
1902	U.S.A.	Baldwin	Improved carry mechanism for Baldwin machine
1906	U.S.A.	DeForrest and others	Vacuum tube
1908	U.S.A.	Hopkins	Burroughs bookkeeping machine
1911	U.S.A.	Monroe, Baldwin	Modified Baldwin machine
1911	U.S.A.	Herman, Coxhead	Mercedes-Euclid machine
1912	U.S.A.	Sundstrand	Improved ten-key adding machine
1913	Switzerland	Jahnz	Improved Rechnitzer machine
1919	U.S.A.	Eccles, Jordan	Flip-flop circuit

TABLE A.1.—Continued

Year	Country	Developer	Development
1920	Spain	Torres y Quevedo	Electro-mechanical typewriter-controlled machine with comparison division
1920	U.S.A.	Monroe, Baldwin	First fully automatic "Monroe" machine
1929	U.S.A.	G. E. Co.	A–C network analyzer
1931	U.S.A.	Bush	Mechanical differential analyzer
1936	U.S.A.	Turing	Ph.D. Thesis at Princeton
1938	U.S.A.	Philbrick	Operational amplifier
1944	U.S.A.	Aiken and IBM	Harvard Mark I—first automatically-sequenced electromagnetic relay-type computer
1947	U.S.A.	Eckert, Mauchly	ENIAC—first high-speed electronic automatic computer
1948	U.S.A.	Bardeen, Brattain, Schockley	Transistor
1949	England	Wilkes	EDSAC—first sorted program computer
1949	U.S.A.	IBM	CPC—first commercially available automatic computer-type machine
1950	U.S.A.	Forrester, Taylor, Everett, Youtz	Whirlwind I—a semibasic type of automatic computer (M.I.T.)
1951	U.S.A.	Remington Rand	UNIVAC I—first large-scale general purpose commercially available automatic computer
1952	U.S.A.	Mauchly, Eckert, Von Neumann	EDVAC—a basic type of automatic computer (U. of Pennsylvania)
1952	U.S.A.	Von Neumann, Goldstine, Burks	IAS—a basic type of automatic computer (Princeton)
1955	U.S.A.	IBM	702—first commercially available automatic computer with variable-block and variable-word length

The middle fifties mark the beginning of the era of the solid-state computers. Rapid developments in the design of magnetic core memories, transistors, and, more recently, integrated circuits; advances in the engineering design of computer systems of high complexity, variety, efficiency, and technical perfection; and great strides in the art of programming, have all resulted in enormous increases in the capacity, speed and reliability, while decreasing the size and cost of digital computers. Above all, they have resulted in the almost universal application of computers in business, science, and in the military. Most promising developments now under way in the computer field are based on technological developments in components and processes, on engineering developments in systems and logic design, and in such concepts as time-sharing, remote access, man-machine interaction, high level languages, and new display systems.

TABLE A.2.
CHRONOLOGICAL TABLE OF AUTOMATIC ELECTRONIC COMPUTERS[a]

Date	Name	Date	Name
1947		*1953 (Cont.)*	
Dec	ENIAC	Apr	CRC-107
1949		Apr	IBM-701
		Apr	OARAC
Aug	BINAC	May	ABC
Dec	IBM-CPC	May	NAREC
		May	CEC-36-101
1950		May	MIDAC
May	SPAC	Jun	ORACLE
June	USAF-Fairchild	Jul	RAYDAC
Dec	WHIRLWIND	Aug	CALDIC
Dec	ERA-1101	Aug	Magnefile
		Aug	Jaincomp
1951		Aug	MINAC
Jan	MARK-III	Sep	FLAC
Feb	UDEC	Sep	Univac Scientific (ERA-1103)
Apr	UNIVAC-I		
May	ONR-Relay	Nov	ERA-1102
Dec	RR-409-2R	Dec	UDEC-II
1952		*1954*	
Jan	LAS		
Jan	SWAC	Jan	OMIBAC
Jan	CRC-102	Mar	JOHNIAC
Mar	ORDVAC	Apr	ORDFIAC (Elecom-200)
Mar	MANIAC	Apr	DYSEAC
Apr	EDVAC	Apr	Elecom-120
Jul	MARK-IV	Jun	Circle
Sep	ILLIAC	Jun	ALWAC
Nov	Elecom-100	Jun	UNIVAC-60 (120)
1953		Jul	MSI-5014
Feb	CRC-105	Jul	BURROUGHS 205
Mar	AVIDAC	Aug	Datatron-204
Mar	Logistics	Sep	WISC
Mar	MONROBOT	Sep	MODAC
		Dec	IBM-650

[a] Chapin, N. *An Introduction to Automatic Computers*, pp. 232–235, D. Van Nostrand, 1957. Mr. Chapin's table was extended to include electronic computers manufactured between 1956 and 1968, as those are reported in "Computer Characteristics Quarterly —Annual Supplement 1968" pp. 18–22, published by Adams Associates.

TABLE A.2.—*Continued*

Date	Name	Date	Name
1955		*1957*	
Feb	NORC	Nov	IBM RAMAC 305 I
Mar	IBM-702	Nov	UNIVAC II
Mar	CRC-106 (Whitesac)	*1958*	
Jun	CDC G-15	Jan	UNIVAC FILE COMPUTER I
Nov	ALWAC III E	Sep	IBM 709
Nov	BURROUGHS E-101	Sep	UNIVAC 1105
Dec	Elecom-125	Dec	BURROUGHS 220
Dec	IBM-704	*1959*	
1956		Jan	IBM 705 III
Jan	NCR-303	*1962*	
Jan	IBM-705 I	Jan	UNIVAC FILE COMPUTER II
Mar	Readix	Mar	IBM RAMAC 305 II
Mar	UNIVAC 1103A		
Sep	LGP-30		
Nov	Univac File		
Dec	Monrobot-MU		
Dec	IBM-705 II		

TABLE A.3.
CHRONOLOGICAL LISTING OF SOLID-STATE COMPUTERS

Date	Name	Date	Name
1958		*1960 (Cont.)*	
Mar	— *STC STANTEC*	Dec	— HONEYWELL 800
Nov	— PHILCO 2000/210	Dec	— RAYTHEON 250
Nov	— RECOMP II	?	— *EMIDEC 1100*
?	— CDC 46	?	— *GAMMA 60*
?	— *EL X1*	?	— *ICT 1101*
		?	— *SEREL 1001*
1959		?	— *SIRIUS*
May	— ITT 025	?	— *ZUSE Z23*
May	— *NEAC 2203*		
Jun	— *FACOM 212*	*1961*	
Jun	— *SIEMENS 2002*	Feb	— RCA 301
Oct	— *GAMMA 300 MCT*	Mar	— B-R 400
Nov	— GE 312	Mar	— *ELLIOTT 803*
Nov	— NCR 304	Mar	— *NEAC 2101*
Nov	— RCA 501	Mar	— *NEAC 2205*
		Apr	— G-20
1960		Apr	— GE 225
Jan	— CDC 1604A	Apr	— GE 255
Jan	— L-3000	Apr	— NCR 310
Jan	— UNIVAC SS 80/90 I, II	May	— DDP-25
Mar	— *MELCOM 1101F*	May	— IBM 7030 STRETCH
Mar	— PHILCO 2000/211	May	— NCR 390
May	— MONROBOT XI	Jun	— *HITAC 201*
May	— UNIVAC LARC	Jun	— RECOMP III
Jun	— IBM 7070	Jul	— CDC 160A
Jun	— IBM 7090	Jul	— GENERAL MILLS AD/ECS 37
Jul	— CDC 160	Aug	— B-R 530
Sep	— B251 (VRC)	Aug	— CDC 136
Sep	— HONEYWELL H290	Sep	— B250
Sep	— IBM 1401	Sep	— *CE 102*
Sep	— RPC 9000	Sep	— *FACOM 222*
Oct	— *AEI 1010*	Sep	— IBM 7080
Oct	— *ELEA 9003*	Sep	— *MADIC IIA*
Oct	— IBM 1620	Oct	— B-R 130
Nov	— GE 210	Nov	— *HIPAC 103*
Nov	— PDP-1	Nov	— IBM 1410
Nov	— RPC 4000		

[a] *Italics* denote foreign manufacture.

TABLE A.3.—Continued

Date	Name	Date	Name

1961 (Cont.)

Dec — HONEYWELL 400
Dec — IBM 7074
Dec — *REGNECENTRALEN GIER*
Dec — UNIVAC 490
? — *CITAC 210B*
? — *EMIDEC 2400*
? — *GAMMA 500*
? — RCA 110

1962

Jan — NCR 315
Feb — *ELEA 6001*
Feb — *ELLIOTT 502*
Feb — *GAMMA 30*
Feb — IBM 1710
Mar — *NEAC 2206*
Mar — *OKITAC 5090D*
Mar — *TOSBAC 4200*
Apr — ASI 210
Apr — *LEO 3*
May — *HITAC 3010*
Jun — CE 55
Jun — IBM 7072
Jun — UNIVAC UIII
Jul — B260
Jul — B270
Jul — B280
Jul — GE 412
Jul — *ICT 1500*
Jul — *NEAC 2204*
Jul — PDP-4
Aug — *KDP-10*
Sep — IBM 7094 I
Sep — *KDN-2*
Sep — SDS 910, 920
Sep — UNIVAC 1107
Nov — RCA 601
Dec — *ATLAS I*
Dec — *DATASAAB D21*
Dec — *FACOM 241*
Dec — *HITAC 3030*

1962 (Cont.)

Dec — *ZUSE Z31*
? — *ICT 1300, 1301*
? — *ICT 1600*
? — PHILCO 2400/410
? — *TELEFUNKEN TR-4*

1963

Jan — ASI 420
Jan — HUGHES H330
Jan — *ORION I*
Feb — B5000
Feb — H-W 15K
Feb — *NEAC 2400*
Feb — PHILCO 2000/212
Mar — COLLINS DATA CENTRAL
Mar — *NEAC 2230*
Mar — *OKITAC 5090H*
Apr — *ELLIOTT 503*
Apr — IBM 7040
Apr — *KDF-9*
Apr — PRODAC 500
Apr — *ZUSE Z25*
May — *FACOM 231*
Jun — B-R 230
Jun — CDC 3600
Jun — DDP-24
Jun — PHILCO 1000
Jul — *FP 6000*
Jul — *GAMMA 305*
Jul — IBM 7044
Sep — GE 215
Sep — *KDF-6*
Sep — *OKITAC 5090M*
Sep — PDP-5
Sep — UNIVAC 1004 I
Sep — UNIVAC 1050 III
Oct — *CAE 510*
Oct — DATANET 30
Oct — IBM 1460
Oct — IBM 7010
Nov — HONEYWELL 1800
Nov — IBM 1440

TABLE A.3.—Continued

Date	Name	Date	Name
1963 (Cont.)		*1964 (Cont.)*	
Nov	*MADIC III*	Jun	*GEC 90/30*
Nov	*NEAC 3800*	Jun	*SDS 925*
Nov	*TOSBAC 3300*	Jun	*SDS 930*
Dec	ASI 2100	Jun	*SETI PALLAS*
Dec	HONEYWELL H610, H620	Jun	UNIVAC 1004 II, III
Dec	H600	Jul	CDC 8090
Dec	ITT 525 VADE	Jul	DMI 610
Dec	L-3055	Jul	GE 205
Dec	PHILCO 4000	Jul	HONEYWELL 200/200
Dec	*SIEMENS 3003*	Jul	*NEAC 2200/200*
?	EPSCO 275	Jul	RCA 3301
?	*GAMMA 10*	Aug	AMBILOG 200
		Aug	B-R 133
1964		Aug	PRODAC 50
Jan	*ATLAS II*	Aug	SEL 820
Jan	B263	Sep	*BULL 415*
Jan	B273	Sep	*BULL 425*
Jan	B283	Sep	CDC 6600
Jan	*MELCOM 1530*	Sep	*ORION II*
Feb	*CE 201*	Sep	*TELEFUNKEN TR10*
Feb	DSI 1000	Sep	UNIVAC 418
Feb	PDS 1068	Oct	*FACOM 230/10*
Mar	IBM 7700	Oct	*KDF-8*
Mar	RAYTHEON 440	Oct	*NEAC 1210*
Apr	B160	Oct	PDP-6
Apr	B170	Nov	B5500
Apr	B180	Nov	CDC 3400
Apr	CDC 160G	Nov	*ELEA 4001*
Apr	GE 235	Nov	*HITAC 4010*
Apr	GE 265	Nov	*ICT 1900*
Apr	GE/PAC 4000	Nov	NCR 315/100
Apr	GE/PAC 4040	Dec	*CAE 90/80*
Apr	IBM 7094 II	Dec	DIGIAC 3080
May	CDC 3200	Dec	*GEC 90/300*
May	GE 415	Dec	*ICT 1905, E, F*
May	GE 420	Dec	*LEO 360*
May	HUGHES H3118	Dec	PDP-7
Jun	BECKMAN 420	Dec	SDS 9300
Jun	B-R 340	Dec	*TOSBAC 3400*
Jun	*CAE 90/40*	Dec	*TOSBAC 4300*
Jun	GE 425	?	CDC 8092
Jun	*GEC 90/25*	?	*MADIC 500*

TABLE A.3.—Continued

Date	Name	Date	Name
1964 (Cont.)		*1965 (Cont.)*	
? — *NEAC 2800*		Sep — DMI 612	
1965		Sep — GE 435	
		Sep — IBM 1130	
Jan — *TOSBAC 5200*		Sep — IBM 360/50	
Feb — *CAE 90/10*		Sep — NCR 500	
Feb — CDC 3100		Sep — *PHILIPS PR 8000*	
Feb — *GEC 90/2*		Sep — *TOSBAC 5400/20*	
Feb — L-2010		Oct — *ELLIOTT 903*	
Feb — SDS 92		Oct — H21, H22	
Mar — EMR ADVANCE 6000		Oct — RAYTHEON 520	
Mar — DDP-224		Ocy — *SIEMENS 4004/15*	
Mar — *EL X8*		Oct — SPECTRA 70/15	
Mar — *FACOM 230/10*		Oct — UNIVAC 491	
Mar — *HITAC 5020*		Oct — UNIVAC 492	
Mar — HUGHES H3324		Nov — *ELLIOTT MCS 920B*	
Mar — LGP-21		Nov — IBM 360/75	
Apr — DDP-116		Nov — *OLIVETTI 115*	
Apr — *ELLIOTT 4120*		Nov — PHILCO 102	
Apr — GE 625		Nov — SCC 660	
Apr — PDP-8		Nov — SCC 670	
Apr — *SIEMENS 303*		Nov — *BULL 435*	
May — GE 635		Dec — CDC 3300	
May — GE 645		Dec — CDC 3800	
May — IBM 360/30		Dec — HONEYWELL 200/2200	
May — IBM 360/40		Dec — SPECTRA 70/25	
May — *ICT 1904, E, F*		Dec — UNIVAC 1108 II	
Jun — *GAMMA M40*		? — *KDF-7*	
Jun — GE/PAC 4050 I		? — *SEREL 505*	
Jun — GE/PAC 4060		? — *STC 8300 ADX*	
Jun — *LEO 326*		*1966*	
Jun — *TOSBAC 5300*		Jan — DDP-124	
Jun — *TOSBAC 5400/10*		Jan — HONEYWELL 200/1200	
Jul — B300		Jan — HUGHES H3118M	
Jul — DMI 620		Jan — IBM 360/20	
Jul — EAI 8400		Jan — *SIEMENS 4004/25*	
Jul — *ICT 1902*		Jan — *ZUSE Z32*	
Jul — *ICT 1903*		Feb — HONEYWELL 200/120	
Jul — NCR 315/RMC-501		Feb — IBM 1800	
Jul — SEL 810A		Feb — UNIVAC 1005 I, II, III	
Jul — SEL 840A		Mar — CDC 1700	
Aug — *ICT 1909*		Mar — *FACOM 230/50*	
Sep — DMI 611			

APPENDIX A 337

TABLE A.3.—Continued

Date	Name	Date	Name
1966 (*Cont.*)		1966 (*Cont.*)	
Mar —	*GAMMA 115*	Dec —	*ICT 1907, E, F*
Mar —	Hughes HM4118	Dec —	*SEA 1500*
Mar —	IBM 360/65	Dec —	*SIEMENS 4004/55*
Mar —	IBM 360/67	Dec —	SIGMA 2
Mar —	UNIVAC 494	Dec —	SIGMA 7
Apr —	CDC 6400	Dec —	*TOSBAC 5100/20*
Apr —	GE 115	Dec —	*TOSBAC 5400/30*
Apr —	SCC 650	? —	*EL X5*
Apr —	SDS 940	1967	
May —	*SEA 4000*		
Jun —	*EL X2*	Jan —	EMR ADVANCE 6130
Jun —	*EL X4*	Jan —	B8500
Jun —	GE/PAC 4050 II	Jan —	C-8500
Jun —	SCC 665	Jan —	*HITAC 8300*
Jul —	*EL X3*	Jan —	*ENGLISH ELECTRIC 4/10*
Jul —	*ELLIOTT 4130*	Feb —	EAI 640
Jul —	LINC-8	Feb —	*GEC S.2*
Jul —	*SIEMENS 4004/45*	Feb —	IBM 360/90
Jul —	*SPECTRA 70/45*	Feb —	*NEAC 1240*
Jul —	*SPECTRA 70/55*	Feb —	*NEAC 2200/300*
Aug —	PDP-9	Feb —	*SIEMENS 4004/35*
Sep —	*FACOM 230/20*	Mar —	CDC 3500
Sep —	*HITAC 8200*	Mar —	*GAMMA 140*
Sep —	*ICT 1901*	Mar —	*ENGLISH ELECTRIC 4/30*
Sep —	*MELCOM 3100/10*	Apr —	DDP-416
Sep —	*MELCOM 3100/30*	Apr —	*GAMMA 145*
Sep —	*MELCOM 3100/50*	May —	B2500
Sep —	PDP-8/S	May —	B3500
Oct —	*CAE 10070*	May —	NCR 315/RMC-502
Oct —	DDP-516	May —	*NEAC 2200/50*
Oct —	*GEC S.7*	May —	*NEAC 3100*
Oct —	GE/PAC 4020	May —	*ZUSE Z26*
Oct —	IBM 360/44	Jun —	CDC 6800
Oct —	*NEAC 2200/400*	Jun —	*TELEFUNKEN TR86*
Oct —	SPECTRA 70/35	Jun —	*TELEFUNKEN TR440*
Nov —	IC 6000/19, 29, 39	Jun —	*TOSBAC 7000/60*
Nov —	*NEAC 2200/100*	Jun —	UNIVAC 920
Nov —	*NEAC 2200/500*	Jul —	SEL 840MP
Dec —	BIT 480	Aug —	SIGMA 5
Dec —	*HITAC 5020E*	Sep —	PDP-10
Dec —	*HITAC 8100*	Sep —	PRODAC 250
Dec —	*ICT 1906, E, F*	Sep —	RAYTHEON 703

TABLE A.3.—*Continued*

Date	Name	Date	Name
1967 (Cont.)		*1968*	
Sep	— SCC 6700	Jan	— B6500
Sep	— *SIEMENS 302*	Jan	— B7500
Sep	— *ENGLISH ELECTRIC 4/50*	Jan	— *CII 10010*
Sep	— *TOSBAC 5100/30*	Jan	— *DATASAAB D22*
Sep	— UNIVAC 9300	Feb	— GE 405
Nov	— *CII 10020*	Feb	— HONEYWELL 200/4200
Nov	— *HITAC 8400*	Mar	— *MELCOM 9100/30*
Nov	— *SIEMENS 305*	Apr	— INTERDATA MODEL 2
Dec	— *HITAC 8500*	Apr	— INTERDATA MODEL 4
Dec	— *ENGLISH ELECTRIC 4/70*	Jun	— *SIEMENS 304*
		Jun	— HONEYWELL 200/8200
		Aug	— *TELEFUNKEN TR84*

Appendix B

Tables of Common Codes

B.1 PURE BINARY

16	8	4	2	1	Dec.
	0	0	0	0	0
	0	0	0	1	1
	0	0	1	0	2
	0	0	1	1	3
	0	1	0	0	4
	0	1	0	1	5
	0	1	1	0	6
	0	1	1	1	7
	1	0	0	0	8
	1	0	0	1	9
	1	0	1	0	10
	1	0	1	1	11
	1	1	0	0	12
	1	1	0	1	13
	1	1	1	0	14
	1	1	1	1	15
1	0	0	0	0	16
1	0	0	0	1	17
←			etc.	etc.

No code limit on number of binary digits (bits)

B.2 4-BIT BINARY CODED DECIMAL

8	4	2	1	Dec.
0	0	0	0	0
0	0	0	1	1
0	0	1	0	2
0	0	1	1	3
0	1	0	0	4
0	1	0	1	5
0	1	1	0	6
0	1	1	1	7
1	0	0	0	8
1	0	0	1	9
1	0	1	0	×
1	0	1	1	×
1	1	0	0	×
1	1	0	1	×
1	1	1	0	×
1	1	1	1	×

4-Bit limit — May be used for other characters

B.3 EXCESS THREE BINARY CODED DECIMAL

				Dec.
0	0	0	0	×
0	0	0	1	×
0	0	1	0	×
0	0	1	1	0
0	1	0	0	1
0	1	0	1	2
0	1	1	0	3
0	1	1	1	4
1	0	0	0	5
1	0	0	1	6
1	0	1	0	7
1	0	1	1	8
1	1	0	0	9
1	1	0	1	×
1	1	1	0	×
1	1	1	1	×

4-Bit limit

B.4 UNIT DISTANCE GRAY CODE

				Dec.
0	0	0	0	0
0	0	0	1	1
0	0	1	1	2
0	0	1	0	3
0	1	1	0	4
0	1	1	1	5
0	1	0	1	6
0	1	0	0	7
1	1	0	0	8
1	1	0	1	9
1	1	1	1	10
1	1	1	0	11
1	0	1	0	12
1	0	1	1	13
1	0	0	1	14
1	0	0	0	15

B.5 UNIVAC II CODE (7 BITS-EXCESS THREE) (SPERRY RAND CORPORATION)

Zone				Detail
0 0	0 1	1 0	1 1	(XS-3)
i	r	t	Σ	0 0 0 0
⊿	,	"	β	0 0 0 1
−	.	|	:	0 0 1 0
0	;	)	+	0 0 1 1
1	A	J	/	0 1 0 0
2	B	K	S	0 1 0 1
3	C	L	T	0 1 1 0
4	D	M	U	0 1 1 1
5	E	N	V	1 0 0 0
6	F	O	W	1 0 0 1
7	G	P	X	1 0 1 0
8	H	Q	Y	1 0 1 1
9	I	R	Z	1 1 0 0
'	#	$	%	1 1 0 1
&	¢	*	=	1 1 1 0
(	@	?	N.U.	1 1 1 1

Construction: 0   0 0   0 0 0 0
               Parity   Zone   Detail
               (Odd)   (XS-3, BCD)

B.6 EXTENDED BINARY CODED DECIMAL INTERCHANGE CODE (EBCDIC)

In EBCDIC, the character has an eight-bit length limit. The four leading bits and the four ending bits are designated "Zone" and "Digit" respectively. The sign of a number is assigned to the Zone with

        1 1 0 0 designating positive
        1 1 0 1 designating negative.

It occupies the Zone of the least significant digit of the number. For example, a four-digit field would have the following format:

Zone   Digit   Zone   Digit   Zone   Digit   Sign   Digit
1 1 1 1 0 0 0 1 1 1 1 1 1 0 0 1 1 1 1 1 0 1 1 0 1 1 0 0 1 0 0 0

and would read + 1968. Care should be taken to avoid confusion between:
"positive" (that is, 1 1 0 0 in the Zone) and "addition" (that is, 0 1 0 0 1 1 1 0)
"negative" (that is, 1 1 0 1 in the Zone) and "subtraction" (that is 0 1 1 0 1 1 0 1)
"Digit" (that is, the end four bits of a character) and "digit" (that is, any one of the positive integers from 0 to 9).

This code is in use with the IBM-360 system. It employs eight-character words, that is, sixty-four-bit words. By using eight bit positions per character, 256 different characters can be coded. This code then permits, for instance, the coding of uppercase and lowercase alphabetic characters, a wide range of special characters, and many control characters that are meaningful to certain input/output devices.

EBCDIC

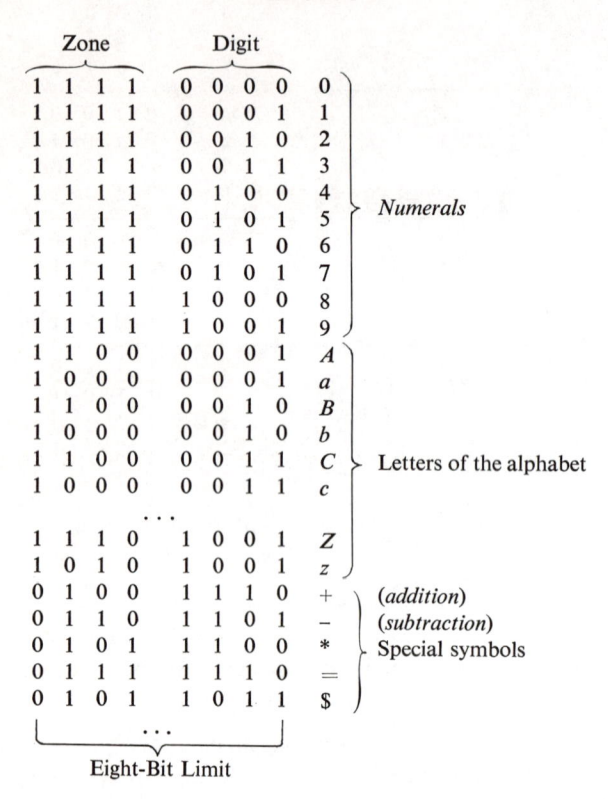

Zone	Digit		
1 1 1 1	0 0 0 0	0	
1 1 1 1	0 0 0 1	1	
1 1 1 1	0 0 1 0	2	
1 1 1 1	0 0 1 1	3	
1 1 1 1	0 1 0 0	4	
1 1 1 1	0 1 0 1	5	Numerals
1 1 1 1	0 1 1 0	6	
1 1 1 1	0 1 1 1	7	
1 1 1 1	1 0 0 0	8	
1 1 1 1	1 0 0 1	9	
1 1 0 0	0 0 0 1	A	
1 0 0 0	0 0 0 1	a	
1 1 0 0	0 0 1 0	B	
1 0 0 0	0 0 1 0	b	
1 1 0 0	0 0 1 1	C	Letters of the alphabet
1 0 0 0	0 0 1 1	c	
. . .			
1 1 1 0	1 0 0 1	Z	
1 0 1 0	1 0 0 1	z	
0 1 0 0	1 1 1 0	+	(addition)
0 1 1 0	1 1 0 1	−	(subtraction)
0 1 0 1	1 1 0 0	*	Special symbols
0 1 1 1	1 1 1 0	=	
0 1 0 1	1 0 1 1	$	

Eight-Bit Limit

Appendix C

**EXAMPLES OF
ASSEMBLY LANGUAGE PROGRAMS
ON THE IBM SYSTEM/360 COMPUTER**

The mnemonics which were used to illustrate assembly language programming in Chapter 2 (Section 2.5, Table 2.1) belong to a small computer, the General Precision RPC 4000. Their simple composition renders them appropriate for tutorial purposes in discussing only the essential features of assembly language programming.

Programming of the larger, modern computers using mnemonic languages is in principle identical to programming the RPC 4000 computer using the mnemonics shown in Table 2.1. However, in practice, assembly languages for larger machines employ mnemonics with somewhat more complex compositions than the ones used in Chapter 2.

Only for the edification of the student a list of mnemonics for use with the IBM System/360 is shown in Fig. C.1. The assembly language programs of the two examples used throughout Chapter 2 (the "temperature conversion" and the "income tax" examples) are shown in Figs. C.2 and C.3, respectively.

It is of interest to note that the compiler would frequently produce the source program in assembly language as an intermediate stage to the final translation into the object (internal machine) language. This feature provides the programmer with the facility to check his program for errors in both the procedure and the assembly languages. The programs shown in Figs. C.2 and C.3, were initially written in FORTRAN IV language (see Chapter 2, Section 2.16) and were translated by the compiler into the machine's assembly language.

In Figs. C.2 and C.3 the hexadecimal numbers in the far left column refer to the memory locations provided for the corresponding instruction words. Each instruction word is composed of the mnemonic part (on the left) and of the operand, which provides essential information for the execution of the command, such as pertinent memory addresses.

NAME	MNEMONIC	TYPE	OPERAND
Add	AR	RR	R1, R2
Add	A	RX	R1, D2(X2, B2)
Add Halfword	AH	RX	R1, D2(X2, B2)
Add Logical	ALR	RR	R1, R2
Add Logical	AL	RX	R1, D2(X2, B2)
AND	NR	RR	R1, R2
AND	N	RX	R1, D2(X2, B2)
AND	NI	SI	D1(B1), I2
AND	NC	SS	D1(L, B1), D2(B2)
Branch and Link	BALR	RR	R1, R2
Branch and Link	BAL	RX	R1, D2(X2, B2)
Branch on Condition	BCR	RR	M1, R2
Branch on Condition	BC	RX	M1, D2(X2, B2)
Branch on Count	BCTR	RR	R1, R2
Branch on Count	BCT	RX	R1, D2(X2, B2)
Branch on Index High	BXH	RS	R1, R3, D2(B2)
Branch on Index Low or Equal	BXLE	RS	R1, R3, D2(B2)
Compare	CR	RR	R1, R2
Compare	C	RX	R1, D2(X2, B2)
Compare Halfword	CH	RX	R1, D2(X2, B2)
Compare Logical	CLR	RR	R1, R2
Compare Logical	CL	RX	R1, D2(X2, B2)
Compare Logical	CLC	SS	D1(L, B1), D2(B2)
Compare Logical	CLI	SI	D1(B1), I2
Convert to Binary	CVB	RX	R1, D2(X2, B2)
Convert to Decimal	CVD	RX	R1, D2(X2, B2)
Diagnose		SI	
Divide	DR	RR	R1, R2
Divide	D	RX	R1, D2(X2, B2)
Exclusive OR	XR	RR	R1, R2
Exclusive OR	X	RX	R1, D2(X2, B2)
Exclusive OR	XI	SI	D1(B1), I2
Exclusive OR	XC	SS	D1(L, B1), D2(B2)
Execute	EX	RX	R1, D2(X2, B2)
Halt I/O	HIO	SI	D1(B1)
Insert Character	IC	RX	R1, D2(X2, B2)
Load	LR	RR	R1, R2
Load	L	RX	R1, D2(X2, B2)
Load Address	LA	RX	R1, D2(X2, B2)
Load and Test	LTR	RR	R1, R2
Load Complement	LCR	RR	R1, R2
Load Halfword	LH	RX	R1, D2(X2, B2)
Load Multiple	LM	RS	R1, R3, D2(B2)
Load Negative	LNR	RR	R1, R2
Load Positive	LPR	RR	R1, R2
Load PSW	LPSW	SI	D1(B1)
Move	MVI	SI	D1(B1), I2
Move	MVC	SS	D1(L, B1), D2(B2)
Move Numerics	MVN	SS	D1(L, B1), D2(B2)
Move with Offset	MVO	SS	D1(L1, B1), D2(L2, B2)
Move Zones	MVZ	SS	D1(L, B1), D2(B2)
Multiply	MR	RR	R1, R2
Multiply	M	RX	R1, D2(X2, B2)
Multiply Halfword	MH	RX	R1, D2(X2, B2)
OR	OR	RR	R1, R2
OR	O	RX	R1, D2(X2, B2)

Fig. C.1 IBM System/360 Instruction Set

NAME	MNEMONIC	TYPE	OPERAND
OR	OI	SI	D1(B1), I2
OR	OC	SS	D1(L, B1), D2(B2)
Pack	PACK	SS	D1(L1, B1), D2(L2, B2)
Set Program Mask	SPM	RR	R1
Set System Mask	SSM	SI	D1(B1)
Shift Left Double	SLDA	RS	R1, D2(B2)
Shift Left Single	SLA	RS	R1, D2(B2)
Shift Left Double Logical	SLDL	RS	R1, D2(B2)
Shift Left Single Logical	SLL	RS	R1, D2(B2)
Shift Right Double	SRDA	RS	R1, D2(B2)
Shift Right Single	SRA	RS	R1, D2(B2)
Shift Right Double Logical	SRDL	RS	R1, D2(B2)
Shift Right Single Logical	SRL	RS	R1, D2(B2)
Start I/O	SIO	SI	D1(B1)
Store	ST	RX	R1, D2(X2, B2)
Store Character	STC	RX	R1, D2(X2, B2)
Store Halfword	STH	RX	R1, D2(X2, B2)
Store Multiple	STM	RS	R1, R3, D2(B2)
Subtract	SR	RR	R1, R2
Subtract	S	RX	R1, D2(X2, B2)
Subtract Halfword	SH	RX	R1, D2(X2, B2)
Subtract Logical	SLR	RR	R1, R2
Subtract Logical	SL	RX	R1, D2(X2, B2)
Supervisor Call	SVC	RR	I
Test and Set	TS	SI	D1(B1)
Test Channel	TCH	SI	D1(B1)
Test I/O	TIO	SI	D1(B1)
Test Under Mask	TM	SI	D1(B1), I2
Translate	TR	SS	D1(L, B1), D2(B2)
Translate and Test	TRT	SS	D1(L, B1), D2(B2)
Unpack	UNPK	SS	D1(L1, B1), D2(L2, B2)

DECIMAL FEATURE INSTRUCTIONS

NAME	MNEMONIC	TYPE	OPERAND
Add Decimal	AP	SS	D1(L1, B1), D2(L2, B2)
Compare Decimal	CP	SS	D1(L1, B1), D2(L2, B2)
Divide Decimal	DP	SS	D1(L1, B1), D2(L2, B2)
Edit	ED	SS	D1(L, B1), D2(B2)
Edit and Mark	EDMK	SS	D1(L, B1), D2(B2)
Multiply Decimal	MP	SS	D1(L1, B1), D2(L2, B2)
Subtract Decimal	SP	SS	D1(L1, B1), D2(L2, B2)
Zero and Add	ZAP	SS	D1(L1, B1), D2(L2, B2)

DIRECT CONTROL FEATURE INSTRUCTIONS

NAME	MNEMONIC	TYPE	OPERAND
Read Direct	RDD	SI	D1(B1), I2
Write Direct	WRD	SI	D1(B1), I2

PROTECTION FEATURE INSTRUCTIONS

NAME	MNEMONIC	TYPE	OPERAND
Insert Storage Key	ISK	RR	R1, R2
Set Storage Key	SSK	RR	R1, R2

Fig. C.1 (Cont.)

```
FORTRAN IV G LEVEL 1, MOD 1           MAIN               DATE = 68130

LOCATION   STA NUM    LABEL    OP      OPERAND              BCD OPERAND
000000                         BC      15,12(0,15)
000004                         DC      06D4C1C9
000008                         DC      D5404040
00000C                         STM     14,12,12(13)
000010                         LM      2,3,40(15)
000014                         LR      4,13
000016                         L       13,36(0,15)
00001A                         ST      13,8(0,4)
00001E                         STM     3,4,0(13)
000022                         BCR     15,2
000024                         DC      00000000             A4
000028                         DC      00000000             A20
00002C                         DC      00000000             A36
0000F8              A36        L       13,4(0,13)
0000FC                         L       14,12(0,13)
000100                         LM      2,12,28(13)
000104                         MVI     12(13),255
000108                         BCR     15,14
00010A              A20        L       15,124(0,13)         IBCOM#
00010E                         LR      12,13
000110                         LR      13,4
000112                         BAL     14,64(0,15)
000116                         LR      13,12
000118      3                  L       15,124(0,13)         IBCOM#
00011C                         BAL     14,0(0,15)
000120                         DC      00000005
000124                         DC      000000B0
000128                         BAL     14,8(0,15)
00012C                         DC      0450D06C
000130                         BAL     14,16(0,15)
000134      5                  L       0,144(0,13)
000138                         ST      0,112(0,13)          I
00013C      6         5        L       0,112(0,13)          I
000140                         A       0,184(0,13)
000144                         ST      0,112(0,13)          I
000148      7                  L       15,124(0,13)         IBCOM#
00014C                         BAL     14,0(0,15)
000150                         DC      00000005
000154                         DC      000000B4
000158                         BAL     14,8(0,15)
00015C                         DC      0470D074
000160                         BAL     14,16(0,15)
000164      9                  LE      0,188(0,13)
000168                         ME      0,116(0,13)          CENTI
00016C                         DE      0,192(0,13)
000170                         AE      0,196(0,13)
000174                         STE     0,120(0,13)          FAHREN
000178     10                  L       15,124(0,13)         IBCOM#
00017C                         BAL     14,4(0,15)
000180                         DC      00000006
000184                         DC      00000089
000188                         BAL     14,8(0,15)
00018C                         DC      0470D074
000190                         BAL     14,8(0,15)
000194                         DC      0470D078
000198                         BAL     14,16(0,15)
00019C     12                  L       0,112(0,13)          I
0001A0                         C       0,108(0,13)          N
0001A4                         L       14,96(0,13)          5
0001A8                         BCR     4,14
0001AA     13                  L       15,124(0,13)         IBCOM#
0001AE                         BAL     14,52(0,15)
0001B2                         DC      05404040
0001B6                         DC      40F0
                               END
     TOTAL MEMORY REQUIREMENTS 0001B8 BYTES
```

Fig. C.2 Programming the temperature conversions example (Chapter II) using the Assembly Language of the IBM System/360 Model 75. (Note that the machine was provided with the FORTRAN program and carried out the translation into the mnemonic language by itself.)

348   APPENDIX C

FORTRAN IV G LEVEL 1, MOD 1   MAIN   DATE = 68130

LOCATION	STA NUM	LABEL	OP	OPERAND	BCD OPERAND
000000			BC	15,12(0,15)	
000004			DC	06D4C1C9	
000008			DC	D5404040	
00000C			STM	14,12,12(13)	
000010			LM	2,3,40(15)	
000014			LR	4,13	
000016			L	13,36(0,15)	
00001A			ST	13,8(0,4)	
00001E			STM	3,4,0(13)	
000022			BCR	15,2	
000024			DC	00000000	A4
000028			DC	00000000	A20
00002C			DC	00000000	A36
000144		A36	L	13,4(0,13)	
000148			L	14,12(0,13)	
00014C			LM	2,12,28(13)	
000150			MVI	12(13),255	
000154			BCR	15,14	
000156		A20	L	15,180(0,13)	IBCOM#
00015A			LR	12,13	
00015C			LR	13,4	
00015E			BAL	14,64(0,15)	
000162			LR	13,12	
000164	3		L	15,180(0,13)	IBCOM#
000168			BAL	14,0(0,15)	
00016C			DC	00000005	
000170			DC	000000E8	
000174			BAL	14,8(0,15)	
000178			DC	0450D0A0	
00017C			BAL	14,16(0,15)	
000180	5		L	0,200(0,13)	
000184			ST	0,164(0,13)	I
000188	6	12	L	0,164(0,13)	I
00018C			A	0,240(0,13)	
000190			ST	0,164(0,13)	I
000194	7		L	15,180(0,13)	IBCOM#
000198			BAL	14,0(0,15)	
00019C			DC	00000005	
0001A0			DC	000000EC	
0001A4			BAL	14,8(0,15)	
0001A8			DC	0470D0A8	
0001AC			BAL	14,16(0,15)	
0001B0	9		LE	0,168(0,13)	GROSS
0001B4			CE	0,244(0,13)	
0001B8			L	14,100(0,13)	4
0001BC			BCR	4,14	
0001BE	10		LE	0,168(0,13)	GROSS
0001C2			CE	0,248(0,13)	
0001C6			L	14,104(0,13)	6
0001CA			BCR	4,14	
0001CC	11		LE	0,168(0,13)	GROSS
0001D0			CE	0,252(0,13)	
0001D4			L	14,108(0,13)	7
0001D8			BCR	4,14	
0001DA	12		L	14,112(0,13)	8
0001DE			BCR	15,14	
0001E0	13	4	LE	0,200(0,13)	
0001E4			STE	0,172(0,13)	TAX
0001E8	14		L	14,116(0,13)	10
0001EC			BCR	15,14	
0001EE	15	6	LE	0,168(0,13)	GROSS
0001F2			SE	0,244(0,13)	
0001F6			ME	0,256(0,13)	
0001FA			STE	0,172(0,13)	TAX
0001FE	16		L	14,116(0,13)	10
000202			BCR	15,14	

Fig. C.3 Programming the income tax example (Chapter II) using the Assembly Language of the IBM System/360 Model 75. (Note that the machine was provided with the FORTRAN program and the compiler carried out the translation into the mnemonic language.)

```
FORTRAN IV G LEVEL 1, MOD 1            MAIN              DATE = 68130

000204      17      7       LE      0,168(0,13)          GROSS
000208                      SE      0,248(0,13)
00020C                      ME      0,260(0,13)
000210                      AE      0,264(0,13)
000214                      STE     0,172(0,13)          TAX
000218      18              L       14,116(0,13)         10
00021C                      BCR     15,14
00021E      19      8       LE      0,168(0,13)          GROSS
000222                      SE      0,252(0,13)
000226                      ME      0,268(0,13)
00022A                      AE      0,272(0,13)
00022E                      STE     0,172(0,13)          TAX
000232      20      10      LE      0,168(0,13)          GROSS
000236                      SE      0,172(0,13)          TAX
00023A                      STE     0,176(0,13)          NET
00023E      21      5       L       15,180(0,13)         IBCOM#
000242                      BCR     0,0
000244                      BAL     14,4(0,15)
000248                      DC      00000006
00024C                      DC      000000F1
000250                      BAL     14,8(0,15)
000254                      DC      0470D0A8
000258                      BAL     14,8(0,15)
00025C                      DC      0470D0B0
000260                      BAL     14,8(0,15)
000264                      DC      0470D0AC
000268                      BAL     14,16(0,15)
00026C      23              L       0,164(0,13)          I
000270                      C       0,160(0,13)          N
000274                      L       14,96(0,13)          12
000278                      BCR     4,14
00027A      24              L       15,180(0,13)         IBCOM#
00027E                      BAL     14,52(0,15)
000282                      DC      05404040
000286                      DC      40F0
                            END
        TOTAL MEMORY REQUIREMENTS 000288 BYTES
```

Fig. C.3 (Cont.)

Appendix D

GLOSSARY OF MOST COMMON COMPUTER TERMS

access time—The time needed to locate data in its storage position and make it available for processing. Also the time needed to return information from a processing unit to a storage location.

accumulator—A temporary storage device (register) for the operands of an arithmetic or logic operation. The accumulator also commonly stores the result of a mathematical operation.

adder—A device or circuit which performs the addition of two or more numbers.

address—An alphanumeric designation of the storage location of data or information in the machine's memory.

AND gate—A device or circuit which has an output, when, and only when, all inputs are present.

binary—A condition in which there are only two alternatives.

binary cell—A storage device which possesses two stable states or conditions.

binary code—A code made up of binary elements.

binary-coded decimal—A decimal number coded by way of binary digits. (A common example, in the NBCD code.)

binary number system—A positional number system with base two.

binary point—Sporatic point in the binary number system.

biquinary code—A seven-bit binary code with the weights 5, 0, 4, 3, 2, 1, 0, respectively. Each combination of the code contains only two 1-bits. For example, the decimal numeral 3 is 0101000.

bit—binary digit.

card feed—A mechanical or numatic device which moves punched cards one at a time into an information processing machine.

card programming—The use of punched cards for the storage of instructions.

card punch—A manually operated device for punching information on cards. Also referred to as a "key punch."

card reader—A mechanism which converts information punched on cards into another form, commonly, electric impulses.

card stacker—A mechanism for stacking cards in a bin or pocket after they have been processed through a card reader.

character—A combination of elementary symbols, usually binary digits, which is used to express information. Also a symbol by itself such as the A, 3, $\pm$, and $-$ signs, and so forth.

check bit—A binary digit which is carried along with coded information to enable verification that the information is correct. (See parity check.)

clock—The source of synchronizing impulses, usually a crystal-controlled oscillator.

code—A symbolic language for the representation of information.

command—A machine instruction.

compiler—A machine program which processes a source program and transforms it into an object program, expressed in the internal machine language.

complement—In the positional notation the base-minus-one complement of a number written in base r is derived by subtracting each digit from $r - 1$. True complement is obtained by adding one to the base-one-complement.

condition of transfer—A machine instruction which discontinues the normal sequencing of machine instructions and orders a different instruction if certain programmed conditions occur.

counter—A device which counts input pulses. Also a device capable of changing state upon receipt of a certain number of input pulses.

cycle—A complete sequence of actions taken by the machine in the course or its normal operation.

data—Numerical or other forms of quantitative information.

data processing—The operations involved in the collection, manipulation, dissemination of data.

data processor—A device or system for data processing. Usually such devices are used primarily for sorting or otherwise manipulating information rather than extensive calculations.

debug—Testing and correcting a machine program.

density—The number of information bits packed per unit space on a storage medium.

digit—A symbol representing an integral quantity.

digital computer—A machine for processing information presented to it in the form of digital numbers.

edit—The rearranging of information involving division of unwanted data, selection of pertinent data, or the insertion of symbols.

END-around carry—A carry digit from the highest-order digit which is added to the lowest-order digit.

erase—Erasing a memory cell involves replacing all bits in the cell by zeros.

excess 3 code—A number code in which each decimal digit, x, is represented by the four-bit binary equivalent of $x + 3$. For example, the decimal 4 is represent by 0111 in the excess 3 code.

fetch phase—The part of the machine cycle wherein the instruction is brought from the memory into the instruction register of the control unit piler to the execute place.

fixed-cycle operation—A machine operation wherein a fixed amount of time allotted for the completion of the operation.

floating-point calculation—Arithmetic operation wherein the operands are written in the floating-point notation.
flow chart—A graphical representation of a program.
gate—A two or more input device or circuit whose output is present when, and only when, some prescribed combination of the inputs is present.
general-purpose computer—An information processing machine capable of solving a variety of problems by external programming.
half adder—A circuit or device which can accept two binary inputs, x and y, and deliver two outputs, s and c, according to the following table

Input		Output	
x,	y	s,	c
0	0	0	0
0	1	1	0
1	0	1	0
1	1	0	1

information—Any fact, data, or opinions which may be communicated or manipulated.
inhibit pulse—A pulse signal which is applied to a magnetic storage cell to prevent flux reversal.
input equipment—Hardware used for accepting information into the machine.
instruction—A command plus one or more addresses in the form of a machine word which prescribes the machine operation to be performed on the indicated data.
instruction register—A register located in the control unit which stores the instruction currently governing the operation of the machine. Also referred to as **program register**.
internal storage—Storage facilities, part of the main body of the machine.
jump—To close the next machine instruction to be selected from a specified storage location.
Karnaugh map—A tabulator arrangement facilitating minimization of logical functions.
line printer—A printer capable of printing an entire line of characters simultaneously.
land program (loader)—A special program which allocates space in the machine's memory and affects the transfer of the object program and the data into the memory in preparation for information processing.
logical design—The mathematical planning of the interrelationships of functional blocks and logic circuits prior to the detailed design of an information processing unit.
logical operation—A machine operation which involves some kind of a decision.
loop—Part of a program in which a group of instruction is repeated on condition.
machine language—The representation of information in a form the machine can interpret.
machine word—A fixed or variable sized sequence of bits which represent a piece of information.
magnetic core—A tiny donut-shaped device in which one bit of information may be stored magnetically.
magnetic drum—A rotating cylindrical drum bearing a magnetic coating for storage of information.

magnetic tape—A metallic paper or plastic tape coated with magnetic material and used for storing information.

memory capacity—The amount of information that can be stored in a memory, usually expressed in words.

message—A group of words bearing an integral piece of information.

NOR element—A gate which produces an output only when all of its inputs are absent.

off-line equipment—Auxillary equipment operating independently of the main units of the machine. Typical off-line equipment include punched cards and magnetic tape units.

on-line equipment—Processing equipment connected directly to the main processing unit and of compatible speed.

operand—A quantity which is used in an arithmetic operation.

operational unit—A combination of electronic, mechanical, or other components which perform a complete machine operation such as storage or input processing.

OR gate—A gate which produces an output when one or more of its inputs are present.

output unit—The portion of the machine which is used for the dissemination of information from the machine to the outside.

overflow—The condition which results when a machine word exceeds the capacity of a register or a machine shell.

paper tape—A strip of paper which bears information in the form of punched holes and blank spaces.

parallel digital computer—See serial digital computer.

parallel transfer—The simultaneous transfer of bits of information.

parity check—A summation check to reveal single-bit errors.

peripheral equipment—See off-line equipment.

precision—The degree of exactness with which a quantity is specified.

program—A complete plan of instructions for the solution of a program by information processing machines.

program register—*see* instruction register.

program tape—A tape, usually magnetic, which bears a program.

programmer—The person responsible for originating a program.

programming—The analysis and synthesis of programs.

punched card—A standard size card which bears information stored in the form of holes and blank spaces.

quantization—A process by which analog information is converted to digital information.

random excess—A method of information retrieval and procurement from any location of the machine's main memory in about equal excess time.

random processing—Processing of randomly accessible data.

read—To copy or sense information, usually from one storge unit into another.

read in—To place information at the specific address of a storage unit.

read out—To copy information from a specific address in the computer's memory.

real-time operation—Processing information while the facts which produced the information take place so that the results of the processing may guide the operation.

register—A temporary storage device or circuit of limited capacity.

reset—To restore a storage device to a prescribed state, usually the zero state.

rig counter—A loop formed of interconnected bistable devices for counting a sequence of input signals.

routine—A sequence of instructions to cause the machine to perform a certain specific operation.

scaling—Shifting the decimal point of a quantity for a specific purpose.

sequential processing—The processing of data that has been operated on prior to its entry into the machine and has been placed in a definite prescribed order.

serial (parallel) digital computer—A machine in which the digits are handled in a serial (parallel) manner.

shift—To move a sequence of bits one or more places to the right or the left.

significant digit—Any digit between the highest known zero digit and the lowest known zero digit.

special-purpose machine—A machine designed to solve specific types of problems, usually operating with a built in (wired) program.

subroutine—A short sequence of instructions for solving a small well-defined part of a larger problem.

tabulator—A machine which copies, types, and brings out specific fields of information.

ternary notation—A number system using the base 3.

transfer—To convey information from one location to another.

translate—To convert the information to one language or code into another.

truth table—A tabular listing of a logical function showing its outcome for all combinations of inputs.

unconditional branch—An instruction which interrupts the normal process of obtaining instructions from an ordered sequence and specifies the address from which the next instruction is to be taken.

variable cycle operation—A machine operation wherein operations may be of varying length.

volatile memory—A storage medium which is incapable of retaining information without continuous power dissipation.

word—A sequence of digits which are treated by the machine as a unit and which bear a piece of information.

word length—The number of bits per word.

write—To record information into a storage medium.

zero suppression—The removal of nonsignificant zeros to the left of the integral part of a quantity before printout.

zone punch—A punch in one of the three top positions (12, 11, 0) of a punched-card column. A zone punch combined with a punch in one of the numeric positions forms an alphabetic character.

Index

Accumulator (see A-register)
Addition, binary, 76, 79, 82, 263–268
   serial, 258–259, 272–276
   parallel, 276–277
Address, 15, 17, 63–64
   modification of, 280
Adjacent cells (see Karnaugh maps)
ALGOL, 49–56
Algorithm, 29
   preparation of an, 34–35
Algebra
   Boolean (see Boolean algebra)
   of propositions, 135–137
   switching, 140–183
All-or-none organs, 250
AND, 136
A-register, 36
Arithmetic, binary, 262
   mechanizing of, 74–76
   operating rules, 257–260
   operations, 260–262
Assemblers, 31, 41, 48–49
Assembly language, 40–43
   IBM System/360 programs in, 344–350
Asynchronous, computers, 10
   operation, 278–280

Binary, addition, 263–268, 311
   arithmetic, 262
   counter, 243
   subtraction, 269, 311
Binary coded numerical systems, 77–82
Binary numerical system, pure, 76–77
Biquinary code, 90–91
Bits, 10
Boole, G., 125, 140
Boolean algebra, 140–183
   axiomatic definition, 142–143
   combinations and functions of, 145–148
   complementation, 150–152
   equations, 149–150
   theorems and operation rules, 143–145
Boolean calculus, 178–179
Buffer, 15, 18

Canonical forms, 152–156
Card, punched, 18
   punches, 21
   readers, 21
Carry flip-flop, 315–317

Character, 12
Class, 67
Classes, calculus of, 127–135
  intersection of, 129
  null, 128
  operation with, 128–130
  union of, 128
  unit, 128
  universal, 127
Clocking, 10, 238, 278, 311
Code converter, 201–202
Codes, 77
  tables of common, 341–343
Code translator, 114–116
Combinational circuits, 184–214
  code converter, 201–202
  half-adder, 202–206, 267–268, 273
  hardware considerations, 190–191
  four-bit register, 200–201
  NAND, NOR, 191–199
Combinational logic, 120
Combinational operations, 185–187
Command (see Instruction)
Compiler, 31, 45, 48–49
  error detection, 54
Complement, of a class, 127
  of a number, 260–261
  of a switching function, 150–152
  of two's, 312
Complementation, number, 260–261
  number, 260–261
  function, 150–152
Complete set of hardware, 188–190
Computers, automatic, chronological table, 333–334
  classification of, 24
  EDP, 5
  glossary, 351–355
  history, 24
  limitations, 306
  logic (see Machine logic)
  scientific, 5
  solid state, chronological table, 335–340
Computing developments, chronological table, 329–332
Contact circuits (see Combinational circuits)

Control, 17
Copy cycle, 283
Counter, binary, 243
  ring, 244
Cycle, 17, 282

Demapping (see Karnaugh maps)
DeMorgan's theorems, 131, 151
Diode, decoding/encoding matrices, 222–224
  electronic AND/OR gates, 220–222
Distance, 93
Don't cares, 185–187, 296–297
DO-statement, 53

End-around-carry, 269
Error correction, 89
  single, 95–96
Error detecting codes, 90–91
Error detection, 88–89
  single, 92–95
Excess-three code, 80–82
Execute cycle, 17, 278–279, 285
Exclusive-or, 136

Fan-in, 190
Fan-out, 190
Fetch cycle, 17, 278–279, 285
Finite state machines (see Machines, finite state)
Flip-flops, 235–241
  D, 238
  J-K, 238
  R-S, 236
  R-S-T, 238
  T, 238
  transistor NOR, 239–241
Flowchart, 35, 37, 39, 50
Flow diagram (see Flowchart)
FORTRAN programming, 56–63
Four-bit register, 200–201
Full-adder, 264–269, 273
  carry output, 266
  sum output, 266

Gates, 107
  diode, 220–222
Glossary, 350–354

Half-adder, 202–206, 267–268, 273
Hamming code, 96–99
Hardware, peripheral, 18
Hollerith, H., 19
Huntington, E. V., 142

Index register, 280–282
Inference, calculus of, 131–135
Information conveyance, 29
Information structure, 10
Inhibit line, 248
Input-output, 18
Instruction, 36, 40, 41
Instruction counter, 18
Instruction word, 36, 41

Karnaugh maps, 161–176
   adjacent cells, 167–170
   demapping, 170–172
   simplification (see Simplification)
Keypunches, 21

Logic design, 103
Logic, symbolic, 125–140
Logical circuits (see Combinational circuits)
Logical diagrams, 116–120
Logical equations, 109–116

Machine, logic, 8
   organization, 14–18, 283–285, 318–319
   programming, 8, 28–66
   timing, 8
Machines, 1
   classification of, 2
   current uses of information-processing, 4
   cybernetic, 6
   finite state, 302–304
   general purpose, 7
   heuristic, 6
   Turing (see Turing machines)
Magnetic core, 245–246
   inhibit line, 248
   matrices, 246
   memory, 245–250
   three-dimensional threading, 248–250

Magnetic tapes, 21
Maps, and Boolean algebraic expressions, 312–315
   composition, 206–208
   decomposition, 206–208
McCluskey, E. J., Jr., 161
Memory, 15–17, 288
   address (see Address)
   address register, MAR, 282
   assembly, 247
   control component, MCC, 283
   cycle, 282
   data register, MDR, 282
   dynamic, 288
   magnetic core, 245–250
      planar array, 250
   static, 288
   word organized, 247–248
Minimizations (see Simplification)
Mnemonics (see Assembly Language)

NAND logic, 191–199
NBCD code, 78–80
NOR logic, 191–199
Number, 68–69
Number representation, floating-point, 82–83
   fixed-point, 82–83
Numeration systems, 70

OR, 136
Overflow, 311, 317–318

Parity, 91
Plugboards, 30–33
Printers, 23
Procedure-oriented language, 45–48
Processing, 17
Processor, 262
Program, branch, 51, 65
   identifiers, 52
   loop, 51, 64
   sentence, 40
   statement, 40, 41, 48
Programming, 28–65
   external, 29, 33–34
   plugboard, 29–33
Propositions, algebra of, 105, 135–137
Punched card (see Card, punched)

Quine, W. V., 161

Radix, 71
  conversion, 72–74
R-drive line, 296
Redundancy, 88
Register, 15, 16
  index (see Index register)
  instruction, I-register, 284
Relays, 224–227
Reliability, 85–102
  interaction, 86
    constructive, 86
    destructive, 86
Reset equations, 315–317
Ring counter, 244
Rounding-off, 99

S-drive line, 296
Secondary excitations, 300
Selector units, 283–285
Sequential circuits (see Sequential machines)
Sequential machines, 287–309
  definitions, 287–288
  logic, 120
Serial addition (see Addition)
Set (see Class)
Set equations, 315–317
Shannon, C. E., 141
Shifting, 261
Shift registers, 16, 241–242
  in multiplication, 270
Simplification, of switching circuits, 156–160
  of Karnaugh maps, 172–176
Source program, 45
State flow diagrams, 290–292, 303, 319–320
  minimized, 293
  primitive, 292
  transitions, 290

Statement brackets, 52
States, 289
  equivalent, 290, 291, 293
  super, 290
State flow diagram, 290–292, 302, 319–320
State transition table, 292–293, 302, 313
  minimized, 293–294
Storage (see Memory)
Store cycle, 282
Symbolic logic, 125–140
Synchronous, computers, 10
  operation, 278–280, 321
Syntax error, 54
Switches, 217–219
Switching algebra, 140–183
Switching devices, 215
  engineering specifications, 216–217

Timing (see Clocking)
Timing chart, 311
Transistor, common emitter logic, 232
  diode-transistor logic, DTL, 234
  emitter-follower logic, 231
  inverter, 229–231
  logic, 228
  NOR flip-flop, 239-241
  resistor-transistor logic, RTL, 233
Transitions, 290
Transition table (see State transition table)
Truncation, 99
Truth table, 146, 312
Turing, A. M., 6, 304
Turing machines, 304–305

Unsolvability, 306

Venn diagram, 127, 133

Word, machine, 13